Science and Engineering Literature

Library Science Text Series

Science and Engineering Literature:
A Guide to Reference Sources

Third Edition

H. Robert Malinowsky
Associate Dean of Libraries
University of Kansas

Jeanne M. Richardson
Science Librarian
University of Kansas

Libraries Unlimited, Inc.   **Littleton, Colorado**
1980

LIBRARIES UNLIMITED, INC.
P.O. Box 263
Littleton, Colorado 80160

Library of Congress Cataloging in Publication Data

Malinowsky, Harold Robert.
  Science and engineering literature.

  (Library science text series)
  Bibliography: p. 279
  Includes index.
  1. Reference books--Science. 2. Science--
Bibliography. 3. Reference books--Engineering.
4. Engineering--Bibliography. I. Richardson,
Jeanne M., 1951-   joint author. II. Title.
Z7401.M28 1980 [Q158.5]   016.5   80-21290
ISBN 0-87287-230-0 (cloth)   0-87287-245-9 (paper)

Libraries Unlimited books are bound with Type II nonwoven material that meets and exceeds National Association of State Textbook Administrators' Type II nonwoven material specifications Class A through E.

Preface to the Third Edition

This edition is completely rewritten and restructured. The audience, however, remains the same—the library school student, the general librarian, and the science librarian specialist. The emphasis is on titles that are current and in print; major bibliographies and abstracting services are given primary attention. In compiling a guide such as this, a certain amount of selection must be made or the book becomes too large for its intended use. As a result, some titles are necessarily omitted. It is our intention to include representative English language titles that are in print; if glaring omissions are discovered, we request that they be brought to our attention.

The text to this edition has decreased, emphasizing the importance of the bibliographic citations. The introductory chapters are shorter and references to individual titles are combined in a chapter called "Multidisciplinary Sources of Information." The book continues to present a progression of material from the general to the specific, using similar form divisions in each chapter for ease of instruction. The "History of Science" chapter precedes the chapters on subject disciplines. The "Mathematical Sciences" chapter has been restructured to combine the sources of information for pure mathematics, statistics, and computer science in one, instead of three separate areas.

The chapter entitled "Geoscience and Environmental Sciences" in the second edition has been split into two chapters—"Geosciences" and "Energy and Environment." The extensive amount of sources on energy and the environment has prompted this separation. The "Biological Science" chapter has been reorganized into four major divisions—General Biology, Cell Biology, Botany/Agriculture, and Zoology. The "Engineering" chapter is divided into five areas—General Engineering, Mechanical and Electrical, Production and Processing, Construction, and Transportation. The chapter in the second edition on "Science Literature and Science of Librarianship" has been eliminated in this edition; these citations are redistributed among the subject disciplines and in the chapter on "Multidisciplinary Sources of Information."

An updated bibliography has been added, as has a section on data bases in chapter 1. The annotations throughout the text are designed to call attention to the scope of the work, its intended audience, and special features that often go unnoticed. The annotations may be one sentence or several paragraphs in length depending on the importance and the type of the source.

Because this bibliographical text is necessarily selective, the reader desiring comprehensive coverage should consult other bibliographic guides. Specifically, we recommend *American Reference Books Annual* (3-23) as a comprehensive source for currently published reference materials, and Walford's *Guide to Reference Material* (3-3) and Sheehy's *Guide to Reference Books* (3-2) for retrospective coverage. When available, literature guides for individual disciplines should also be consulted.

The second edition, which was co-authored by Dorothy A. Gray and Richard A. Gray, and the present third edition are based on Malinowsky's *Science and Reference Sources* (Littleton, CO: Libraries Unlimited, 1967). The first edition covered 435 reference sources, the second edition covered 1,096, and this edition covers 1,273.

There is naturally room for improvement to any bibliographic guide. The authors and the publisher welcome comments, corrections, and additions that should be incorporated into the fourth edition.

May, 1980   H. Robert Malinowsky
                                                                   Jeanne M. Richardson

Table of Contents

1

Introduction

Science, engineering, and technology encompass a vast body of knowledge that relates to our being, surviving, and creating. Science and technology have existed from the early times of man. Man's quest for answers has continued throughout history and is responsible for today's new discoveries and technologies. This book systematically examines the disciplines in science and engineering, alerting the reader to the variety of printed sources necessary for research and discovery. It covers mathematical science, astronomy, physics, chemistry, geoscience, environment and energy, biological science, biomedical science, and the many areas of engineering. An introduction to each discipline precedes the bibliographic text.

CURRENT VERSUS RETROSPECTIVE COLLECTIONS

When talking about the literature of science and engineering, current sources are usually emphasized over the older items that are in a library. However, older materials, as well as current materials, are needed in research, because the past is the key to the present. Rarely does a scientist begin research without first surveying what has been published; no one wants to reinvent the wheel or rediscover penicillin. It is because of this that libraries exist today. If the information were expendable and of no use after a certain period of time, then libraries would not be faced with the dilemma of housing all the published and unpublished materials that are needed in research.

Current information is vital to the practitioner. No doctor would use a 1930 procedure or diagnosis when a 1980 method is better and safer. Use of current information is, of course, important to all researchers. Keeping abreast of the literature is essential to avoid duplication of efforts and discoveries. Libraries therefore attempt to subscribe to the latest journals. If they do not, they fail to provide important, current research data. Still, retrospective information is extremely important. For example, taxonomic studies depend on older materials because specimens are named only once; the first publication of the name remains official. Collecting and maintaining older literature in these areas is an important task for the librarian. Even more important is ensuring that accessibility to older literature is adequate. Often important pieces of information are lost because the indexing is inadequate. For the librarian there is no distinction between the importance of current and retrospective materials. Libraries are expected to have the information when it is needed and librarians are expected to gather the information.

COST OF MATERIALS

Maintaining a major collection of science and engineering materials is big business. Anyone who has had an active hand in acquiring the books and periodicals that are needed in research can report how quickly their budgets are spent. The spiraling costs increase every year. Monographic materials are becoming more and more expensive to purchase. Books that used to cost $10.00 to $15.00 ten years ago now cost $40.00 to $60.00. The cost of reference materials is even higher. It is not unusual for a handbook in the sciences or engineering to cost over $100.00. Medical books lead the list in price and chemistry books are a close second. The price of indexes and abstracts can cause even the seasoned acquisitions librarian to take a second look. For example, in 1980, *Chemical Abstracts* is priced at $4,200.00 a year with an institutional discount of $400.00; *Biological Abstracts* is $1,360.00; *Mathematical Reviews* is $1,050.00; *Physics Abstracts* is $840.00; *Excerpta Medica* is $127.00 to $332.00 per section; and *Engineering Index* costs $695.00. These sources are mandatory for any research library in the sciences.

Soaring costs are not limited to books, indexes, and abstracts; journal subscriptions are also unusually high. Most large academic libraries are spending at least half of their acquisitions budgets on subscriptions each year. The decrease in the buying power of the dollar overseas also limits libraries' attempts to purchase foreign journals. Foreign titles are often cut when libraries need extra monies in the budget. Even the standard journals have increased in price. For example, in 1980, part A of *Physical Review* is $70.00 and part B is $40.00; *Journal of Chemical Physics* is $195.00; *Journal of Soviet Mathematics* is $365.00; *Journal of Geophysics* is $152.00; and the *Journal of Biological Chemistry* is $200.00.

There is no easy solution to this problem. The cost of producing printed material has risen drastically and this cost is passed on to the consumer. Increasing acquisition costs are forcing libraries to be more selective and prompting libraries to participate in cooperative buying plans. Libraries must continue to analyze the situation and develop new methods to cope with the problem.

BIBLIOGRAPHIC CONTROL

Science and engineering librarians have always been proud of the excellent bibliographic control maintained within their collections. Librarians attempt to catalog collected materials regardless of type or format. As a result, the printed catalogs of several research libraries include fully cataloged books and periodicals, as well as reprints, chapters, technical reports, and documents.

Because libraries strive to have as much bibliographic control as possible, it is only natural that they produce numerous printed catalogs and union lists. Production of these research tools has a two-fold purpose. First, they alert users to collection strengths of other libraries and, second, they allow the research community to know what is collected, indexed and available for public use. These tools are invaluable for interlibrary loan purposes and promote cooperation of resources in the science and engineering communities.

With the concept of bibliographic utilities that has developed over the past ten years, the control of bibliographic literature becomes all-encompassing. Cooperative cataloging accelerates the cataloging process and cuts down the expense of original cataloging. OCLC, Inc. is an example of a bibliographical utility that has benefited the library community. It is a fully online cataloging system that allows libraries direct access to a data base of over 4 million records. OCLC is only one of several available utilities. Others may have different features and programs, but all enable the library to catalog online and produce machine-readable records. Machine-readable records are an important aspect of bibliographic control of today's literature. For example, through manipulation of these records, libraries are developing new alternatives to the present card catalog. Large research libraries are increasingly concerned about the costs needed to maintain large and complicated card catalogs. This concern is growing with the proposed implementation by the Library of Congress of the new *AACR* in January of 1981. Libraries will undoubtedly "start over" and create new catalogs that are more accurate and easier to use. More COM and online catalogs will probably result.

Another advantage of online cataloging networks is the creation of union catalogs for each library that belongs to a utility. By calling up any one record, it becomes apparent which libraries own copies of the item in hand. As budgets become more inflexible, librarians will use these utilities when deciding to purchase expensive items. If other nearby libraries own copies of the item, they may decide not to purchase. For interlibrary loan activities, these online catalogs locate the availability of material. As the utilities expand, additional uses can be made of the machine-readable records. In acquisitions, for example, programs exist that create order information and actually order the items that a library may want. This information is stored and becomes part of the online data base, showing the status of each order and thus eliminating the time of filing and refiling order slips, letters, cards, etc. The accounting of budgetary information is also automatically processed so that on a moment's notice one can see how much money has been spent to date, a process that is quite cumbersome when performed manually. Online serials and circulation systems are also being developed through the use of machine-readable records.

LOCATION OF SCIENCE AND ENGINEERING INFORMATION

Most any library can boast of some kind of collection in the sciences or engineering. Many prominent libraries acquire private collections that individuals have developed with their own funds. These collections are purchased, kept intact. Acquisition techniques such as this result in collections that are unique in a given field.

Browsing through the G. K. Hall catalog quickly reveals major library collections: *Catalog of the John Crerar Library*; *Catalog of the Library of the Academy of Natural Sciences of Philadelphia*; *Catalog of the Scripps Institution of Oceanography*; *Catalog of the Royal Botanic Gardens*; and *Dictionary Catalog of the Blacker-Wood Library of Zoology and Ornithology*. These collections are important, and researchers are aware of what they hold. Many libraries have gone to great expense to provide as many access points as possible and have gone far beyond standard cataloging practices found in the general collection.

Other science and engineering libraries attempt to preserve printed research. Libraries such as the Center for Research Libraries (CRL), Linda Hall Library, National Lending Library (NLL), National Library of Medicine, and the National Agricultural Library stress preservation as well as accessibility to the scientist. The CRL maintains what is termed "Rarely Held Scientific Journals." Its purpose is to preserve items that are rarely used in research so that individual libraries need not cope with the expense and time of maintaining them. Several libraries have formed a consortium so that this plan can be carried out; in return, the participating libraries can borrow, without charge, material from CRL. The materials at CRL are either purchased or placed on permanent loan by one of the member libraries. This arrangement has been successful and will improve as more libraries realize the difficulties of maintaining obscure materials that are seldom used.

The Linda Hall Library in Kansas City, Missouri, collects materials in the physical sciences. Its periodical literature is especially strong because it attempts to collect complete runs of international journals. Through its interlibrary loan policy, the Linda Hall Library furnishes copies of the requested material in relatively quick turn-around time. The National Lending Library in Great Britain is another library that collects current scientific materials. It began as a service to the scientific community in Great Britain by supplying photocopies of journal articles, technical reports, proceedings and translations. With the success of this service in Great Britain and with no opposition from the British publishers, the NLL inaugurated an overseas program. CRL subscribed to this program in 1975 and termed it the Journals Access Service (JAS). Its purpose is to furnish materials from 1970 that are not held by CRL. Libraries that use this service find it useful.

Currently, attempts are being made to establish a National Periodicals Center in the United States similar to Great Britain's. Most librarians believe that if British publishers can accept NLL's photocopying center, so should U.S. publishers. The concept is being discussed at the national level, and funding has been requested from the federal government. In the meantime, libraries continue to analyze collections, to buy and house materials cooperatively, and to give prompt, reliable service.

LOCATING WHAT YOU WANT – LITERATURE SEARCHING

It can be difficult to find a specific fact or list of articles in the library if the task is not approached in the proper manner. Locating books in most libraries is relatively simple because librarians spend a great deal of time cataloging individual books and provide more than one access point per title. If the library does not own a particular book, there are numerous tools available to determine the availability of the item in other libraries.

Locating facts or data in a library is a bit more difficult, but can be done with germane reference sources. Reference librarians are trained to locate all kinds of information, including the most obscure. In addition to encyclopedias, dictionaries, and handbooks, hundreds of specialized books provide factual information in the sciences. These include guides to the literature, bibliographies of bibliographies, indexes, abstracts, directories, atlases, field guides, treatises, and manuals. Many tools are easy to use, but others, such as the *Science Citation Index*, initially require guidance from someone who is familiar with the source. Never underestimate the ability of a librarian to locate what is needed. In fact, the librarians are often responsible for developing the source that finds the fact.

Locating information in journal and technical report literature is probably the most difficult. Some journals and technical reports are not indexed or abstracted even in such publications as *Chemical Abstracts* and *Biological Abstracts*, which strive to provide comprehensive international coverage of the literature. (The list of journals indexed by these services runs into hundreds of pages.) For the researcher, searching the literature is serious business. Hours of valuable time can be wasted if the search is conducted improperly. Even with today's sophisticated online data bases, one must carefully determine what kind of information is needed.

Unfortunately, some scientists approach literature searching in a rather haphazard manner. All too often the false assumption is made that nothing has been published on the topic. When trying to convince library users that this is not necessarily true, researcher-librarian relationships are sometimes placed in jeopardy. Literature searching must be performed with proper experience or guidance. Many researchers are uncertain of their exact needs and should critically ask themselves whether they know enough about the subject they are searching, whether there are any restrictions on the search, whether the material is needed immediately, and what the proper bibliographical format should be.

According to Parke, in his *Guide to the Literature of Mathematics and Physics*, there are eight important traits that a literature searcher should have before attempting any systematic search, whether manual or online.[1] First, imagination is needed. It is necessary to list all places that could possibly have the desired information. A second trait is mental flexibility. In the process of conducting the search, one must allow for rapid adjustment to new ideas and possibilities. Thoroughness is a third trait. Having the patience to be thorough guarantees that no information is overlooked. A fourth trait is orderliness. It is necessary to keep a record of exactly what was searched and where the information was found, to prevent any back tracking. In order not to give up too soon when nothing is found, a fifth trait is needed, persistence. Persistence goes hand in hand with imagination enabling the searcher to find what is needed or to determine that nothing is available on the topic. The power of observation is a sixth trait. Terminology changes, for instance, require observation skills when performing retrospective searches. During the search, judgment, the seventh trait, is needed so that time is not wasted following a wrong path. Accuracy is the eighth and final trait. Copying an incorrect citation can cause trouble when trying to 'locate the item. With these eight traits a good, thorough search can be conducted.

Before starting any literature search, the searcher should review the subject. This serves as a review of terminology and background, and helps determine any subject limitations. General reviews are usually found in encyclopedias, treatises, textbooks, or monographs. Reviews acquaint the searcher with the subject and give an idea of how the topic relates to other topics. Once the review is made, a more detailed study is conducted to ensure a thorough understanding of the subject. If the searcher is performing a search for someone else, this is the time to make certain that both parties understand the subject equally well.

The next step is to become familiar with the sources that will be used in the search. In addition to the handbooks and encyclopedias, bibliographies, indexes, and abstracts should be consulted. When using indexes and abstracts, it is necessary to have a firm grasp of subject terminology, because terms may change from year to year. It is also true that indexed terms found in different tools are seldom identical. The best plan is to prepare a list of all expected terms and refer to that list while searching the indexes and abstracts, noting any new terms not on the list. When searching online data bases it is necessary to compile this list of terms so that the computer can search all possible terminology. Usually the data bases have printed thesauri that help identify the proper terms.

---

[1] Nathan Grier Parke, *Guide to the Literature of Mathematics and Physics*, 2nd rev. ed. (New York: Dover, 1958), p. 37.

Searching the indexes and abstracts is the next step. This can be done with the printed copy or with computer data bases. When performing manual searches record the information exactly as it is found. Also, record the location of the citation in the index so that if there is a question, the information is easily verified. In most cases primary sources are consulted, requiring the use of library catalogs or union lists.

The rewards of any literature search are the resulting bibliographies. When compiling a bibliography, keep in mind that it is worthless unless all notations are easily understood; nothing is more exasperating than bibliographic hieroglyphics. The information in the bibliography must be complete, listing author, article title, and periodical title spelled out in full. If abbreviations are used, list only standard abbreviations accepted by major abstracting services. The volume, year, number, months and pages of each journal article must also be clearly stated. For reports, all information on report number, accession numbers, corporate authors, and personal authors must be given. For books, all imprint information should be included. If annotations are not included in the bibliography, abstract citations are useful to determine the pertinence of the article.

Another way of searching the literature is to trace the bibliographies of one or two key, recent papers in a particular field. Because these articles are important their bibliographies should also contain pertinent citations. By checking each entry in the bibliography, additional titles are identified, and checking their bibliographies should provide yet more titles. This procedure is called "fan searching," since each article contains a bibliography and the bibliographies multiply like a fan. The method works well with current research. Current issues of major periodicals are scanned. If a key article is discovered, then a fan search can be conducted.

Citation searching is another recent innovation in science and engineering, although it has existed for a long time in the legal profession. Instead of using subject terms, the names of key authors are used. The names are checked in a citation index, which lists all papers that cite particular authors. The authors to these papers can also be checked in the citation index to locate more information. Doing this two or three times helps compile a broad bibliography. Another approach to a citation index occurs when authors look up their own names and determine who is citing their articles. This enables them to identify who is working on similar research problems.

DATA BASES

With the creation of machine-readable data, numerous data bases have been developed in the science and engineering fields. These data bases are constantly changing in coverage and cost but are extremely important to anyone doing research. They save countless hours of searching manually and are much more thorough than anyone could be manually. Each data base is different, and its use is dictated by a detailed manual that the librarian must be familiar with. Some libraries provide data base searching as a free service and others provide it at cost.

The following list of data bases includes the most important ones in the various science and engineering fields. There are others and users should consult with their librarians as to data base availability.

AGRICOLA is produced by the National Agriculture Library of the U.S. Department of Agriculture, and covers worldwide literature in agriculture and related subject fields. It is the machine-readable version of *Bibliography of Agriculture* and the *National Agriculture Library Catalog*. Available from 1970 through BRS, Lockheed, and SDC.

Agricultural Abstracts. See CAB ABSTRACTS.

Air Pollution Abstracts. See APTIC.

ALCOHOL USE/ABUSE. See BRS/DRUG.

Alloys Index. See METADEX.

Animal Breeding Abstracts. See CAB ABSTRACTS.

Antarctic Bibliography. See COLD.

APILIT covers literature relating to petroleum refining and the petrochemical industry, and is produced by the Control Abstracting and Indexing Service of the American Petroleum Institute. Available through SDC from 1964 to date.

APIPAT is published by the Central Abstracting Service of the American Petroleum Institute and covers patents related to petroleum refining and the petrochemical industry. Available through SDC from 1964 to date.

APTIC is produced by the Manpower and Technical Information Branch of the U.S. Environmental Protection Agency and covers all aspects of air pollution. The data base includes the abstracts that appeared in *Air Pollution Abstracts*, which is no longer published. Available through Lockheed from 1966-1978. There are no updates after 1978.

AQUACULTURE is produced by the National Oceanic and Atmospheric Administration and covers information on marine, brackish, and freshwater organisms. It is available through Lockheed from 1970 to the present.

Aquatic Sciences and Fisheries Abstracts. See ASFA.

ASFA is a comprehensive data base on the science, technology, and management of marine and freshwater environments. This data base and the corresponding printed version, *Aquatic Sciences and Fisheries Abstracts*, are produced and published by Information Retrieval, Ltd. under contract to the Food and Agriculture Organization of the United Nations. Available from 1978 through Lockheed.

BHRA FLUID ENGINEERING provides citations to international information on all aspects of fluid engineering. It is produced by British Hydromechanics Research Association and is available from 1974 through Lockheed.

Bibliography and Index of Geology. See GEOREF.

Bibliography and Index of Geology Exclusive of North America. See GEOREF.

Bibliography of Agriculture. See AGRICOLA.

Bibliography of Thesis in Geology. See GEOREF.

Bibliography on Cold Regions Sciences and Technology. See COLD.

Biological Abstracts. See BIOSIS.

Biological Abstracts/RRM. See BIOSIS.

Bioresearch Index. See BIOSIS.

BIOSIS provides comprehensive coverage of research in the life sciences. It is the machine-readable version of *Biological Abstracts, Bioresearch Index* (1967-1979) and *Biological Abstracts/RRM*, the major publications of Bio-Science Information Service of Biological Abstracts. BIOSIS is available through BRS, Lockheed, and SDC; each vendor divides the tapes into several distinct files.

BRS/DRUG contains citations from the Drug Information Service Center at the College of Pharmacy, University of Minnesota, and the Hazeldon Foundation. It is composed of two data bases: DRUG INFO, which focuses on alcohol and drug use/abuse, and ALCHOL USE/ABUSE, which deals with the evaluation of treatment and family therapy of alcohol use and abuse. Available through BRS from 1968; there is no printed version.

CA Formula Index. See CHEM ABS.

CA Index Guide. See CHEM ABS.

CA Registry Handbook. See CHEM ABS.

CA Substance Index. See CHEM ABS.

CAB ABSTRACTS is a file of agricultural and biological information published by the Commonwealth Agricultural Bureaux. The following printed sources are included in CAB ABSTRACTS: *Animal Breeding Abstracts*; *Agricultural Abstracts*; *Dairy Science Abstracts*; *Field Crop Abstracts*; *Forestry Abstracts*; *Helminthological Abstracts (A and B)*; *Herbage Abstracts*; *Horticultural Abstracts*; *Index Veterinarius*; *Nutrition Abstracts and Reviews (A and B)*; *Plant Breeding Abstracts*; *Protozoological Abstracts*; *Review of Applied Entomology (A and B)*; *Review of Medical and Veterinary Mycology*; *Review of Plant Pathology*; *Soils and Fertilizers*; *Veterinary Bulletin*; *Weed Abstracts*; and *World Agricultural Economics and Rural Sociology Abstracts*. Available through Lockheed from 1973.

CHEM ABS is the machine-readable version of the publications produced by Chemical Abstracts Service (CAS). BRS, Lockheed, and SDC provide a variety of integrated files from the CAS tapes. *Chemical Abstracts, CA Substance Index, CA Registry Handbook, CA Index Guide*, and the *CA Formula Index* are the printed publications of this chemistry data bases. Lockheed provides citations from 1967; BRS and SDC provide citations from 1970.

Chemical Abstracts. See CHEM ABS.

CLAIMS. There are three CLAIMS data bases available through Lockheed. CLAIMS/CHEM 1950-1970 contains chemical and chemically related patents issued from 1950 to 1970. CLAIMS/U.S. PATENTS 71-77 and CLAIMS/U.S. PATENTS 78- provide access to U.S. patents classified by the U.S. Patent and Trademark Office and announced in the U.S. Patent and Trademark office *Official Gazette* since 1971.

COLD covers source material on all disciplines dealing with Antarctica, the Antarctic Ocean, and subantarctic islands. It is produced by the Cold Regions Research and Engineering Laboratory and is available from 1962 through SDC. The printed versions of this data base are *Antarctic Bibliography* (1962-) and *Bibliography on Cold Regions Science and Technology* (1964-).

COMPENDEX, produced by Engineering Index, Inc., covers the world's engineering and technological literature. COMPENDEX is the machine-readable version of *The Engineering Index* and is available from 1970 through Lockheed and SDC.

Computer and Control Abstracts. See INSPEC.

Computer and Information Systems Abstracts Journal. See ELCOM.

Current Contents. See SCISEARCH.

Dairy Science Abstracts. See CAB ABSTRACTS.

DRUG INFO. See BRS/DRUG.

Drug Research Report (The Blue Sheet). See PHARMACEUTICAL NEWS INDEX (PNI).

ELCOM is a file of electronics and computer information published by Cambridge Scientific Abstracts, Inc. It is the machine-readable version of *Electronics and Communications Abstracts* (August, 1977-) and *Computer and Information Systems Abstracts* (1978-). Available through SDC.

Electrical and Electronics Abstracts. See INSPEC.

Electronics and Communications Abstracts Journal. See ELCOM.

Energy Information Abstracts. See ENERGYLINE.

ENERGYLINE is produced by the Environment Information Center, Inc., and includes citations that appear in *Energy Information Abstracts* and energy materials from *Environment Abstracts*. It provides primary source material relating to all aspects of energy. Available from 1971 through Lockheed and SDC.

Engineering Index. See COMPENDEX.

ENVIROLINE is produced by the Environment Information Center, Inc., and covers the world's environmental information. It is the machine-readable version of *Environment Abstracts*. Available through Lockheed and SDC.

Environment Abstracts. See ENVIROLINE.

ENVIRONMENTAL IMPACT STATEMENTS (EIS) is produced by the Information Resources Press and includes citations to environmental impact statements from federal, state, county, and municipal agencies. Available through BRS since 1977.

EXCERPTA MEDICA contains biomedical and pharmaceutical information that is published in the 42 abstract journals of *Excerpta Medica*. It is available from 1974 through Lockheed.

FDC Reports (The Pink Sheet). See PHARMACEUTICAL NEWS INDEX (PNI).

Field Crop Abstracts. See CAB ABSTRACTS.

Fluidex. See BHRA FLUID ENGINEERING.

Food Science and Technology Abstracts. See FSTA.

Forestry Abstracts. See CAB ABSTRACTS.

FSTA provides access to the literature in food science and technology. It is produced by International Food Information Service, England, and is the machine readable version of *Food Science and Technology Abstracts.* Available from 1969 through Lockheed and SDC.

Geophysical Abstracts. See GEOREF.

GEOREF is supplied by the American Geological Institute and provides comprehensive coverage to geology literature. The following printed sources are included in GEOREF: *Bibliography and Index of North American Geology* (1961-1970); *Bibliography and Index of Geology Exclusive of North America* (1967-1968); *Geophysical Abstracts* (1966-1971); *Bibliography of Theses in Geology* (1965-1966); and *Bibliography and Index of Geology* (1969-). Available through SDC.

Government Reports Announcements. See NTIS.

GPO MONTHLY CATALOG is the machine readable version of the printed *Monthly Catalog of United States Government Publications* published by the Government Printing Office. It contains information issued by all U.S. federal agencies and is available through Lockheed from July 1973, and through BRS from 1976.

Helminthological Abstracts. See CAB ABSTRACTS.

Herbage Abstracts. See CAB ABSTRACTS.

Horticultural Abstracts. See CAB ABSTRACTS.

Index Veterinarius. See CAB ABSTRACTS.

INSPEC, produced by the Institution of Electrical Engineers (IEE), provides international coverage of the literature in physics and electrical and electronics engineering. The data base includes abstracts that appear in *Physics Abstracts, Electrical and Electronics Abstracts*, and *Computer and Control Abstracts*. Available since 1969 through Lockheed and SDC, and from 1970 through BRS.

INTERNATIONAL PHARMACEUTICAL ABSTRACTS (IPA) is published by the American Society of Hospital Pharmacists and provides information on all phases of drugs and professional pharmaceutical practice. It is the machine-readable version of the printed *International Pharmaceutical Abstracts* and is available from 1970 through Lockheed.

IPA. See INTERNATIONAL PHARMACEUTICAL ABSTRACTS.

ISMEC is the machine-readable file of the printed *Ismec Bulletin*. ISMEC covers information on mechanical engineering, production engineering, and engineering management. It is available from 1973 through Lockheed and SDC.

Ismec Bulletin. See ISMEC.

Medical Devices, Diagnostic and Instrumentation Reports (The Gray Sheet). See PHARMACEUTICAL NEWS INDEX (PNI).

METADEX is produced by the American Society for Metals and the Metals Society, and provides international coverage of metallurgy literature. METADEX is the machine-readable version of *Review of Metal Literature* (1966-1967), *Metals Abstracts* (1968-), and *Alloys Index* (1974-). It is available only through Lockheed.

Metals Abstracts. See METADEX.

Monthly Catalog of United States Government Publications. See GPO MONTHLY CATALOG.

National Agricultural Library Catalog. See AGRICOLA.

NIMH is the data base produced by the National Institute of Mental Health and covers all aspects of mental health and related information. There is no printed version of this data file. BRS and Lockheed plan to supply NIMH from 1969 in the near future.

NTIS consists of U.S. government-sponsored research and development technical reports in the physical, biological, and social sciences. It is the machine-readable version of *Weekly Government Abstracts* and *Government Reports Announcements*, which is supplied by the National Technical Information Service. Available from 1964 through Lockheed, and from 1970 through BRS and SDC.

Nutrition Abstracts and Reviews. See CAB ABSTRACTS.

OCEANIC ABSTRACTS covers international literature on marine-related subjects. It is the machine readable version of *Oceanic Abstracts*, published by Data Courier, Inc. Available through SDC, which refers to the data base as OCEANIC, and Lockheed from 1964.

Official Gazette. See CLAIMS.

P/E NEWS includes references to articles from major suppliers of petroleum and energy business news. It is supplied by the Central Abstracting and Indexing Service of the American Petroleum Institute and is available through SDC from 1975 to date.

Petroleum Abstracts. See TULSA.

PHARMACEUTICAL NEWS INDEX (PNI) covers current news about pharmaceuticals, drug legislation, and related health fields. It is the machine-readable version of *FDC Reports (The Pink Sheet)*; *Drug Research Reports (The Blue Sheet)*; *Medical Devices, Diagnostic and Instrumentation Reports (The Gray Sheet)*; *Weekly Pharmacy Reports (The Green Sheet)*; *Quality Control Reports (The Gold Sheet)*; *PMA Newsletter*; and *Washington Drug and Device Letter*. Available through BRS, Lockheed and SDC.

Physics Abstracts. See INSPEC.

Plant Breeding Abstracts. See CAB ABSTRACTS.

PMA Newsletter. See PHARMACEUTICAL NEWS INDEX (PNI).

POLLUTION ABSTRACTS provides information on environmentally related literature on pollution, its sources, and its control. POLLUTION ABSTRACTS and the printed version, also entitled *Pollution Abstracts*, are produced by Data Courier, Inc. Available through BRS, Lockheed, and SDC from 1970.

Protozoological Abstracts. See CAB ABSTRACTS.

Quality Control Reports (The Gold Sheet). See PHARMACEUTICAL NEWS INDEX (PNI).

Review of Applied Entomology. See CAB ABSTRACTS.

Review of Medical and Veterinary Mycology. See CAB ABSTRACTS.

Review of Metal Literature. See METADEX.

Review of Plant Pathology. See CAB ABSTRACTS.

SAE, produced by the Society of Automotive Engineers, provides access to technical papers on the technology of the automotive industry. There is no printed publication to the SAE data base. Available through SDC from 1966.

SAFETY is the machine-readable file of the printed *Safety Science Abstracts Journal*. SAFETY covers information on a variety of safety topics including industrial and occupational safety, transportation safety, general safety, and environmental safety. Available from June 1975 through SDC.

Safety Science Abstracts Journal. See SAFETY.

Science Citation Index. See SCISEARCH.

SCISEARCH is a multidisciplinary index to the literature of science and technology. It is prepared by the Institute for Scientific Information and contains all the records in the printed *Science Citation Index* and *Current Contents.* Available through Lockheed from 1974.

Soils and Fertilizers. See CAB ABSTRACTS.

SPIN, Searchable Physics Information Notes, is supplied by the American Institute of Physics and provides coverage to all areas of physics, including mathematical and statistical physics, astronomy, astrophysics and geophysics. Available through Lockheed from 1975.

SSIE is a data base that contains reports on research in progress and recently completed research sponsored by both government and privately funded scientific projects. It includes all areas of the life, physical, social, behavioral, and engineering sciences. SDC offers SSIE from fiscal year 1974 to date; Lockheed and BRS retain the last two years in their files.

TSCA INITIAL INVENTORY is a non-bibliographic dictionary file listing the chemical substances on the initial inventory of the *Toxic Substances Control Act (TSCA) Chemical Substance Inventory* compiled by the Environmental Protection Agency. Available through Lockheed.

TULSA provides international coverage of literature and patents dealing with the exploration, development and production of oil and natural gas. TULSA and its printed version, *Petroleum Abstracts*, are produced by the Information Services Department at the University of Tulsa. Available from 1965 through SDC.

Veterinary Bulletin. See CAB ABSTRACTS.

Washington Drug and Device Letter. See PHARMACEUTICAL NEWS INDEX (PNI).

Weed Abstracts. See CAB ABSTRACTS.

Weekly Government Abstracts. See NTIS.

Weekly Pharmacy Reports (The Green Sheet). See PHARMACEUTICAL NEWS INDEX (PNI).

WELDASEARCH is produced by the Welding Institute of England and provides coverage to the international literature on welding. Available through Lockheed from 1967.

World Agricultural Economics and Rural Sociology Abstracts. See CAB ABSTRACTS.

WORLD ALUMINUM ABSTRACTS provides coverage of the world's technical literature on aluminum. It is supplied by the American Society of Metals and is available through Lockheed from 1968 to the present.

THE FUTURE

It is obvious that libraries cannot continue to manage spiraling costs of materials and lack of storage space. In fact, many of the larger research libraries have already cut acquisitions budgets. When this happens in a library, the research community becomes uneasy and frustrated. Often the user does not fully understand the problem and how it will affect research. When asked to supply unnecessary book or periodical titles, users often answer that everything is needed and that nothing can be eliminated. That is a natural reaction; librarians should not despair. It is difficult to give up materials that have been handy over the years. Nonetheless, the problem still exists and the librarian must cooperate with the researcher and the publisher.

If a three pronged problem exists among the librarian, publisher, and researcher, what has been done and what can be done to help eliminate it? Because librarians have been aware of the problem longer than the others, they have tried to develop strategies to offset the decrease in purchasing power. So far, these strategies have had little effect on the availability of materials; some have had slight effect on publishers. Librarians have probably exhausted their resources without causing stir in the research and publishing communities.

One course of action by librarians is the elimination of multiple copies. If there is a fixed amount of money to spend, many librarians believe it should be spent to purchase something not already owned. However, this can have somewhat of a backlash effect because publishers often charge more for a title when sales are decreasing to meet publication costs. Consequently, the full amount of savings is unknown. Elimination of multiple copies may decrease loan periods for users since others may be waiting to use the item. Periodicals may not be routed if the only copy must remain in the library. Users will probably spend more time in the library reading materials than before. Departments may share journal subscriptions. In the academic institutions, branch libraries will depend on larger libraries since they cannot afford subscriptions to every desired journal. Users may be forced to consult several libraries before locating all needed items. This is having a positive effect on branch libraries because many smaller libraries are combining into one larger library with more efficient use of staff and monies. Although the researcher may initially resent having to travel across campus to another building, the advantages of housing materials in one place rather than in three or four will quickly be realized. For the institutions faced with building problems, it is of course more economical to build one facility than to build three or four. In another ten years, decentralization of libraries on academic campuses will be a thing of the past.

Cooperation between institutions and libraries has increased over the past five years. As libraries cut subscriptions they look to other institutions in their areas for the titles. Agreements are being made to determine journal titles cancelled by one library but retained by a second library, and vice versa. There are hazards to this approach, especially if a cancelled journal becomes an essential tool in a research field. Copyright restrictions may also interfere and create difficulties.

Interlibrary services will most likely increase in the next few years. This is one area on which librarians have worked for quite some time. Several bibliographical utilities promote the exchange of interlibrary loan requests by use of computers. Some of the more specialized libraries have developed into holding libraries. Interlibrary loan activity with these libraries will continue to be heavy. More holding libraries will most likely develop in the next ten years and may be funded by the federal government. Publishers are watching this development. They realize the problem and hope to aid in developing solutions. However, most publishers are profit organizations and must consider their costs if libraries continue to purchase cooperatively and to use interlibrary loan networks. Subscription prices will probably rise and use of royalties may become common. The publisher and librarian must attempt to solve the dilemma or the researcher will be on the losing end.

To help solve library space problems, publishers have taken the reins by providing microfilming and computer data bases. It is evident that housing a library in microform takes up less space than housing material in hard form. This does not mean, however, that the cost of running a microform installation is any less. Equipment is needed to read the microforms and to access the data bases. Many libraries have committed themselves to microforms and have a vigorous program of converting older materials to film and purchasing new materials only on film. Publishers have helped by providing titles in both film and hard copy. Some large microform publishing firms are spending millions of dollars filming back copies of periodicals so that libraries can purchase film and discard hard copy. Data bases are a way of paying on demand for information located in printed indexes and abstracts. Most suppliers are still furnishing hard copy and machine-readable copy; however, some publishers are eliminating hard copy format. Special libraries no longer are required to allot space for long runs of indexes and abstracts. Instead, terminals can take their place and data bases are accessed only when needed. Access to some data bases is too costly for many libraries. Hopefully new technologies will lessen tape and terminal costs.

Publishers are beginning to consider ways of cutting the cost of publishing periodicals. They are aware that libraries cannot continue to spend more and more each year on journals. Some publishers have merged journals in an effort to cut costs; other publishers have discontinued overlapping publications. This sounds fine to the librarians, but means there will be fewer journals to publish in for the researcher. In the academic community, fewer places to publish causes alarm for those who must "publish or perish." However, there are solutions. One possibility is to publish technical reports that are available through the government from federal grants. Another way is to promote more oral reporting at conferences and symposia and to publish the reports in transactions or proceedings. A third idea would be to have a central collecting agency in each subject discipline where researchers deposit reports that do not get published in the journals. These depositories would then index the reports, accession them, and create machine-readable data bases for public access. When using the data base, scientists could obtain offset copies from the clearinghouse or depository.

It may look bleak to the librarian, the publisher, and the researcher, but there is hope. New technologies and ideas will help solve these problems. All three will continue to exist in the future and all three will have to cooperate.

2

Forms of Literature

PRIMARY FORMS OF LITERATURE

In the scientific and engineering disciplines the most important kinds of information are obtained from primary sources. Primary information sources present new facts and help guide research efforts in the proper direction. They appear in several forms, the most important being the periodical. In addition to the periodical, other forms of primary information include patents, technical reports, notebooks, diaries, archives, proceedings, transactions, theses, dissertations, manuscripts, computer output, government documents, and oral recordings. To qualify as a primary source, the information must be original. It must not be analyzed, reviewed, or abstracted by anyone prior to its publication. Primary sources contain raw data used to produce additional information so that most information sources usually contain a mix of primary and secondary information.

Scientific Periodicals

In 1665 the *Journal des Scavants* began publication as the first scientific journal. Two years later the Royal Society's *Philosophical Transactions* appeared, which continues to be published today. By 1800 the number of scientific periodicals approached 100, and by 1900 the figure was at 10 thousand. This figure has steadily increased; estimates today place the figure at 50,000 titles. Librarians are aware that the periodical budgets are growing at a fast rate due to inflation of current subscriptions and to the increase of new titles. The literature includes many articles discussing the journal explosion. These figures vary from a low of 25,000 to as many as 120,000 titles per year. It is difficult to validate these figures because the method of counting varies and because few authors incorporate cessation of publication rates into their figures.

Because not every journal article is indexed or abstracted, it can be difficult to locate primary sources of information. At the most, a third to a half of the current journal articles are abstracted or indexed. This becomes a major obstacle for the scientist. Articles that are published in unindexed periodicals may be cited in later articles that are abstracted or indexed. As a result, the original article becomes available to the research community at a much later date. The librarian's job is to organize the material in such a manner that it is quickly retrieved. Through the use of machine-readable data bases, this organization is becoming easier. More and more of the primary materials are being indexed and abstracted at a faster rate.

Periodicals are only one form of literature within a larger category called serials. A serial is any publication that is indefinitely issued in parts on a regular or irregular basis. In addition to periodicals, there are other forms of serials. The "letter" was originally intended to promote prompt publication of information in unedited form and to alert the scientist to new research. The idea was to create correspondence between researchers that provided information in a format that was not intended to be kept for future reference. Now, however, letters contain information that never appears in print again, and is important for retrospective searching. Today, most letters are abstracted and indexed. *Physics Letters* and *Tetrahedron Letters* are two examples of letter publications. Other types of serial are "Advances," "Progresses," and "Annual Reports." These items are usually published annually and contain review literature. The material is usually a mixture of primary and secondary sources of information. Examples include *Advances in Chemical Physics, Advances in Medicine, Progress in Astronautical Sciences*, and *Annual Reviews of Chemistry*.

The house journal or organ is another type of serial that can have research value. As its name implies, it is produced by an industrial, commercial, or research organization to provide a vehicle of communication among its employees and to present a favorable image of itself to the public. The public relations journals are less significant than internal communication publications. The *IBM Journal of Research and Development* is an example of a house organ. It is indexed and abstracted by the leading services in biology, chemistry, physics, engineering, and mathematics as one of the house organs that has research value. The public relations house organs are not indexed and may or may not be retained by libraries. Some libraries, however, are beginning to recognize the research value of these items and are collecting them for future use.

Technical Reports

Technical reports are written to describe and report results of ongoing research. Since most research is supported by grants and contracts issued by the government, written reports must be submitted to the sponsoring governmental agency and made available to other researchers. At one time, these reports were unique and easily identified; now, however, they may appear in the form of proceedings, transactions, translations, annual reports, bibliographies, indexes, or abstracts. There are well over one million technical reports, covering every aspect of science and engineering, and ranging in length from one page to several volumes.

Every governmental, state, national, and international agency, and every research and development organization produces technical reports. Many of the reports are indexed by governmental agencies, but some are picked up by commercial firms and others are indexed in-house by the library owning the report. The usefulness of technical reports depends on the purpose of the report. Some reports are merely status reports containing little or no data; others are detailed accounts of specific research projects. Because of the timeliness of the information, technical reports are usually indexed in great detail.

The National Aeronautics and Space Administration (NASA) is one of the major producers of technical report literature. NASA's indexing and abstracting service, STAR, makes these reports available to the public and provides microfiche and hard copies to depository libraries. Probably the largest depository of technical reports is the National Technical Information Services (NTIS). Its mission is to make available all unclassified results of federally sponsored research and development. Its indexing publication is the *Government Reports and Announcements Index*.

The shelving arrangement of technical reports differs in each library. Libraries usually devise unique schemes to meet the needs of their research communities. Many reports have three and four different report numbers that are applicable in citing the report. Other reports are the proceedings and transactions of symposia or translations of foreign literature. The rule of thumb regarding technical report literature is to provide as much access as possible so that it can be retrieved regardless of how it is cited in the literature. The author citing a technical report should cite all report numbers, agencies, and authors. In the past few years, data base searching has aided technical report retrieval.

Patents

Over four million patents have been issued in the United States covering everything from simple machines to botanical plants. Patents are an important part of the research literature and need to be considered when searching that literature. Almost all developed countries issue patents. Unfortunately, patents are honored only in the country of issue. If an inventor wants to secure patent rights, patents must be obtained from all industrialized countries.

The patent is a document representing an agreement between a national government and an inventor, granting the inventor the right to exclusive exploitation of his product for a period of 15 to 20 years. In the United States, the period is 17 years except for patents on ornamental designs, which are granted for terms of 3½, 7, and 14 years. Each patent is given a number. These numbers are issued in consecutive order in four separate categories: patents, design patents, plant patents, and reissue patents. In the United States the Patent and Trademark Office issues patents every Tuesday. When a patent is submitted to the patent office, it is assigned a serial number and an application date, and is legally described as "patent pending." All patents must have a title or name descriptive of the patentable idea, defined as any new and useful process, machine, manufacture, composition of matter, or new and useful improvement thereof; any distinct and new variety of plant, other than a tuber-propagated plant which is asexually reproduced; or any new, original, and ornamental design for an article of manufacture. The inventor is called the patentee, and the person or firm to which the inventor assigns his or her patent, if necessary, is called the assignee. In the United States, no firm may apply for a patent. Each patent is assigned a subject classification by the Patent and Trademark Office. In submitting an application, the inventor must give specifications or arguments for the patent. This is called the claim on which all legal determinations are based. A person who challenges the validity of a patent is called a disclaimer.

It is important to search the patent literature to determine if an invention or innovation is patentable, if it would infringe on a patent owned by someone else, or if an existing patent is valid. Patent literature also documents what others have done or described, and summarizes the developments in a particular line. Patent searching is complex and takes a considerable amount of time. Large research organizations routinely employ a professionally qualified person such as a patent attorney or agent. In some subject areas patent literature is abstracted and appears in abstracting and indexing sources. For example, Chemical Abstracts Services includes chemical patents or patents relating to chemical processes in its publications.

The best way to identify patents in the United States is to go to the U.S. Patent and Trademark Office where all patents are arranged by classification number. The Library of the Patent and Trademark Office receives three copies of each patent issued by countries that produce printed patents. These are arranged by the issuing country's classification scheme, number of the patent, and corresponding subject classification in the U.S. system. Science librarians in this country are, of course, most directly concerned with the U.S. patents.

SECONDARY SOURCES OF INFORMATION

Secondary sources of information include all organized information that has been compiled from original sources and arranged according to some definite plan. Secondary sources cover a wide range of categories of materials. In fact, if one wanted to be specific, everything that a library may have on its shelves could be included, because even some of the primary sources of information include information that is secondary.

Monographs

The monograph is the bridge between older collected encyclopedic material and newer periodical literature. At one time each monograph was unique and not related to other monographs. Today, however, there are many series of monographs that are related by a common subject title. Selecting research monographs for any library is a difficult task, due to the inflation rate and budgetary restrictions. Before librarians spend hundreds of dollars on specialized monographs, they must be certain of their content by reading review journals and publisher information and by contacting other libraries.

Textbooks

Textbooks are a special kind of monograph. They provide background information, since research monographs often bypass introductory material. Textbooks are meant to be read from cover to cover, chapter by chapter, since each new chapter depends on the material that was explained in the preceding chapter. They are not, however, expected to discuss recent developments. Because there are so many textbooks, librarians must be very selective and consult with researchers when determining which texts should be kept or purchased.

Treatises

Treatises are historical summaries. They usually provide complete and authoritative compilations and summaries of the subject in question. Treatises often suggest areas of future research. There is no sharp distinction between an advanced textbook and a treatise, but a treatise usually does not include introductory material that may be needed by the non-subject specialist. An important characteristic of the treatise is that it is well documented providing citations to the original articles.

Manuals and Guides

Manuals are sources that explain procedures or outline in detail how a piece of equipment operates. They can be general or detailed depending on their purpose. Guides, on the other hand, are intended to suggest where to go for further information. Current guides are important to any library. Unfortunately, in science and engineering, there are only a few up-to-date, excellent guides that cover the literature of individual disciplines. Many guides were excellent at the time of publication but are now dated. Another kind of guide that should be mentioned is the field guide, common in the biological sciences. These are usually guides to identification of plants, animals, rocks, and fossils. They are detailed and need not be updated as often as literature guides.

Encyclopedias

Encyclopedias provide general informative articles to a variety of subject topics. Most science and engineering encyclopedias follow a logical order — definition, relation to other topics, history, discussion and bibliography. Researchers use encyclopedias to acquaint themselves with unfamiliar general topics. Because the information in encyclopedias is usually outdated at the time of publication, many encyclopedias in science and engineering have supplements that are issued in serial format. Libraries may subscribe to the set and receive volumes as they are published. Encyclopedias are expensive reference sources in any library and great care should be taken to select the proper titles.

Dictionaries

Dictionaries usually supply the spelling, meaning, etymology, pronunciation, and usage of words. However, it is not unusual for pronunciation and etymology to be omitted. Language dictionaries, which may be bilingual or multilingual, rarely list word meanings. Encyclopedic dictionaries include short discussions, which become useful when terms have various meanings in different subject areas.

Handbooks

Handbooks are compact reference sources that contain data and descriptive information. They are logically arranged and provide charts, tables, graphs, glossaries, and detailed discussions. Most handbooks are compiled by subject specialists who know what kinds of information are needed the most. Scholarly handbooks that are used extensively are kept up to date with new editions. Engineering and the biomedical sciences have more handbooks than the other disciplines discussed in this book.

Bibliographies

A bibliography is a list of information sources pertinent to a particular subject or topic. Theoretically, all abstracts, reviews, and indexes are bibliographies. The card catalog of any library is a bibliography of the library's holdings. When background information is needed, the availability of a bibliography is important and can be time saving. Important features to note when using a bibliography are scope and periods of coverage. The date of publication does not indicate the date of the bibliography. Articles providing bibliographies are also important. If a periodical article contains a notable bibliography, the abstract or index entry will note the fact. The G. K. Hall Publishing Co. understands the importance of comprehensive bibliographies. It films card catalogs of specialized collections and sells them to the public in book format.

Indexes

An index is an alphabetically arranged table connecting keywords to page numbers. Without the index, there would be no systematic method of locating detailed information in printed sources. Cumulative indexes covering books and periodicals are focal points for literature searching and, consequently, must be complete to ensure adequate coverage. Most indexes are available in printed and machine-readable form. Some indexes are continuing publications that maintain access to current information. Other indexes provide access to individual books, closed serial sets, and special works. The card catalog, serves as an index by providing stack locations to library materials.

Abstracts

Abstracting publications are among the most important secondary sources. Because few libraries house every article and book published in a given field, abstracts are needed to provide summaries of journal articles. In many cases enough information is provided in the abstract to determine whether or not the entire article is needed. Because abstracting services offer international coverage, translating articles into English often causes abstracts of foreign language material to be late. In some cases this can be one or two years.

Directories

A directory is an organized listing of products, people, firms, services, etc. The kind of information listed depends on the kind of directory. Product directories list names of products, producers, and the necessary information needed to obtain the product. Directories of firms and organizations provide pertinent information about the firm, such as its purpose, owners, directors, address, telephone numbers, and net worth. Biographical directories may list living or deceased persons, and usually provide biographical sketches about each entry including birth and death dates, full name, academic degrees, publications, affiliations with asociations, and positions. Current directories are important library sources that answer a wealth of questions.

Atlases

Atlases are bound collections of maps, charts, plates, or tables illustrating any subject. There are two types of atlases. Geographical atlases are collections of maps or charts depicting various kinds of information. Star atlases are included in this category; they locate stars and planets at certain times of the year and in the different sky quadrants. Biomedical atlases are collections of charts and plates that locate body position in three dimensions. For example, a brain atlas depicts cross sections of the brain. There are atlases of the eye, heart, etc., which are important aids in the study of the body.

Union Lists

Union lists are location guides to books and periodicals around the country. They may be of a specialized nature, listing only the locations of books and periodicals in a special subject, or they may be much broader, listing all types of material in science and engineering. Each entry in a union list is usually entered in full bibliographic form followed by a list of libraries that own the item. Union lists are vital to interlibrary loan librarians. With the increasing number of cooperative programs across the country, there are more and more union lists. While some union lists are small and only locally useful, others are large, such as the *National Union Catalog*, produced by the Library of Congress, and the *Union List of Serials*, produced by cooperative institutions. OCLC also serves as a union catalog, as do other bibliographic utilities.

One could discuss numerous other secondary sources of information, but those that have been mentioned are the more important ones. In the following chapters, the various disciplines are divided according to the type of reference source. Note that some subject disciplines do not have examples of every type. Some sources could be placed under more than one type but in this book each source has been placed under only one type; the index provides further access to particular items.

3

Multidiscipline Sources of Information

The citations in this chapter are to works that are multidisciplinary. These works may cover just science and engineering, or they may be even more broad and include the humanities and social sciences. As a result, this chapter includes more citations within more divisions than most of the other chapters. Many of the titles listed here are key titles to several disciplines, and could easily have been listed in separate chapters. It was decided to list them here as an aid to those wanting to cover more than one subject at a time. The various divisions are Guides to the Literature and Librarianship, Abstracts and Indexes, Bibliographies, Encyclopedias and Treatises, Dictionaries, Handbooks, Serials, Directories, Biographical Directories, Catalogs, Theses and Dissertations, Meetings, Translations, Copyright and Patents, and Government Documents and Technical Reports.

GUIDES TO THE LITERATURE AND LIBRARIANSHIP

3-1 **A Brief Guide to Sources of Scientific and Technical Information.** By Saul Herner. Washington: Information Resources Press, 1970. 102p. illus. index.

A readable guide to major sources of information, particularly relating to directories and source guidance, ongoing research and development, current or recent research and development results, major American research collections, organization of personal index files, and relationship of the scientist and engineer to his or her information tools and mechanisms.

3-2 **Guide to Reference Books.** Comp. by Eugene P. Sheehy. 9th ed. Chicago: American Library Association, 1976. 1015p.

The purpose of this work is to list reference books that are basic to both general and special research. It is intended to serve as a reference manual for the library user, a selection aid for the librarian, and a textbook for the student. Some 10,000 titles in all areas and many languages are included. For all practical purposes the cutoff date is 1973. The portion that concerns science and technology is part E, Pure and Applied Sciences. Various disciplines constitute sub-parts and are assigned double-letter designations (e.g., Chemistry is ED, Earth Sciences is EE). Entries are arranged within each sub-part under various categories (e.g., guides and manuals, bibliographies, serial publications). Within categories, entries are arranged alphabetically by main entry and are numbered consecutively throughout a sub-part.

3-3 **Guide to Reference Material.** v. 1- . Ed. by A. J. Walford. 4th ed. London: Library Association; distr. New York: R. R. Bowker, 1980- .

Only volume 1, Science and Technology, is applicable for science reference work. This new edition continues the tradition of previous editions by covering UDC classes 5/6 with main entries for over 4,000 items, plus subsumed entries for roughly a further 700. Most annotations are primarily descriptive and not critical; they refer infrequently to similar works, and occasionally they contain references to a selected number of published reviews.

3-4   **Guide to the Research Collections of the New York Public Library.** Comp. by Sam P. Williams. Chicago: American Library Association, 1975. 336p. index.

This important guide is arranged in four major categories: general materials, humanities, social sciences, and pure and applied sciences. These categories are divided into subject classes and subclasses and include a statement of scope and collection policy and a brief history of the development of the library's holdings followed by a detailed description of the collection. The data are accurate as of December 1969 with supplements appearing in the *Bulletin* of the NYPL.

3-5   **Information Sources in Science and Technology.** By C. C. Parker and R. V. Turley. Boston: Butterworths, 1975. 223p. illus. index.

Mainly written for those using materials published in Britain. The main part of the book covers people, organizations, the literature, and information services and libraries.

3-6   **Introduction to Librarianship.** By Jean Key Gates. 2nd ed. New York: McGraw-Hill, 1976. 288p. bibliog. index. (McGraw-Hill Series in Library Education).

A good supplementary text for undergraduate courses in library science. Part 1 is called the "Story of Libraries" and gives the historical development of libraries. Part 2, "Librarianship as a Profession," covers professional organizations and library education. Part 3, "Kinds of Libraries and Library Services," covers such topics as state responsibility for library service, the municipal public library, academic libraries, research libraries, etc.

3-7   **Introduction to Library Science: Basic Elements of Library Service.** By Jesse H. Shera. Littleton, CO: Libraries Unlimited, 1976. 208p. illus. bibliog. index. (Library Science Text Series).

A good text that provides a point of entry for those for whom librarianship, as an anticipated career, is only beginning. Each of the eight chapters covers a specific area or problem of librarianship. There are notes and selected readings for each chapter.

3-8   **Science and Technology: An Introduction to the Literature.** By Denis Grogan. 3rd ed. rev. London: Clive Bingley; Hamden, CT: Linnet Books, 1976. 343p. index.

An effective text for library schools in Great Britain that covers the various literatures of the sciences and technology. Most of the information is up to date as of 1973.

3-9   **Scientific and Technical Libraries: Their Organization and Administration.** By Lucille J. Strauss, Irene M. Shreve, and Alberta L. Brown. 2nd ed. New York: Becker and Hayes, 1972. 450p. illus. index. (A Wiley-Becker-Hayes Publication).

This manual of science library practice for library school students, beginning special librarians, managers, and library consultants discusses scientific and technical libraries, budgets, staffs, buildings, materials, services, processes, and acquisitions.

3-10   **Standards and Specifications; Information Sources: A Guide to Literature and to Public and Private Agencies Concerned with Technological Uniformities.** By Erasmus J. Struglia. Detroit, MI: Gale Research Co., 1965. 187p. (Management Information Guide, 6).

Lists sources of information for all types of standards and specifications. It is divided by types: general sources and directories, bibliographies and indexes to periodicals, catalogs and indexes of standards and specifications, government sources, associations and societies, international standardization, and periodicals. Most book and periodical entries are annotated.

3-11   **University Science and Engineering Libraries.** By Ellis Mount. Westport, CT: Greenwood Press, 1975. 214p. index. (Contributions in Librarianship and Information Science, no. 15).

A good text that provides a general overview of major issues faced by most academic libraries that have departmental units or special collections.

ABSTRACTS AND INDEXES

3-12   **Bulletin Signaletique.** v. 1- . Paris: Centre National de la Recherche Scientifique, 1948- . monthly; some sections are quarterly. Title varies; formerly called *Bulletin Analytique.*

An excellent abstracting service covering all fields of science and technology. It is divided into many separate sections, each of which covers a specific subject and is available separately. All titles of articles not in French are translated into French with a notation of the original language of the title. Each entry gives a complete citation and descriptive abstract. There are author indexes to each issue. Although the periodicals abstracted overlap those of other abstract-indexes, the *Bulletin* is in some cases more current. The following numbered sections are pertinent to the sciences:

101.   Sciences de l'Informati, documentation.
110.   Informascience.
120.   Astronomie. Physique spatiale. Geophysique.
130.   Physique. Mathematique.
140.   Electrotechnique.
145.   Electronique.
160.   Physique de l'etat condense.
161.   Cristallographie.
165.   Atomes et molecules. Physique des fluides et plasmas.
170.   Chimie.
Bibliographie des sciences de la terre.
    220.   Mineralogie, geochimie, geologie extraterrestre.
    221.   Gisemonts metalliques et non metalliques.
    222.   Roches cristallines.
    223.   Roches sedimentaires, geologie marine.
    224.   Stratigraphie, geologie regionale et general.
    225.   Tectonique.
    226.   Hydrologie, geologie de l'ingenieur, formationes.
    227.   Paleontologie.

310. Genie biomedical. Informatique biomedicale.
330. Sciences pharmacologiques. Toxicologie.
340. Microbiologie. Virologie – Immunologie.
346. Ophtalmologie.
347. Oto-rhino-laryngologie, stomatologie, pathologie cerricofaciale.
348. Dermatologie – Venereologie.
349. Anesthesie. Reanimation.
351. Revue bibliographique cancer.
352. Maladies de l'appareil respiratoire du coeur et des vaisseaux – Chirurgie thoracique et vasculaire.
354. Maladies de l'appareil digestif chirurgie abdominale.
355. Maladies des reins et des voies urinaires – Chirurgie de l'appareil urinaire.
356. Maladies du systeme nerveux myopathies-neurochirurgie.
357. Maladies des os et des articulations – Chirurgie orthopedique – Traumatologie.
359. Maladies du sang.
360. Biologie animale. Physiologie et pathologie des protozoaires et des invertebres. Ecologie.
361. Endocrinologie et reproduction.
362. Diabete. Maladie metaboliques.
363. Genetique.
365. Physiologie des vertebres.
370. Biologie et physiologie vegetables.
380. Agronomie-Zootechnie-Phytopathologie-Industries alimentaires.
390. Psychologie-Psychopathologie-Psychiatrie.
522. Histoire de sciences et des techniques.
730. Combustibles-Energie.
740. Metaux-Metallurgie.
761. Microscopie electronique-Diffraction electronique.
780. Polymeres. Peintures. Bois. Cuirs.
880. Genie chimique-Industries chimique et parachimique.
885. Nuisances.
890. Industries mecaniques-Batiment-Travaux public-Transports.

3-13 **Current Contents.** Philadelphia, PA: Institute for Scientific Information, 1961- . weekly.
Not an index but a listing of the contents of the current issues of many leading journals in the sciences. Available in five sections: physical and chemical sciences; engineering, technology and applied sciences; social and behavioral sciences; life sciences; and agricultural, biology, and environmental sciences.

3-14 **Index to Book Reviews in the Sciences.** v. 1- . Philadelphia, PA: Institute for Scientific Information, 1980- . monthly with semiannual cumulations.
This publication indexes over 35,000 recently published scientific book reviews each year gathered from some 3,000 leading science and technology journals. The reviews are arranged by author or editor of the book reviewed. Each entry contains complete bibliographic information of the book review, as well as the book reviewed. The bibliographic information includes primary author or editor; full book title and subtitle; and publisher information. Each entry also includes a

complete citation for the source in which the book review appeared, the review language, and reviewer's name. A *Permuterm Subject Index to Book Titles* provides keyword subject access to the book titles.

3-15 **Index to Scientific and Technical Proceedings.** v. 1- . Philadelphia, PA: Institute for Scientific Information, 1978- . monthly with semiannual cumulations.

This is the only publication that indexes conference literature in science and technology to the article level. Nearly 90,000 papers, published in 3,000 volumes of proceedings are indexed annually. Material in the ISTP is indexed in six different ways: author's name, conference sponsor, subject category, meeting location, title words, and author's organization. This allows the user a variety of ways to find complete descriptions of proceedings contents and individual papers.

3-16 **Index to Scientific Reviews.** v. 1- . Philadelphia, PA: Institute for Scientific Information, 1974- . semiannually with annual cumulation.

"An international interdisciplinary index to the review literature of science, medicine, agriculture, technology, and the behavioral sciences." Includes all articles with keywords such as Advances, Review, Progress, etc. in their titles; all articles with 40 or more references; and all articles coded "R" (review) in the *Science Citation Index* (3-18) data base. ISR's arrangement and use parallel the *Science Citation Index*; citation and keyword searching are possible by using ISR's Citation and Permuterm Subject Indexes.

3-17 **Referativnyi Zhurnal.** Moscow: Akademiia Nauk SSSR, 1953- . monthly.

This is probably the world's most comprehensive current abstracting service covering all areas of science/technology. The *Zhurnal* is printed in Russian, but the abstract entries are in the language of the article. It appears on a monthly basis in over 65 sections, each covering a separate subject. It has international coverage of periodical articles, books, and bibliographies. Dissertations and patents are included in some sections. About 25% of the abstracted material is from literature published behind the Iron Curtain. The abstracts are informative, and in some cases they provide direct quotes from the abstracting services of other countries. There are author indexes for all sections in both Cyrillic and Latin alphabets, but subject indexes for all sections have not been issued. The indexes are generally late. Abstracts appear approximately six months after the publication of the abstracted work. Following are the *Zhurnal* sections that pertain to the sciences:

Astronomiya (Astronomy).
Aviatsionnye i raketnye dvigateli (Aircraft & rocket engines).
Avtomatika, telemekhanita i vychislitel'naya (Automation . . .).
Avtomobil'nye dorogi (Motor roads).
Avtomobil'nyi i gorodskoi transport (Motor & municipal transport).
Biofizika (Biophysics).
Biologiya (Biology).
Dvigateli vnutrennogo sgoraniya (Internal combustion engine).
Ekonomika promyshlennosti (Industrial economics).
Elektronika i ee primenenie (Electronic engineering).
Elekrosvyaz (Electric communication).
Elektroteknika i energetika (Electrical & power engineering).

Farmakologiya . . . (Pharmacology).
Fizika (Physics).
Fotokinoteknika (Photography & cinematography).
Geodeziya i aeros'emka (Geodesy & aerial surveying).
Geofizika (Geophysics).
Geografiya (Geography).
Geologiya (Geology).
Gornoe delo (Mining).
Gornoe i neftlepromyslovoe machinostroenie (Mining & oil industry
    machines).
Informatika (Informatics).
Issledovanie kosmicheskogo prostranstva (Space research).
Khimicheskoe, neftepererabatyvayushchee i polimernoe machinostroenie
    (Chemical, oil-refining & polymer machinery).
Kimiya (Chemistry).
Kibernetika (Cybernetics).
Kommulan'noe, bytovoe i torgovoe oborudovanie (Municipal, household
    and trading equipment).
Koroziya i zashchita ot korrozii (Corrosion . . .).
Kotlostroenie (Boiler-making).
Legkaya promyshlennost (Textile industry).
Lesovedenie i lesovodstvo (Forestry).
Mashinostroitel'nye materialy (Engineering materials . . .).
Matematika (Mathematics).
Meditsinskaya geografiya (Medical geography).
Mekhanika (Mechanics).
Metallurgiya (Metallurgy).
Metrologiya i izmeritel'naya tekhnika (Metrology & measuring instruments).
Nasostroenie . . . (Pumps).
Oborudovanie pishchevoi promyshlennosti (Food industry machinery).
Obshchie voprosy patologii (Pathology . . .).
Okhrana prirody i vosproizvodstvo prirodnykh (Natural resources).
Onkologiya (Oncology).
Organizatsiya upravleniya promyshlennost'yu (Industrial management &
    organization).
Pochvovedenie i agrokhimiya (Soil science & agricultural chemistry).
Pozharnaya okhrana (Fire protection).
Promyshlennyi transport (Industrial transport).
Radiatsionnaya biologiya (Radiation biology).
Radiotekhnika (Radio engineering).
Raketostroenie (Rocket engineering).
Rastenievodstvo (Plant breeding).
Stroitel'nye i dorozhyne machiny (Building & road machines).
Svarka (Welding).
Tekhnologiya i oborudovanie tsellyuloznobumazhnogo i poligraficheskogo
    proizvodstva (Technology & machinery of paper-making & printing).
Tekhnologiya machinostroeniya (Mechanical engineering).
Teploenergetika (Thermal power engineering).
Traktory . . . (Tractors).
Truboprovodnyi transport (Pipelines).
Turbostroenie (Turbine engineering).
Vodnyi transport (Water transport).

Voprosy teckhnicheskogo progressa (Problems of technical progress).

Vozdushnyi transport (Air transport).

Vzaimodeistvie raznykh vidov transport i konteinernye perevozki (Coordination of transport containers).

Yadernye reaktory (Nuclear reactors).

Zheleznodorozhnyi transport (Rail transport).

Zhivotnovodstvo i veterinariya (Animal husbandry & veterinary science).

3-18   **Science Citation Index.** v. 1- . Philadelphia, PA: Institute for Scientific Information, 1961- . quarterly with annual cumulations.

This tool is based on the premise that because authors cite the works of other authors, the cited and original works are related in subject. The *Index* is divided into three major sections: *Citation Index, Source Index*, and *Permuterm Subject Index.* The *Citation Index* is an alphabetical listing by first author of all the cited or referenced works in a given year. Each entry is followed by a list of current authors and their works who have cited the original paper. Thus, all authors and their works cited in footnotes and bibliographies are listed in the *Citation Index* and are followed by the author and source of the article in which the footnote or bibliography appeared. About 85% of the citations are from periodicals. The remainder are from patents, government reports, books, dissertations, contracts, and personal communications. The *Source Index* is a bibliographic list of all the works containing the footnotes or bibliographies cited in the *Citation Index.* It is arranged alphabetically by author and lists co-author, full title, publication source, volumes, issue, pages, year, type of item, and the number of reference in the bibliography. Corporate and anonymous source listings are also provided. The *Permuterm Subject Index* is an alphabetical keyword index that refers to the works listed in the *Source Index.* One primary keyword from the document title is paired with a co-term and followed by the author; the author and respective work must be located in the *Source Index.* The *Science Citation Index* is international in scope and covers all scientific disciplines. Five year cumulations are available.

BIBLIOGRAPHIES

3-19   **The AAAS Science Book List: A Selected and Annotated List of Science and Mathematics Books for Secondary School Students, College Undergraduates, and Nonspecialists.** By Hilary J. Deason. Washington: American Association for the Advancement of Science, 1970. 439p.

This edition contains 2,441 titles. All bibliographical descriptions are complete, including prices and LC numbers; annotations are descriptive and frequently offer evaluative comments. The material is arranged under broad DDC subject categories. Although dated, it is still a good basic list and should be checked for newer editions.

3-20   **ASLIB Book List: A Monthly List of Recommended Scientific and Technical Books, with Annotations.** London: ASLIB, 1935- . monthly.

Arranged by UDC categories with a broad subject index, this monthly list contains critical notes on the books.

3-21   **American Book Publishing Record.** New York: R. R. Bowker, 1960- . monthly.

Best known as *BPR*, this is a monthly cumulation of the *Weekly Record*, formerly contained in *Publishers Weekly* and now (since 1974) issued as a separate journal. Whereas the latter is arranged alphabetically by main entry, *BPR* presents the entries rearranged by subject according to DDC. It does not include governmental publications, subscription books, second or subsequent printings, dissertations, periodicals, pamphlets of fewer than 49 pages, or specialized advertising or transitory publications. Price is included where possible. Adult fiction, juvenile fiction, paperback books and "below-the-line" entries (i.e., reprints, foreign works that lack an exclusive U.S. distributor, etc.) are arranged by main entry and presented in separate listings. Author and title indexes conclude each issue.

3-22　**American Library Resources: A Bibliographical Guide.** By Robert B. Downs. Chicago: American Library Association, 1951. 428p.

3-22a　**Supplement** 1950-1961. 1962. 226p.

3-22b　**Supplement** 1961-1970. 1972. 256p.
Contains bibliographies of bibliographies listing resources in American libraries. The term "bibliography" in this context covers union lists, surveys, checklists, and catalogs both of particular libraries and of special collections within them. Since the basic arrangement is by DDC, science librarians should check those sections of the basic volume and of the supplements that contain the 500 and 600 classes. Fully indexed by subject and author.

3-23　**American Reference Books Annual.** Ed. by Bohdan S. Wynar. Littleton, CO: Libraries Unlimited, 1970- . annual.
The purpose of *ARBA* is to provide reviews of all reference books published in the United States (and foreign imprints with an exclusive U.S. distributor) from January through December of each year. Coverage is as complete as possible. A total of 18,813 titles have been reviewed in the first 11 volumes. Each annual volume reviews new books published during the past calendar year and does not cumulate previous titles. All subjects are covered at every level with a selection of reference books published by government agencies. Reviews, which vary from 75 to 300 words, are written by over 375 specialists and the staff of *ARBA*. To aid those searching for complete information and various critical opinions about a particular title, citations to reviews in major journals are appended to each *ARBA* review. In the eleventh volume of *ARBA* (1980), 305 out of a total of 1,635 reviews were of science reference books, distributed over the entire range of science. It is safe to assert that no other source provides an equally current and comprehensive account of science reference book publishing on an annual basis. Essential for the science librarian working in reference, book selection, and acquisitions.

3-24　**Annotated Bibliography of Automation in Libraries and Information Systems, 1972-1975.** Comp. by Maxine MacCafferty. London: ASLIB, 1976. 147p. index.
This compilation of 813 items updates Tinker's *Annotated Bibliography of Library Automation 1968-1972* (London: ASLIB, 1973. 85p.). Includes topics on information processing and management, housekeeping routines, bibliographic control, and retrieval and search.

3-25　**A Basic Collection for Scientific and Technical Libraries.** By Effie B. Lunsford and Theodore J. Kopkin. New York: Special Libraries Association, 1971. 274p. index.

An annotated guide to 2,409 books and periodicals prepared for community colleges, technical institutes, and vocational schools developing new or enlarging existing collections. Even though a number of the titles listed are out of date and some are irrelevant, this is on the whole a well-balanced list. Those using the list should be aware that a great number of the entries will have newer editions.

3-26 **Bibliographic Index: A Cumulative Bibliography of Bibliographies.** New York: H. W. Wilson, 1938- . biannual with annual cumulations.
A classified list of both separately published and concealed bibliographies of which many are pertinent to the sciences.

3-27 **Bibliographische Berichte, im Auftrag des deutschen bibliographischen Kuratoriums.** Bearb. von Erich Zimmerman. Frankfurt am Main: Kloster-mann, 1959- . quarterly.
Recent bibliographies presented in classified order. Includes concealed bibliographies as well as those that are separately published, unlike Besterman, which is limited to the latter.

3-28 **A Bibliography of Scientific, Technical and Specialized Dictionaries, Polyglot, Bilingual, Unilingual.** By C. W. Rechenbach and E. R. Garnett. Washington: Catholic University of America Press, 1969. 158p.
This guide helps to identify specialized foreign language science dictionaries. Since it is somewhat dated, the entries need to be checked for newer editions.

3-29 **Books in Print: An Author-Title Index to the Publishers' Trade List Annual.** New York: R. R. Bowker, 1948- . annual.
This annual index lists in-print and forthcoming books. The 1978-1979 edition listed some 496,000 titles from 6,900 publishers. The author index volume lists for each entry: author (editor, translator, compiler), title, publisher, price, series, whether illustrated, edition statement, year of publication and collation, if supplied by publisher. Title index volume lists for each entry: title, author, price, publisher. A publishers' directory, listing addresses for every known U.S. publisher whether or not it is listed in PTLA, is also included. *BIP* and its companion, *Subject Guide to Books in Print* (3-40), are the basic tools for determining availability and price of a book. *Forthcoming Books* (New York: R. R. Bowker, 1966- . bimonthly.), a supplement to *BIP*, provides a bimonthly cumulative index to new announcements and to books published after the summer closing date for the current *BIP*. *BIP* is updated by the *Books in Print Supplement* (New York: R. R. Bowker, 1972/73- . annual.). Listings not found in the main volume of *BIP* are supposed to be recorded here, as well as new books published between July and September.

3-30 **British Library Resources: A Bibliographical Guide.** By Robert B. Downs. London: Mansell Information; distr. Chicago: American Library Association, 1973. 332p. index.
Contains bibliographies of bibliographies listing resources in British libraries. Includes union lists, surveys, checklists, and catalogs both of particular libraries and of special collections within them. The basic arrangement is by DDC.

3-31 **Cumulative Book Index.** New York: H. W. Wilson, 1898- . monthly (except August) with annual bound cumulations.

Since 1929, it has strived to include all English language imprints of the year of coverage, with the exception of several categories (i.e., government publications, maps, sheet music, miniature editions, pamphlets, tracts, and miscellaneous local and ephemeral publications). *CBI*'s subject analysis (based on Sears) is adequate to identify English language imprints on particular scientific subjects. In addition, it serves as an indispensable tool for verification of titles inasmuch as it provides full data needed for complete bibliographic descriptions.

3-32 **Dictionaries of English and Foreign Languages: A Bibliographical Guide to Both General and Technical Dictionaries with Historical and Explanatory Notes and References.** By Robert L. Collison. 2nd ed. New York: Hafner, 1971. 303p. bibliog. index.
A good guide to check for the standard dictionaries. Need to check entries for newer editions.

3-33 **Directory of Directories.** Detroit, MI: Gale Research Co., 1979. ca900p.
Listings are under nearly 900 detailed subject headings with numerous cross references; includes those directories produced by industries and professional societies. It is similar in format to the *Encyclopedia of Associations* (3-110) and lists accurate titles and subtitles, full publisher addresses, description of coverage, analysis of information in a typical entry, arrangement, type of indexes, number of pages, frequency, editor's name, former title, price, and whether data base is machine-readable. It updates and revises the entries that are found in the interedition supplements, *Directory Information Service*, and adds detailed entries on directories not covered in the supplements.

3-34 **Guide to Microforms in Print: Author, Title: Incorporating International Microforms in Print.** Westport, CT: Microforms Review, 1961- . annual.
This is a cumulative annual listing of books, journals, newspapers, government publications, archival material and collections, but not of theses or dissertations. Each entry gives the following information: author, title, volumes, date, price, publisher code, and type of microform code. For subject access to this listing there is a *Subject Guide to Microforms in Print.*

3-35 **Guide to Reprints.** Kent, CT: Guide to Reprints, 1967- . annual.
An annual list of books, journals, and other materials available in reprint (full size, hardbound) from publishers in the United States and some foreign countries. The 1978 edition covered about 400 publishers. There is a directory of publishers represented in the guide. This annual is supplemented by *Announced Reprints* in which prices are frequently not included and many "announced" titles fail to materialize.

3-36 **Handbooks and Tables in Science and Technology.** Ed. by Russell H. Powell. Phoenix, AZ: Oryx Press, 1979. 184p.
This is a bibliography of well over 2,000 handbooks and tables in the science and technology field. They are listed alphabetically by title with separate sections for medical handbooks and National Bureau of Standards publications. Subject and author indexes are included.

3-37 **Scientific and Technical Books and Serials in Print, 1978.** New York: R. R. Bowker, 1978. 2343p.

An outgrowth of *Books in Print* (3-29) and *Subject Guide to Books in Print* (3-40), this lists titles available from some 1,990 publishers, providing access by subject, author, and title. LC subject headings are used. Entries include the following bibliographic information: author, co-author, or editor, if any; title; price; imprint; and year of publication. Also included is a list of 15,500 serial titles under 270 subject headings.

3-38 **Scientific, Engineering, and Medical Societies Publications in Print 1978-1979.** By James M. Kyed and James M. Matarazzo. New York: R. R. Bowker, 1979. 571p. index.

This book is intended to provide some bibliographic control and sales information for material not normally covered in standard publication-finding tools. The 365 organizations that submitted publications lists (513 were contacted) are listed in alphabetical order. Under the name of each society, publications are listed in one of three categories—books, periodicals, and nonprint materials. Book titles are listed alphabetically under subject headings supplied by the society (an exception is monographic series, which are listed in numerical order). Where applicable, author, edition, publication date, order number, and price are listed. Periodicals and nonprint materials are also listed alphabetically by title. Author, subject and periodical indexes complete the volume.

3-39 **Scientific, Medical, and Technical Books Published in the United States of America: A Selected List of Titles in Print with Annotations.** By Reginald Robert Hawkins and prepared under the direction of the National Academy of Sciences—National Research Council's Committee on Bibliography of American Scientific and Technical Books. 2nd ed. (books published to December 1956). Washington, 1958; distr. New York: R. R. Bowker, 1958. 1491p.

Despite its age, this list of 8,000 books published through 1956 and in print in that year is still serviceable for selection of basic titles.

3-40 **Subject Guide to Books in Print.** New York: R. R. Bowker, 1957- . annual.

The *Subject Guide,* based on LC subject headings, is as complete as the conditions and circumstances of compilation of the data base permit. It provides subject access to the 496,000 titles in *Books in Print* (3-29).

3-41 **A World Bibliography of Bibliographies and of Bibliographical Catalogues, Calendars, Abstracts, Digests, Indexes and the Like.** By Theodore Besterman. 4th and final ed. rev. and greatly enlarged. Lausanne: Societas Bibliographica, 1965-1966. 5v.

The single most important bibliography of separately published bibliographies in all fields. Arranged by subjects, including those in science and technology. The first four volumes consist of the main part of the bibliography with volume 5 being the index. Includes only those bibliographies that appear as separates, excluding those that are concealed within monographs or serials. The reprint publisher, Rowman and Littlefield, has issued many of the Besterman subject sections as separates.

ENCYCLOPEDIAS AND TREATISES

3-42 **Abstracting Scientific and Technical Literature: An Introductory Guide and Text for Scientists, Abstractors, and Management.** By Robert Maizell, Julian F. Smith, and T. E. R. Singer. New York: Wiley-Interscience, 1971. 297p. bibliog. glossary. index.

This book teaches the techniques for achieving clarity and brevity in the art of abstracting. Flow charts supplement the sections on organizing and administering an abstracting service as a small business enterprise with or without computer assistance. The book serves well to introduce prospective users to the composition and standards of *Chemical Abstracts* (8-10), *Biological Abstracts* (9-4), *Metals Abstracts* (12-108), the *Official Gazette* (3-164) of the U.S. Patent Office, and NTIS services. Illustrations consist of reprints of guidelines and actual abstracts.

3-43 **Encyclopedia of Library and Information Science.** v. 1- . Ed. by Allen Kent and Harold Lancour. New York: Marcel Dekker, 1968- .

Ten volumes have been published to date. The purpose of this project is clearly stated in the preface to the first volume: "The emphasis has been, throughout, on depth of treatment. While encouraged to express their evaluative opinions as well, wherever possible, to suggest and indicate future trends as they saw them . . . the editors are . . . committed to a 'one world' concept of their science. To this end the approach has been strongly international. . . . A more accurate description of the basic editorial policy would be that this work is not so much inter-national as it is non-national, although, admittedly this has not been easy to accomplish."

3-44 **Information Retrieval, British and American, 1876-1976.** By John Metcalfe. Metuchen, NJ: Scarecrow Press, 1976. 243p. index.

A valuable historical treatment of information retrieval arranged in chronological order. Actually begins with 3000 B.C.

3-45 **Information Storage and Retrieval Systems for Individual Researchers.** By Gerald Jahoda. New York: Wiley, 1970. 135p. index. (Information Sciences Series).

This book is addressed primarily to the librarian and the researcher interested either in improving the index to a documents collection or in starting one. It covers all important aspects of indexing, including controlled index vocabularies, conventional indexes, coordinate indexes, keyword-form-title indexes, citation indexes, as well as other types. There is also a separate chapter on the planning, design, and evaluation of personal indexes.

3-46 **Information Systems.** By B. C. Vickery. Hamden, CT: Archon Books, 1973. 350p. illus. charts. graphs. index.

Consists of 12 chapters discussing basic channels of information exchange, components of information systems, reference retrieval, conceptual mathematical models, retrieval language models, and the design of the system.

3-47 **Library Automation Systems.** By Stephen R. Salmon. New York: Marcel Dekker, 1975. 291p. (Books in Library and Information Science, v. 15).

A good summation of the state-of-the-art on the automation of technical services and circulation systems. Computerized information services have been omitted as have all non-operational or theoretical systems.

3-48   **McGraw-Hill Encyclopedia of Science and Technology.** 4th ed. New
York: McGraw-Hill, 1977. 15v. illus. (part col.). index.
A one-of-a-kind encyclopedia, for high school and undergraduate students with
general coverage of all areas of science and technology. It is also useful for
advanced researchers who may need some information on a topic not in their
field. This edition contains some 7,800 articles written by more than 2,500 scientists and engineers. All major articles are signed. Most longer articles begin with a
definition of the subject, followed by a detailed discussion of certain aspects of a
given topic, with a brief bibliography appended. There are many helpful cross
references in the text, but, unfortunately, pronunciation of scientific terms is not
indicated. This encyclopedia is an authoritative and up-to-date work for libraries
of all types. Supplemented by *McGraw-Hill Yearbook of Science and Technology*
(3-49).

3-49   **McGraw-Hill Yearbook of Science and Technology.** Comp. by the staff
of McGraw-Hill Encyclopedia of Science and Technology. New York:
McGraw-Hill, 1972- . annual.
One of the best and most comprehensive encyclopedia yearbooks, with
authoritative, signed articles and excellent format. In addition to covering the key
advances, discoveries, and developments in science and technology during the
year, it contains major review articles on topics of broad interest and growing
significance.

3-50   **On Documentation of Scientific Literature.** By Thomas P. Loosjes. 2nd
ed. London: Butterworths; distr. Hamden, CT: Archon Books, 1973.
187p. illus. index.
Concentrates on theoretical problems of bibliographic control and information
retrieval, discussing bibliographic control of articles in periodicals, serials,
monographs; general design of retrieval systems; etc. It also contains discussions
of bibliographic aids, the use thereof and the problems of machine translation.

3-51   **Technical Reporting.** By Joseph N. Ulman and Jay R. Gould. 3rd ed.
New York: Holt, Rinehart and Winston, 1972. 419p. illus. index.
An outstanding text, which provides help in writing formal and informal reports,
laboratory reports, theses, instructions, proposals, technical papers, and articles.
Includes a review of style requirements, grammar, punctuation, and mechanics,
as well as instructions on tables and other visual presentations.

3-52   **Understanding Scientific Literatures: A Bibliometric Approach.** By
Joseph C. Donohue. Cambridge, MA: MIT Press, 1974. 101p. illus.
graphs. bibliog. index.
This book is important to science libraries in that it describes bibliometric techniques and their application to definite subject literature, including Bradford's
law of scattering, Zipf and Booth's relationships on word frequency,
bibliographic coupling, citation tracing, and epidemic theory.

3-53   **Van Nostrand's Scientific Encyclopedia.** 5th ed. Princeton, NJ: Van
Nostrand Reinhold, 1976. 2370p. illus. (part col.).
One of the best one-volume encyclopedias of science and technology, it is
authoritative and well edited. Definitions are given first in simple terms, followed
by more detailed and technical explanations. Over 2,450 line drawings and small
diagrams illustrate the text. Science in this work includes earth and space

sciences, life sciences, mathematics and information sciences, energy technology, materials science, and physics and chemistry. In the area of energy, terms associated with the topical subjects of chemical fuels, fossil fuels, geothermal energy, hydropower, nuclear energy, solar energy, and tidal and wind energy are covered. A good source for references to named laws and effects. Within definitions, words that are defined elsewhere in the book appear in boldface. There are frequent cross references, but no biographies.

DICTIONARIES

3-54 **Abbreviations Dictionary.** By Ralph De Sola. New International 5th ed. New York: Elsevier North Holland, 1978. 654p.

This compact volume of over 160,000 entries gives in one alphabet abbreviations, acronyms, anonyms, eponyms, appellations, contractions, geographical equivalents, historical and mythological characters, initials, nicknames, short forms, slang shortcuts, signs, and symbols, with corresponding full identification.

3-55 **Acronyms, Initialisms, and Abbreviations Dictionary: A Guide to Alphabetic Designations, Contractions, Acronyms, Initialisms, Abbreviations and Similar Condensed Apellations.** Ed. by Ellen T. Crowley. 6th ed. Detroit, MI: Gale Research Co., 1978. 1103p.

3-55a **New Acronyms, Initialisms, and Abbreviations.** Detroit, MI: Gale Research Co., 1979- .

This is the most complete book of its type with over 178,000 entries. It is alphabetical by acronym, initialism, or abbreviation with corresponding full name. The latest usage has been checked and obsolete terms have been identified with cross references when necessary. The main volume is published every three years with *New Acronyms, Initialisms, and Abbreviations* published in the intervening years. These two volumes are considered volumes 1 and 2 of a three-volume set with *Reverse Acronyms, Initialisms, and Abbreviations Dictionary* (3-69).

3-56 **Bioscientific Terminology: Words from Latin and Greek Stems.** By Donald M. Ayers. Tucson, AZ: University of Arizona, 1972. 425p.

Divided into two major sections, the first of which deals with bioscientific words derived from Greek and the second with those from Latin. The book is organized so that a student can proceed through a series of lessons introducing the classical languages and their use in scientific terminology. It is not a dictionary of Greek and Latin prefixes and suffixes but rather an organized presentation of the use of these languages in forming scientific terms. Hundreds of examples are given, with definitions. There are interesting sections also on such topics as scientific terms from classical mythology and how the meanings of words were derived from what are now known to be erroneous scientific theories.

3-57 **Chambers Dictionary of Science and Technology.** Edinburgh: W & R Chambers, 1975; distr. Totowa, NJ: Littlefield, Adams, 1976. 2v. illus. (Chambers Paperback Reference Books).

A completely revised and expanded edition of a classic work. The terms are arranged alphabetically in letter-by-letter order with compound or modified words grouped under the first component, although a few biological terms have been inverted. Each definition is labeled to indicate its field. If a term is used in more than one field, the definitions are arranged alphabetically by the label. Trade names, when used, are also noted.

3-58 **Dictionary of Physical Sciences.** Ed. by John Daintith. New York: Pica Press; distr. New York: Universe Books, 1977. 333p. illus.
A dictionary of the more commonly used scientific terms in the main branches of physics, chemistry and astronomy, as well as relevant terms from the related fields of electronics, computer science, and mathematics.

3-59 **Dictionary of Science and Technology, English-French.** Comp. by A. F. Dorian. Amsterdam, NY: Elsevier Scientific Publishing, 1979. 1486p.
A very comprehensive English-French dictionary covering some 150,000 terms from acoustics to zoology. A short definition is added in many cases to prevent misinterpretation.

3-60 **Dictionary of Science and Technology: English-German.** 2nd ed. rev. Amsterdam, NY: Elsevier Scientific Publishing, 1978. 1401p.

3-60a **Dictionary of Science and Technology: German-English. Handworterbuch der Naturwissenschaft und Technik: Deutsch-Englisch.** Amsterdam, NY: Elsevier, 1970. 879p.
A dictionary of terms from virtually every branch of science and technology as well as mathematics and medicine. Terms are adequately defined, and abbreviations indicate association with a particular field; gender is also given.

3-61 **Dictionary of Scientific Units, Including Dimensionless Numbers and Scales.** By H. G. Jerrard and D. B. McNeill. New York: Barnes and Noble, 1972. 212p.
Alphabetical listing of over 400 units with concise definitions, appropriate historical references, and additional sources of information. In this specialized dictionary one will find a number of lesser known units; better known units will be found in most general scientific dictionaries.

3-62 **Dictionary of Technical Terms.** By Frederic Swing Crispin. 11th ed. rev. New York: Bruce Publishing Co., 1970. 455p. illus.
Intended primarily for public and high school libraries. There are over 10,000 clear precise definitions. More specialized terms are classified into one or more of 89 categories, ranging from advertising to mechanics to woodworking. The dictionary's greatest value, however, is that the accent, syllable division, hyphenation, and capitalization are shown in the vocabulary entry.

3-63 **Eponyms Dictionaries Index: A Reference Guide to Persons, Both Real and Imaginary, and the Terms Derived from Their Names.** Ed. by James A. Ruffner. Detroit, MI: Gale Research Co., 1977. 730p.
An index containing some 33,000 entries covering 20,000 eponyms, together with the 13,000 personal names upon which they are based. Each eponym entry provides the word or phrase, the person's name on which the eponym is based, one of 60 broad subject categories, and citations of sources. The biographical entries are

arranged in the same alphabetic sequence, with names, dates of birth and death, nationality, occupation, citations giving biographical details and eponym(s) based on the person's name. Several categories of eponyms are excluded by design: trade names, legal cases, geographical features, buildings, streets, and astronomical bodies and features.

3-64 **French-English Science and Technology Dictionary.** By Louis de Vries. 4th ed. Revised and enlarged by Stanley Hochman. New York: McGraw-Hill, 1976. 683p.

An excellent language dictionary with the terms derived from the fields of aeronautics, agronomy, astronautics, astronomy, astrophysics, botany, chemistry, data processing, electronics, entomology, genetics, geology, geophysics, horticulture, mathematics, nuclear science, oceanography, physics, radar, radio, television, and zoology.

3-65 **German-English Science Dictionary.** By Louis de Vries. Updated and expanded by Leon Jacolev. 4th ed. New York: McGraw-Hill, 1978. 628p.

A total of 65,000 terms are included from the fields of nuclear science and engineering, computer science and data processing, solid state physics, molecular biology, genetics, automation, soil and environmental sciences, electronics, chemistry, physics, biology, and agriculture.

3-66 **McGraw-Hill Dictionary of Scientific and Technical Terms.** Daniel N. Lapedes, Editor in chief. New York: McGraw-Hill, 1978. 1771p. illus.

A comprehensive dictionary of some 100,000 terms in all areas of science but especially in electronics, computer science, physics, and chemistry. Includes a section containing over 1,000 biographical entries on many of the individuals associated with the laws, reactions, and phenomena that are defined in the dictionary.

3-67 **Ocran's Acronyms: A Dictionary of Abbreviations and Acronyms Used in Scientific and Technical Writing.** By Emanuel Benjamin Ocran. Boston: Routledge and Kegan Paul, 1978. 262p.

The entries in this dictionary are arranged under one of 47 different subject areas in the sciences, engineering, and related technical areas. Each entry is defined and in some cases an indication of which country uses that term is included followed by a listing of the subject fields in which the abbreviation or acronym is used.

3-68 **The Penguin Dictionary of Science.** Ed. by E. B. Uvarov and D. R. Chapman. 4th ed. rev. by Alan Isaacs. New York: Schocken Books, 1972. 443p. illus. tables.

An alphabetical, word-by-word listing of approximately 5,000 entries in the fields of astronomy, biology, biochemistry, cosmology, mathematics, physics, and, in response to contemporary demands, the space sciences. Entries average two or three lines. There are cross references, and italicized words in a definition denote those which are defined in the dictionary. There are occasional figures and tables. The material is directed to the student, rather than the specialist.

3-69 **Reverse Acronyms, Initialisms, and Abbreviations Dictionary.** Ed. by Ellen T. Crowley. 6th ed. Detroit, MI: Gale Research Co., 1978. 1103p.

This is a companion volume to Gale's *Acronyms, Initialisms and Abbreviations Dictionary* (3-55) and its supplement, *New Acronyms, Initialisms, and Abbreviations* (3-55a). It re-alphabetizes the entries in the first two titles by the first word of the meaning, or translation of the acronym so that user can locate the acronym, initialism, or abbreviation. It is considered as volume 3 of the three-volume set covering acronyms, initialisms, and abbreviations.

3-70   **Russian-English Scientific and Technical Dictionary.** By M. H. T. Alford and V. L. Alford. New York: Pergamon, 1970. 2v.

Contains some 100,000 entries drawn from nine different disciplines, technologies, and other professional subjects. In addition to scientific terminology, this work adds the basic vocabulary needed by scientists who are learning Russian. One of the most comprehensive dictionaries in this important area.

3-71   **Scientific Words: Their Structure and Meaning.** By Walter Edgar Flood. New York: Duell, Sloan, and Pearce, 1960; repr. Westport, CT: Greenwood Press, 1974. 220p.

An alphabetical listing of word-elements (roots, prefixes, and suffixes) that enter into the formation of scientific terms, with a definition of each element and an illustration of how that element is used in a word.

3-72   **Thesaurus of Information Science Terminology.** By Claire K. Schultz. Completely rev. and expanded ed. Metuchen, NJ: Scarecrow Press, 1978. 288p.

A thesaurus consisting of an alphabetic list of terms, a hierarchical arrangement, and an index to multi-word terms. There are several chapters covering thesaurus building and use.

3-73   **Trade Names Dictionary: A Guide to Trade Names, Brand Names, Product Names, Coined Names, Model Names, and Design Names, with Addresses of Their Manufacturers, Importers, Marketers, or Distributors.** Ed. by Ellen T. Crowley. 2nd ed. Detroit, MI: Gale Research Co., 1979. 2v.

This is a compilation of over 135,000 trade names, brand names, and company names arranged in one alphabet and covering apparel, appliances, automobiles, beverages, candy, cosmetics, decorative accessories, drugs, fabrics, food, furniture, games, glass products, hardware, jewelry, paper products, pet supplies, tobacco products, toys, and other consumer-oriented items. Each entry includes a brief description of the product, name of the company that manufactures, imports, or distributes it, and a code indicating the directory from which the information was taken. Company listings give the company's name and address. A companion volume to this dictionary is the *Trade Names Dictionary: Company Index*, which arranges all of the listed products under the company name. For each product listed a description and a code indicating the trade directory that provided the information is given.

3-74   **The Vocabulary of Science.** By Lancelot Thomas Hogben. New York: Stein and Day, 1970. 184p.

The purpose of the book is to acquaint librarians and others with the Latin and Greek base words used in current scientific and technical terminology. A major part of the book lists English-Greek and English-Latin vocabulary by parts of speech and categories of nouns. Hogben's is one of the few works of this kind dealing solely with science.

3-75 **World Guide to Abbreviations of Organizations.** By F. A. Buttress. 5th ed. Detroit, MI: Gale Research Co., 1974. 470p.
Contains approximately 18,000 abbreviations with almost 5,000 of the entries relating to the European Economic Community.

HANDBOOKS

3-76 **Handbook of Chemistry and Physics: A Ready-Reference Book of Chemical and Physical Data.** Cleveland, OH: Chemical Rubber Co., 1913- . annual.
An invaluable source of tabular data in physics and chemistry. It includes such sections as the elements, atomic weights, organic compounds and physical constants. Since it is revised annually, it can and does revise its data to incorporate the latest verified discoveries.

3-77 **Handbook of Data Processing for Libraries.** By Robert M. Hayes and Joseph Becker. 2nd ed. Sponsored by the Council on Library Resources. Los Angeles: Melville, 1974. 688p. illus. bibliog. index. (A Wiley-Becker-Hayes Series Book; Information Sciences Series).
Undoubtedly the best handbook available on the application of data processing techniques to library operations. Its four sections (Introduction, Management of Library Data Processing, Data Processing Technology, and Library Subsystems) offer a clear exposition of the current state of the art. For the science librarian, the single most revealing chapter may well be number 20, Mechanized Information Services. This chapter, in surveying representative operational systems of mechanized bibliographic retrieval, describes a number of excellent models that lend themselves to replication in a wide variety of situations.

3-78 **Handbook of Technical Writing Practices.** Ed. by Stella Jordan. New York: Wiley, 1971. 2v. illus. index. (Wiley Series on Human Communication).
The four parts of this comprehensive reference handbook cover: part 1 — various documents and publications produced by the technical writer, such as equipment instruction manuals, parts catalogs, management reports, proposals, sales literature, etc.; part 2 — support activities as editing, illustrating, and data processing; part 3 — the management of technical writing; and part 4 — guides and references, including style manuals and data sources. The work includes an extensive glossary.

3-79 **International Critical Tables.** By the National Research Council. New York: McGraw-Hill, 1926-1933. 7v.
Although dated, this set still is useful as a compilation of chemical and physical tables.

3-80 **Tables of Physical and Chemical Constants, and Some Mathematical Functions.** By G. W. C. Kaye and T. H. Laby. 14th ed. rev. Ed. by E. Bailey, and others. New York: Longman, 1973. 386p.
An excellent listing of pertinent tables arranged in three main sections: physics, chemistry, and mathematical functions.

SERIALS

3-81 **Abstracts and Indexes in Science and Technology: A Descriptive Guide.**
By Dolores B. Owen and Marguerite M. Hanchey. Metuchen, NJ:
Scarecrow Press, 1974. 154p. index.
Provides a description of the materials one would encounter in conducting a
literature search in scientific and technological subjects. The abstracts and indexes are listed under the following headings: general, mathematics and statistics,
astronomy, chemistry and physics, nuclear science and space science, earth
sciences, engineering and technology, geological sciences, agricultural sciences,
health sciences, and environment. Each title is described in outline form, with the
following topics covered: arrangement, coverage, scope, locating material,
abstracts, indexes, other material, and periodicals scanned.

3-82 **Ayer Directory: Newspapers, Magazines and Trade Publications.** Ed. by
Leonard Bray. Philadelphia, PA: Ayerpress, 1880- . annual.
The subtitle of this directory is *A Guide to Publications Printed and Published in
the United States and its Territories, the Dominion of Canada, Bermuda, the
Republic of Panama and the Philippines; Descriptions of the States, Provinces,
Cities and Towns in Which They Are Published; Classified Lists; and 70 maps.*
This is a bibliography of currently published serial publications. It has several
classified indexes, the most important of which for science librarians is the one
entitled "Trade, technical and class publications, by subject."

3-83 **Books in Series in the United States: Original, Reprinted, In-Print, and
Out-of-Print Books, Published or Distributed in the U.S. in Popular,
Scholarly, and Professional Series.** 2nd ed. New York: R. R. Bowker,
1979. 3273p.
Includes all entries in the first edition and its supplements plus new entries based
on the definition of a monographic series according to the *Anglo-American
Cataloguing Rules.* Exclusions are series intended for children, primary and
secondary school texts, and U.S. government publications. Generally, publisher's
series are included. Divided into three sections: series index with listings by series
name and then numerically or alphabetically if the series is not numbered; author
index; and title index. It includes access to some 113,154 titles issued by some
1,270 publishers and distributors in over 10,837 series.

3-84 **Current Serials Received by the N.L.L., March 1971.** London:
H. M. S. O., 1971. 540p.
This listing is produced by the British Library Lending Division which formerly
was the National Lending Library. Part 1, Current Titles except for Cyrillic; part
2, Cyrillic titles; part 3, Cover-to-cover translations of Cyrillic journals. A major
bibliographic tool for British science libraries and, since August of 1975, for
those American science libraries that are members of the Center for Research
Libraries.

3-85 **Directory of Published Proceedings, Series SEMT — Science/Engineer-
ing/Medicine/Technology.** v. 1- . White Plains, NY: InterDok Corp.,
1965- .
A bibliographic directory of preprints and published proceedings of congresses,
conferences, symposia, meetings, seminars, and summer schools that have been
held worldwide from 1964 to date. With volume 6, annual cumulative

supplements have been issued. Arrangement is chronological; entries give place, name of meeting, principal sponsors, publisher or distributor, address, order numbers, price, plus, if needed, title and series or journal of publication, editor, and pagination. LC card number is given if known. Sponsor and subject indexes.

3-86 **Gebbie House Magazine Directory.** Sioux City, IA: House Magazine Publishing Co., 1946- . triennial.

Consists of several sections: alphabetical list by company name giving full bibliographic details; a geographic listing; title list; printers' list; and an arrangement by SIC (Standard Industrial Classification) numbers.

3-87 **Guide to Magazine and Serial Agents.** By Bill Katz and Peter Gellatly. New York: R. R. Bowker, 1975. 239p. index.

This guide is designed to help librarians determine the best periodicals or serials subscription agent for their particular library by "giving enough background information concerning serials and their management to provide an understanding of the agent-library relationship and by providing facts, details, and descriptions of the services and procedures of the major and selected smaller domestic and foreign serials subscription agents." The first three parts of this guide describe the essentials of ordering serial publications, record maintenance, agency services, selecting an agent, claims and claiming, etc. The last section includes a helpful checklist of the typical agency services and a well-annotated list of 172 agencies that handle serials.

3-88 **Guide to Special Issues and Indexes of Periodicals.** Ed. by Charlotte M. Devers, Doris B. Katz, and Mary Margaret Regan. 2nd ed. New York: Special Libraries Association, 1976. 289p. index.

This guide provides easy access to the specialized contents of selected American and Canadian trade, technical, and consumer journals. Arranged in alphabetical order by periodical title, the main text contains a descriptive listing of annual special issues, features, supplementary issues, and/or sections appearing on a recurring basis, as well as information on editorial and advertiser indexes. A comprehensive subject index aids in locating the contents of the specials.

3-89 **History of Scientific and Technical Periodicals: The Origins and Development of the Scientific and Technical Press, 1665-1790.** By David A. Kronick. 2nd ed. Metuchen, NJ: Scarecrow Press, 1976. 336p. bibliog. index.

This is the only comprehensive study of the origins and developments of the scientific journal. It includes data on definitions and history of the scientific periodical, discussions on substantive journals, society proceedings, abstract and review journals, collections, almanacs and annuals, the general periodical, and bibliographic control of the literature. A comprehensive 24-page bibliography and separate subject, name, and title indexes are also included.

3-90 **Irregular Serials and Annuals: An International Directory.** 4th ed. 1976-1977. New York: R. R. Bowker, 1976. 1068p.

The directory covers such diverse types of materials as proceedings, transactions, advances, progresses, yearbooks, and handbooks published on an irregular basis. Entries, arranged by title under 251 subject headings, provide full title, frequency, price, publisher's name and address, and DDC number. This edition also provides a listing of over 1,450 serials that have ceased or suspended publication since the last edition.

3-91   **Magazines for Libraries.** By Bill Katz and Berry Gargal. 2nd ed. New York: R. R. Bowker, 1972. 822p. bibliog. index.

3-91a   **Supplement.** 1974. 328p. index.
An excellent source for descriptive and evaluative information on periodicals. Although directed to libraries at large, its coverage of science journals is unexpectedly thorough. The authors' judgments on the most important science journals are often astute. In every instance they are worth reading.

3-92   **Monographic Series.** Washington: Library of Congress, 1976- . quarterly; with annual cumulations.
This is a compilation of LC printed catalog cards representing all monographs cataloged by the Library as parts of series.

3-93   **New Serials Titles: A Union List of Serials Commencing Publication after December 31, 1949.** 1950- . Washington: Library of Congress, 1951- . Cumulations include 1950-1970 (New York: R. R. Bowker), 1971-1975, 1976-1977, and 1976-1978.
The third edition of the *Union List of Serials* (3-101), published in five volumes in 1965, contained 156,499 entries for serial titles held by 956 libraries in the United States and Canada. The information was current to 1949. In 1950, the Joint Committee of the *Union List of Serials* decided against a new edition of the *List* because of the cost and the burden it placed on cooperating libraries. In the same year, the Library of Congress conceived a plan for putting records of serial holdings on punched cards. The compilation of *New Serial Titles* was issued in 1951. There are more than 800 cooperating libraries in the United States and Canada. The serials are arranged alphabetically according to title and are listed under entries prepared in accordance with the rules in the ALA Cataloging Rules . . . (2nd ed., 1949) or the *Anglo-American Cataloguing Rules*, following generally the form used in the *Union List of Serials*. It is indicated in the introduction that "since the entries for inclusion in this publication often may be prepared prior to their official establishment for cataloging records, new corporate entries appearing in this list may not always agree with the established form adopted subsequently. In such cases the entries which are affected are revised."

3-94   **New Serials Titles: 1950-1970 Subject Guide.** New York: R. R. Bowker, 1975. 2v.
Arranged by slightly modified Dewey Decimal class numbers. DDC classes in the 500s and 600s will obviously be of principal interest to science librarians. For those not familiar with DDC, the work provides the following helpful indexing aids: subject headings in numeric sequence, subject headings in alphabetic sequence, index to subject headings, and a table correlating Dewey numbers to subject headings. Since 1975, the following uncumulated series has diminished in importance: *New Serial Titles—Classed Subject Arrangement* (Washington: Library of Congress, 1955- . monthly.); but it retains value for the years after 1970.

3-95   **Proceedings in Print.** Mattapan, MA: Special Libraries Association, 1964- . bimonthly.
Although originally restricted to aerospace technology, it is now inclusive of "the various fields of science, technology and medicine." Includes full citations for

proceedings. Cumulative index for each annual volume. Unlike the chronological arrangement of other works, this one is arranged by unique conference titles. Indexed by sponsors and by corporate authors.

3-96 **Scientific Periodicals: Their Historical Development, Characteristics and Control.** By Bernard Houghton. London: Clive Bingley; Hamden, CT: Linnet Books, 1975. 135p. illus. index.
This is a good history of the scientific periodical including selective but useful lists of science periodicals, arranged by subject, an anlysis of abstracting and indexing services, and a concise discussion of citation studies as a tool for identifying the periodicals most likely to be needed in libraries.

3-97 **Serials Automation in the United States: A Bibliographic History.** By Gary M. Pitkin. Metuchen, NJ: Scarecrow Press, 1976. 148p. index.
A source "to all information published on the automation of serials control functions in the United States and cited in Library Literature" between 1951 and 1974. Excluded are articles on union lists or other cooperative projects such as OCLC or the National Serials Data Program. Most of the articles are annotated.

3-98 **Sources of Serials: An International Publisher and Corporate Author Directory.** New York: R. R. Bowker, 1977- . index. (A Bowker Serials Bibliography).
This very important book is basically an international name authority file for all serial publishers and corporate authors included in *Ulrich's International Periodicals Directory* (3-100), *Irregular Serials and Annuals* (3-90), and *Ulrich's Quarterly*. Some 63,000 publishers and corporate authors are arranged under 181 countries with some 90,000 listings giving complete address, all serial titles published or sponsored, and ISSN.

3-99 **Titles in Series: A Handbook for Librarians and Students.** By Eleanora A. Baer. 3rd ed. Metuchen, NJ: Scarecrow Press, 1978. 4v.
This work now includes 69,657 monographic works in series published prior to January, 1975. The work's contribution lies in its identification of monographs as parts of series, information that is often essential for locating them in libraries.

3-100 **Ulrich's International Periodicals Directory: A Classified Guide to Current Periodicals, Foreign and Domestic.** New York: R. R. Bowker, 1932- . biennial.
The seventeenth edition, 1977-1978, contains classified listings of approximately 62,000 currently published periodicals, predominantly those of the United States and the British Commonwealth. The classification system allows rapid identification (with complete bibliographic description) of science periodicals. For example, biology as a broad field is covered first, followed immediately by 11 subfields beginning with biological chemistry and ending with zoology. Chemistry, physics, and engineering, among others, are similarly structured. A valuable feature of Ulrich's is the citation of indexing or abstracting services and microform availability for each periodical. *Ulrich's Quarterly* provides a continuous source of information about new serial titles, title changes, and cessations. This replaces *Bowker Serials Bibliography Supplement*.

3-101 **Union List of Serials in Libraries of the United States and Canada.** Ed. by Edna Brown Titus. 3rd ed. New York: H.W. Wilson, 1965. 5v.

A record of holdings for libraries in the United States and Canada with its principal value being for interlibrary loan. The successor to this important work is *New Serials Titles* (3-93).

3-102 **World List of Scientific Periodicals Published in the Years 1900-1960.** 4th ed. New York: Butterworths, 1963-1965. 3v.

The most nearly complete listing of scientific publications ever compiled, this is a union list of the scientific periodicals held by British libraries, plus some titles that are not available in any British library. Each entry gives full title, accepted abbreviation, date of first issue, date it ceased publication if applicable, and symbols indicating which libraries in Great Britain have files. The work is arranged by convertible or amalgamated forms of standard title elements, thus making it not necessary to have an accurate citation in order to locate a file of the serial in question. Identification is possible even from a highly truncated citation. Despite its location of files in British libraries, this work is essential for American science librarians, for it is the only one-stop source that helps with the truncated citation problem. Check here first if the citation seems at all suspect; then, after finding an authoritative full title, check *ULS* (3-101) or *NST* (3-93) for a U.S. location.

3-103 **Yearbook of International Congress Proceedings.** Brussels: Union of International Associations; distr. New York: International Publications Service, 1969- . annual.

Provides bibliographical coverage of proceedings, reports, symposia, and other documents emanating from international congresses. Keyword and author indexes. Of historical interest is the following: *International Congresses and Conferences, 1840-1937: A Union List of Their Publications Available in Libraries of the United States and Canada* (New York: H. W. Wilson, 1938. 229p.).

DIRECTORIES

3-104 **American Library Directory.** Ed. by Jacques Cattell Press. 31st ed. New York: R. R. Bowker, 1978- . annual.

The 31st edition of this standard reference source provides current details on the holdings, personnel, services, income, and expenditures of 35,125 public, college, special, U.S. government and armed forces, law, medical, and religious libraries in the United States and Canada. Arranged geographically by state and city, the entries provide such data as name of library, address, telephone number, names and titles of key personnel, number of volumes, microform holdings, income and expenditures, subject interests, and special collections. Additional details include (if applicable): circulation figures, branch locations, memberships in regional systems and book processing centers, amount of federal and state funding, salaries paid professionals and student assistants, and book, microform, and AV budgets. There is an index to all libraries listed. Additional updating is through a bimonthly *American Library Directory Updating Service.*

3-105 **Annual Register of Grant Support: A Guide to Grant Support Programs of Government Agencies, Foundations, and Business and Professional Organizations.** Los Angeles: Academic Media, 1969- . annual.

One of the four main sections of this annual is devoted to science, appropriately subdivided by discipline. Entries describe the grant programs, sums of money allowable, and conditions and requirements for acceptance of applications. Subject, organizational, and geographic indexes are included. Supersedes *Grant Data Quarterly*.

3-106 **Computer-Readable Data Bases: A Directory and Data Sourcebook.** Ed. by Martha E. Williams. Washington: American Society for Information Science, 1979. 1v. (various paging).

Covers over 500 data bases. The information that is provided for each data base includes: acronym and full name; issuance; correspondence with printed source; producer, distributor and/or generator; subject matter and scope; indexing; tape specifications; data base services; producer user services, if offered, and the address and telephone number of the person to be contacted for further information. There is also a listing of data bases that are now outmoded, that have been replaced by more current data bases, or whose existence cannot be verified. Indexes are included for broad subject categories, data base name, producer, and processor. The directory is compiled from the data base on data bases maintained at the University of Illinois' Information Retrieval Research Laboratory.

3-107 **Directory of Research Grants 1976-1977.** Comp. by William K. Wilson, ed. by Betty L. Wilson. Phoenix, AZ: Oryx Press, 1976. 235p. index.

A directory of research grants awarded to academic departments and arranged by 84 academic disciplines. Each entry includes a brief description of the grant, the amount, application date, and sponsor. Indexes cover grant names, sponsoring organizations, and sponsoring organizations by type.

3-108 **Directory of Scientific Directories: A World Guide to Scientific Directories Including Medicine, Agriculture, Engineering, Manufacturing and Industrial Directories.** Comp. by Anthony P. Harvey. Guernsey, Channel Islands: F. Hodgson; distr. New York: International Publications Service, 1972. 513p.

Lists about 1,800 directories published from 1950 to 1971 and includes those published separately as well as in periodicals. Arrangement is by continent, then by country, and within each country by subject.

3-109 **Directory of Special Libraries and Information Centers.** Ed. by Margaret Labash Young, Harold Chester Young, and Anthony T. Kruzas. 5th ed. Detroit, MI: Gale Research Co., 1979. 3v.

This directory covers all types of special libraries, in the broadest sense of the word—e.g., special collections in university or public libraries, governmental libraries, company libraries, specialized libraries of associations and institutions, etc. Volume 1 provides an alphabetical approach to over 14,000 special libraries. Each listing contains: name of sponsoring organization; name of library; address and telephone number; name of person in charge; founding date; size of staff; important subjects in the collection; special collections; size and composition of holdings; number of regularly received serials; services available to outside inquirers; publications of the library; special catalogs and indexes maintained; and names of supervisory staff plus automated library operations and memberships in networks and/or consortia. A subject index concludes the volume. Volume 2 serves as a geographic-personnel index. The geographic section lists by state or province all the institutions included in the first volume. The second section provides an alphabetical listing of the names of library personnel with professional title and affiliation. Volume 3 is an inter-edition supplement called *New Special Libraries* that keeps subscribers up to date on new libraries or important changes.

3-110 **Encyclopedia of Associations.** 13th ed. Detroit, MI: Gale Research Co., 1979. 3v.

Strives to cover all national, non-profit membership organizations, foreign groups, international groups, some local and regional groups, and citizen action groups. Each entry is listed under 1 of 17 broad categories. Information given includes name, acronym, keyword, address, telephone number, chief official and title, founding date, number of members, staff, state and local groups, description, committees, sections and divisions, publications, affiliated organizations, mergers and changes of names, and conventions/meetings. Volume 1 includes an alphabetical keyword index and the listings of the national oganizations of the United States; the geographic-executive index makes up volume 2. Volume 3 contains periodic issues of *New Associations and Projects*, which keeps the encyclopedia updated between editions.

3-111 **Encyclopedia of Information Systems and Services.** Ed. by Anthony T. Kruzas. 2nd international ed. Ann Arbor, MI: Anthony T. Kruzas Associates; distr. Ann Arbor, MI: Edwards Brothers, 1974. 1271p. index.

Provides information on "1,750 organizations concerned with new forms, new media, and new methods for providing information services. Among these organizations are publishers, computer software and timesharing companies, micrographic firms, libraries, information centers, government agencies, clearinghouses, research centers, professional associations, and consultants." Not covered are printed and legal services, traditional academic and special libraries, public information offices, hardware manufacturers and distributors in the computer and micrographics field, conventional indexing and abstracting services, and most library automation programs, which are limited to such housekeeping functions as circulation, serials control, and acquisitions. The information was obtained by questionnaire or other means. The first of two sections contains one-page listings for each major organization; the second provides a more condensed description for organizations of lesser importance or for entities that failed to supply all needed information. Each entry provides: name, date established, executive and equivalent full-time staff, description of system or service, input, holdings, publications, and microform applications and services. Among the 13 supplementary indexes are a geographic index and indexes of commercially available data bases, of micrographic associations, of serials publications, etc.

3-112 **European Research Index.** Ed. by Colin H. Williams. 4th ed. St. Peter Port, Guernsey: Francis Hodgson Ltd.; distr. New York: International Publications Service, 1977. 2v.

A guide to establishments in Europe, including Eastern Europe, that conduct, promote, or encourage research in science, technology, agriculture, and medicine. Covers government and independent establishments, university research departments and research laboratories of industrial firms, and all international science bodies with European headquarters. Listings provide name, address, parent body (if any), director of research, and brief details of the research carried out or promoted.

3-113 **Federal Library Resources: A User's Guide to Research Collections.**
Comp. by Mildred Benton. New York: Science Associates/International,
1973. 111p. index.

The following information is provided for 163 libraries: name and address, person in charge, agency affiliation, hours the library is open, description of services, and brief information on resources. Appended is a subject index to library collections, an index to names of administrative personnel, and a list of libraries to which questionnaires were sent, with indication of changes and non-respondents.

3-114 **The Foundation Directory.** Ed. by Marianna O. Lewis. 6th ed. Comp. by
the Foundation Center. New York: The Foundation Center; distr. New
York: Columbia University Press, 1977. 661p. index.

This edition lists 2,818 foundations. Included are foundations that possessed assets of $1 million or more, or made total contributions of $100,000 or more in the year of record. The foundations must be "nonprofit, with funds and program managed by its own trustees or directors, and established to maintain or aid social, educational, charitable, religious, or other activities serving the common welfare, primarily through the making of grants." The arrangement is by state, with the following data provided: address, founding date, purpose and activities, financial data (including assets and expenditures), and a list of officers. Four indexes are included: index of field of interest; cities within states; donors, trustees, and administrators; and foundation name.

3-115 **Industrial Research Laboratories of the United States.** Ed. by Jaques
Cattel Press. 15th ed. New York: R. R. Bowker, 1977. 828p.

Limited to 10,028 non-governmental laboratories devoted to fundamental and applied research, including development of products and processes. The laboratories belong to 3,241 organizations. Information, based on a questionnaire, provides all the essential facts about each company, its subsidiary laboratories, its chief administrative and research executives, its personnel, and a brief statement of its chief research and development activities. Alphabetically arranged, it concludes with three indexes — personnel, geographical, and subject.

3-116 **National Directory of Newsletters and Reporting Services.** 2nd ed.
Detroit, MI: Gale Research Co., 1978. 109p. index.

Issued in four parts, this directory describes newsletters issued on a regular basis by businesses, associations, societies, clubs, government agencies, and other groups. Each entry includes title, name and address of the sponsoring organization, editor, description of scope and purpose, frequency, subscription cost, circulation, former or alternative titles, and advertising information. The entries are listed by title with a publisher and subject index.

3-117 **National Trade and Professional Associations of the United States.**
Washington: Columbia Books, 1966- . annual.

Each entry gives the address, the chief paid executive, number of members, size of staff, telephone, publications, annual budget, and annual meetings. Arranged alphabetically by organization with category index. Its more concise format and paper binding make this directory less expensive than the *Encyclopedia of Association* (3-110), though they are highly duplicative.

3-118 **Research Centers Directory.** Ed. by Archie M. Palmer. 6th ed. Detroit, MI: Gale Research Co., 1979. 1121p. index.

This edition, with 6,268 entries, retains the structure of the previous edition. The material is arranged in 14 broad subject categories from agriculture, home economics, and nutrition to social sciences, humanities, and religion. The volume concludes with three indexes: institutional index, alphabetic index of research centers, and subject. Information provided for each research center includes: name, address, director's name, year founded, sources of support, size and type of staff, annual budget, principal fields of activity, research facilities, publications, and library facilities. All scientific research centers of consequence are thoroughly covered. Updating the sixth edition is *New Research Centers*, an inter-edition series of paperbound supplements, published periodically.

3-119 **Scientific, Technical, and Related Societies of the United States.** 9th ed. rev. Washington: National Academy of Sciences, 1971. 213p. index.

Information is based on a questionnaire that defines the scope of coverage. Entries are limited to membership societies devoted to particular scientific, engineering, and other technical disciplines. Excluded are trade associations, institutions composed primarily of paid staff, small local groups, and fund-raising organizations. All entries are in one alphabet by the name of the association, and the following information is provided: name of association, address, executive officers, a brief paragraph on history and organization, statement of purpose, membership, meetings, professional activities, and publications.

3-120 **Subject Collections: A Guide to Special Book Collections and Subject Emphases as Reported by University, College, Public and Special Libraries in the United States and Canada.** By Lee Ash. 4th ed. rev. and enlarged. New York: R. R. Bowker, 1974. 908p.

This is a standard guide to special book collections in all areas including those in the sciences and technology. The entries are listed under one or more of some 1,400 subject headings.

3-121 **Subject Directory of Special Libraries and Information Centers.** Ed. by Margaret Labash Young, Harold Chester Young, and Anthony T. Kruzas. 5th ed. Detroit, MI: Gale Research Co., 1979. 5v.

This is a subject index to the entries in the *Directory of Special Libraries and Information Centers* (3-109). The five volumes cover: v. 1—Business and Law Libraries, Including Military and Transportation Libraries; v. 2—Education and Information Science Libraries, Including Audiovisual, Picture, Publishing, Rare Book, and Recreational Libraries; v. 3—Health Sciences Libraries, Including All Aspects of Basic and Applied Medical Sciences; v. 4—Social Sciences and Humanities Libraries, Including Area/Ethnic, Art, Geography/Map, History, Music, Religion/Theology, Theater, and Urban/Regional Planning Libraries; v. 5—Science and Technology Libraries, Including Agriculture, Environment/Conservation, and Food Sciences Libraries.

3-122 **World Guide to Technical Information and Documentation Services. Guide mondial des centres de documentation et d'information techniques.** 2nd ed. rev. and enlarged. Paris: Unesco; distr. New York: Unipub, 1975. 514p.

This guide describes 476 centers in 93 countries and territories which responded to a questionnaire concerning their history, address and telephone, staff, subject specialists, library collection, and bibliographic, abstracting, reprographic, and translating services. The fields most adequately represented are agriculture, architecture, building, chemical engineering, chemistry, electrical engineering, forestry, geology, industrial management, mechanical engineering, metallurgy, nuclear physics and chemistry, nutrition, physics, textiles, and water.

BIOGRAPHICAL DIRECTORIES

3-123 **American Men and Women of Science. Physical and Biological Sciences.**
Ed. by Jaques Cattell Press. 14th ed. New York: R. R. Bowker, 1979. 8v.
An outstanding biographical directory whose criteria for inclusion are: 1) achievement, by reason of experience and learning, of a stature in scientific work equivalent to that required for a doctoral degree, coupled with currently ongoing activity in scientific research; or 2) research activity of high quality in sciences as evidenced by publications in reputable scientific journals; or, for those whose work cannot be published because of government or industrial security, research activity of high quality in science as evidenced by the judgment of the person's peers; or 3) attainment of a position of substantial responsibility requiring scientific training and experience to the extent described for 1) or 2). Gives name, birth date and place, marital status, degrees, major field of interest, positions held, memberships, and address of each of 130,500 scientists representing 79 broad disciplines. The last volume serves as an index to the set and contains a discipline index and geographical index. A section for the engineering sciences will soon be published. Each section is now on a three-year revision cycle.

3-124 **Asimov's Biographical Encyclopedia of Science and Technology: The Lives and Achievements of 1,195 Great Scientists from Ancient Times to the Present Chronologically Arranged.** By Isaac Asimov. New rev. ed. New York: Avon Books, 1976. 805p.
This is a paperback reprint of the Doubleday edition that was published in hardcover in 1972. Its purpose is to provide a history of science through a chronology of biographies of great scientists. Although it contains 1,195 entries there are omissions.

3-125 **Biographical Dictionary of American Science: The Seventeenth through the Nineteenth Centuries.** By Clark A. Elliott. Westport, CT: Greenwood Press, 1979. 360p. index.
This work has been designed as a retrospective companion to *American Men and Women of Science* (3-123). Each of the some 600 scientists has a sketch that includes standard biographical data. There are five appendixes, arranged by year of birth, place of birth, education, occupation, and fields of science. There is an index that provides subject, publication, and other kinds of references as well as a list of abbreviations.

3-126 **A Biographical Dictionary of Scientists.** Ed. by Trevor I. Williams. 2nd ed. New York: Wiley, 1974. 641p.
Covers deceased scientists, technologists, inventors, and explorers from ancient times. Biographical sketches vary in length and are written and signed by some 50 contributors. At the end of each sketch bibliographical references are provided.

3-127 **Biographical Memoirs.** Ed. by the National Academy of Sciences. v. 1- . Washington: National Academy of Sciences, 1877/79- .

An annual providing biographies of deceased members of the Academy. Each entry includes a complete biographical sketch, portrait, list of publications, and chronology of events.

3-128 **Biographical Memoirs of Fellows of the Royal Society, Volumes 1-9, 1932-1954.** London: The Royal Society of London, 1955. 9v.

Each volume has portraits and biographies of those members who have died since the publication of the last volume. A complete bio-bibliography of the member's papers is included with each biography. For obituary notices prior to 1932, consult the *Proceedings of the Royal Society.* Volume 75 of the *Proceedings* contains obituaries for the period 1898-1904, with a general index of obituary notices from 1860-1899. Science librarians should be aware that the Royal Society has many foreign membership hence, its *Biographical Memoirs* can be profitably consulted for scientists who were not British nationals.

3-129 **Biographisch-Literarisches Handworterbuch zur Geschichte der exacten Wissenschaften.** By Johoun Christine Poggendorff. Berlin: Akademie Verlag, 1863- . Publisher varies. Leipzig: Barth, 1863-1904; Leipzig: Verlag Chemie, 1925-1940. Series resumed in 1953 with Band 7a.

The authoritative German-language handbook of information on the lives and published works of mathematicians, physicists, chemists, mineralogists, and other physical science scientists of all countries. Information on each entry includes full name, academic and honorary degrees, positions held, dates and places of birth and death, principal articles about the scientist including obituary notices and principal publications. Titles of articles are cited in the language in which they were published.

3-130 **Dictionary of Scientific Biography.** Charles Coulston Gillispie, editor in chief. New York: Scribner's, 1970-1978. 15v.

The first 14 volumes include some 5,000 names selected by the editorial board and its consultants under the auspices of the American Council of Learned Societies. Volume 15 is supplement 1 covering additional biographies A-Z and topical essays. It has well-documented biographical and evaluative sketches of deceased scientists. The controlling criterion for admission was whether the scientist's contributions "were sufficiently distinctive to make an identifiable difference to the profession or community of knowledge." Unlike the *DAB* and *DNB*, to which it bears close affinities, the present work has no national biases. Thoroughly international in scope, the work includes scientific thinkers from the Greeks down to those of the very recent past. In only one respect do the editors concede that their work may embody a bias. They concede that scientists outside the European-Western tradition — i.e., those of ancient China, India, and other parts of the Far East — may have suffered comparative neglect. This dictionary is destined to be the definitive critical biographical dictionary in the history of science for many years to come.

3-131 **ISI's Who Is Publishing in Science: An International Directory of Scientists and Scholars in the Life, Physical, Social and Applied Sciences.** Philadelphia, PA: Institute for Scientific Information, 1971- . annual.

This continues the annual *International Directory of Research and Development Scientists* (ISI, 1968-1970). It provides the addresses and organizational affiliations of authors of articles published in journals covered by the *Science Citation Index* (3-18). The listings are arranged by author, organization, and geographical location.

3-132 **An Index to Biographical Fragments in Unspecialized Scientific Journals.** By E. Scott Barr. University, AL: University of Alabama Press, 1974. 294p.

An index to biographical information in seven English-language journals of general science. Although the time span varies with the journal, it ranges from 1798 to 1933. The journals are: *American Journal of Science* (1819-1920); *Nature* (1869-1918); *Philosophical Magazine* (1798-1902); *Popular Science Monthly* (1872-1915); *Royal Society of Edinburgh. Proceedings* (1832-1920); *Royal Society of London. Proceedings* (1800-1933); and *Science* (1883-1919). There are approximately 15,000 citations to 7,700 obscure as well as noted scientists, and to non-scientists as well, plus references to about 1,500 portraits. Each entry gives, where possible, journal citations, the full name of each person, his or her dates, nationality, and specialties. Tables at the end of the book show the year or years that correspond to each volume of each journal indexed.

3-133 **Index to Scientists of the World from Ancient to Modern Times: Biographies and Portraits.** By Norma Olin Ireland. Boston: Faxon, 1962. 662p. (Useful Reference Series, No. 90).

A layman's index to collected and composite works of biographical reference.

3-134 **Scientists and Inventors.** By Anthony Feldman and Peter Ford. New York: Facts on File, 1979. 336p. illus. (part col.). index.

Over 150 scientists and inventors have been selected for inclusion in this useful book. The entries are presented in chronological order and each receives a two-page spread that includes a portrait, a picture or diagram of the scientific principle or the equipment or product discovered or invented, and in some cases a photograph of a contemporary application.

3-135 **Who Was Who in American History — Science and Technology.** Chicago: Marquis Who's Who, 1976. 688p.

A selective biographical reference work comprising abbreviated sketches of 10,000 engineers, inventors, and scientists now deceased whose contributions to the advance of science and/or technology during the past 350 years are considered notable. Data consist of facts on birth and death, parentage, education, marriage and offspring, academic/business positions and accomplishments, major publications, honors, and awards.

3-136 **Who's Who in Science in Europe: A Reference Guide to European Scientists.** 3rd ed. Guernsey: Hodgson, 1978. 4v.

Lists approximately 50,000 scientists from Eastern and Western Europe working in the biological, physical, medical, and agricultural sciences. Includes typical "who's who" information for each entrant.

3-137 **Who's Who of British Scientists, 1971-72.** Athens, OH: Ohio University Press, 1971. 1022p.

In addition to biographical information, a list of scientific societies and their journals, a list of independent or commercially produced scientific journals, research establishments, and a classified index by area of specialization are included. Its purpose is to list the "top 10,000 of British men and women in the biological and physical sciences or who are enjoying careers of distinction after basic training in a scientific discipline."

3-138 **World Who's Who in Science: A Biographical Dictionary of Notable Scientists from Antiquity to the Present.** Chicago: Marquis Who's Who, 1968. 1855p.

This biographical directory contains about 30,000 entries, half of which are of historical interest only. Most entries include information on the families, careers, and scientific interests and contributions of the biographical subjects.

CATALOGS

3-139 **Catalog of the John Crerar Library, Chicago.** Boston: G. K. Hall, 1967. 78v.

The catalog is divided into the following component subsections: Author/Title Catalog, 35v.; Classified Subject Catalog, 42v.; Subject Index to the Classified Subject Catalog, 610p. Crerar ranks among the greatest scientific and technical libraries in the world with collections in excess of 1.1 million volumes. This set of volumes is a printing in book form of the actual catalog cards. The collections, which include holdings of the Illinois Institute of Technology, are of research strength in the basic sciences, such as physics, chemistry and biology; medicine, including anatomy, physiology, biochemistry and pharmacology; agriculture, especially agricultural engineering and chemicals; and technology, including all branches of engineering. Science librarians should note also that Crerar is one of the few remaining libraries with a classified subject catalog.

3-140 **Catalog of the Library of the Academy of Natural Sciences of Philadelphia.** Boston: G. K. Hall, 1972. 16v.

First supplement in preparation. A dictionary catalog of 150,000 volumes including many analytic entries in the fields of systematics, paleontology, and stratigraphic geology.

3-141 **Center for Research Libraries Catalogue: Monographs.** Chicago: Center for Research Libraries, 1969-1970. 5v.

This is a main entry catalog of monographs held by the center. It omits such categories of materials as archival materials; children's books; Chinese books; foreign, state, and federal government documents; and dissertations.

3-142 **Center for Research Libraries Catalogue: Serials.** Chicago: Center for Research Libraries, 1972. 2v.

The center is noted for its holdings of serials that are not held in many of the larger research libraries. This two volume set lists by main entry those items that were formerly listed in the center's earlier *Rarely Held Scientific Serials* (1963).

3-143 **National Union Catalog. Pre-1956 Imprints. A Cumulative Author List Representing Library of Congress Printed Cards and Titles Reported by Other American Libraries.** v. 1- . Comp. and ed. with the cooperation of the Library of Congress and the National Union Catalog Subcommittee of the Resources and Technical Services Division, American Library Association. London: Mansell, 1968- .

Cumulates the Library of Congress *Catalog of Books* and its supplement, 1942-1947; the *Library of Congress Author Catalog, 1948-1952*; and *National Union Catalog, 1952-1955 Imprints*; the *National Union Catalog . . . 1953-1957*; and the entries from the *Union Catalog* card file at the Library of Congress. It is thus "a repertory of the cataloged holdings of selected portions of the cataloged collections of the major research libraries of the United States and Canada, plus the more rarely held items in the collections of selected smaller and specialized libraries."

3-144 **National Union Catalog, 1956 through 1967. A Cumulative Author List Representing Library of Congress Printed Cards and Titles Reported by Other American Libraries.** Totowa, NJ: Rowman and Littlefield, 1970-72. 125v.

"A new and augmented twelve-year catalog being a compilation into one alphabet of the fourth and fifth supplements of the *National Union Catalog* with a key to additional locations through 1967 and with a unique identifying number allocated to each title."

3-145 **Published Library Catalogues: An Introduction to Their Contents and Use.** By Robert Collison. London: Mansell Information; distr. New York: R. R. Bowker, 1973. 184p. index.

This volume, which limits itself to the "English-speaking world," lists circa 600 library catalogs and narrates in a series of survey chapters 11 areas that they cover: general catalogs, auction sale catalogs, book industries, philosophy and religion, social sciences, science and technology, arts and architecture, literature, history and biography, music, and geography.

THESES AND DISSERTATIONS

3-146 **Comprehensive Dissertation Index, 1861-1972.** Ann Arbor, MI: University Microfilms International, 1973. 37v.

This comprehensive inventory of doctoral dissertations, with a total of 417,000 entries, provides a nearly complete listing of the impressive output of graduate schools in the United States, plus some foreign universities. It is an expanded version of its predecessor, *Dissertation Abstracts International Retrospective Index* (3-147), published by University Microfilms in 1970. Other sources were consulted as well—namely, *American Doctoral Dissertations, Index to American Doctoral Dissertations, Dissertation Abstracts International, Dissertation Abstracts*, and *Microfilm Abstracts*. In addition to these University Microfilms sources, other reference tools were also used, especially the Library of Congress list of *American Doctoral Dissertations Printed in 1912-1932* and H. W. Wilson's *Doctoral Dissertations Accepted by American Universities 1933/34-1954/55*. The new work also contains listings of doctoral dissertations from some 70 universities not completely covered by any of the above-mentioned works.

3-147 **Dissertation Abstracts International. B: The Sciences and Engineering.** Ann Arbor, MI: University Microfilms International, 1969- . monthly.

The abstracts of dissertations in question are those available from University Microfilms either on microfilm or as xerographic reproductions. The number of cooperating institutions (i.e., universities that require dissertations to be sent to University Microfilms) has increased steadily, making *DAI* and related services of great importance. A list of these cooperating institutions is provided, including the year that each began cooperation. *DAI* appeared previously as *Microfilm Abstracts* (v. 1-11, 1935-1951) and later as *Dissertation Abstracts* (v. 12-29, 1952-1969). The current title reflects the inclusion of dissertations from Europe with most from Western Europe. Beginning in 1966, *DA* appeared in two sections—A: The Humanities, and B: The Sciences and Engineering. In 1977 section C: European Abstracts was added. Keyword, title, and author indexes appear monthly.

3-148 **Guide to Theses and Dissertations: An Annotated, International Bibliography of Bibliographies.** By Michael M. Reynolds. Detroit, MI: Gale Research Co., 1975. 599p. index.

A classified bibliography of listings of completed dissertations (and masters' theses) and announcements of those in progress. The first two sections are "universal" and "national." The latter is arranged by country, listing all general bibliographies with broad national inclusions. The remainder of the book consists of subject lists arranged by discipline; among these are lists of both completed dissertations, sometimes with abstracts, and dissertations in progress or preparation. This work is confined to separately published lists only. Check the *Bibliographic Index* (New York: H. W. Wilson, 1937- .) for concealed lists of dissertations in progress.

3-149 **Index to Theses Accepted for Higher Degrees in the Universities of Great Britain and Ireland.** London: ASLIB, 1953- . annual.

Dissertations and theses (simply "theses" in British usage) emanating from science departments of universities in the United Kingdom of Great Britain as well as the Republic of Ireland are ably represented in this important annual series. Theses are arranged by degree-awarding school, then by discipline, and then by author. Author index.

MEETINGS

3-150 **International Congress Calendar.** Brussels: Union of International Associations, 1961- . annual.

Prepared by the Congress Department of the Union of International Associations from primary sources of information, it is divided into a geographical and a chronological list with an international organization index and an analytical index. The calendar is complemented by monthly supplements appearing in the magazine *Transnational Associations* (1977-).

3-151 **Scientific Meetings.** v. 1- . New York: Special Libraries Association, 1956- .

An alphabetical listing of scientific, technical, medical, health, engineering, and management organizations and universities and colleges that are sponsoring future national, international, and regional meetings, symposia, colloquia, and institutes. It is arranged by association, university, or society. Each issue contains a chronological record of all meetings that are listed, plus a subject index. An excellent source for meetings of national and regional associations held within the United States.

3-152 **World Meetings: United States and Canada.** Newton Centre, MA: Technical Meetings Information Service, 1963- . quarterly.

3-152a **World Meetings: Outside USA and Canada.** Newton Centre, MA: Technical Meetings Information Service, 1968- . quarterly.

Both series provide information on dates and places of meetings that will take place during a two-year period following the date of issue.

TRANSLATIONS

3-153 **Consolidated Index of Translations into English.** Comp. by the National Translations Center, The John Crerar Library. New York: Special Libraries Association, 1969. 948p. index.

Provides information on the availability of translations that have appeared in a number of different lists issued by many agencies. Included are translation bibliographies, selective translation journals, and collections of translations. Cover-to-cover translations of journals are not listed article by article. Items listed in this volume include citations from listings published through 1966.

3-154 **Guide to Scientific and Technical Journals in Translation.** Comp. by Carl J. Himmelsbach and Grace E. Brociner. 2nd ed. New York: Special Libraries Association, 1972. 1v. (various paging).

According to the compilers, little has been done to illuminate the twisted paths of the "translation jungle"—a situation that this compilation remedies. They list some 278 "Cover to Cover Translation Journals" and 53 "Selections, Collections, and Other 'Translation Journals.'" Most of them are from the Soviet Union. Titles have been transliterated and are arranged alphabetically. Part III gives cross references from translated titles to the originals. The last three sections provide "Some Frequently Encountered Russian Abbreviations of Soviet Journals and Their Full Titles," a "Key to Addresses of Publishers and Distributors," and finally, the "Russian Alphabet and Transliteration." Every effort has been made to identify each journal precisely, and to tell what is and what is not available, which ones have merged with others or ceased publication or changed names, and from where they may be ordered.

3-155 **Translation and Translators: An International Directory and Guide.** Comp. and ed. by Stefan Congrat-Butlar. New York: R. R. Bowker, 1979. 241p.

The register of translators and interpreters is arranged in two categories: literary and industrial, and scientific and technical. There is an index by language so that you can locate the right person. There is also a list of translation associations and centers; and training programs and guidelines for translators. A good reference source for any scientific library that has to handle much foreign materials.

3-156 **Translations Register-Index.** New York: National Translations Center, 1967- . twice monthly.

In 1967, the Special Libraries Association, in collaboration with the John Crerar Library, began issuing a periodical listing of all new translations added to the

Crerar depository, the SLA Translations Center. This periodical succeeds *Technical Translations* (Washington: Clearinghouse for Federal Scientific and Technical Information, 1959-1967. 18v.).

COPYRIGHTS AND PATENTS

3-157 **Complete Guide to the New Copyright Law.** By the New York Law School Law Review. Dayton, OH: Lorenz Press; distr. Port Washington, NY: Independent Publishers Group, 1977. 448p.

This guide should be the nearest approximation of a compliance handbook that librarians and educators will have. Part 1 covers aspects of the proposed legislation that did not change as a result of the final Senate debates, and part 2 provides commentaries on the law as passed. Topics include works made for hire; duration of copyright; notice, deposit, and registration; analysis of infringement sections; fair use; CATV copyright; sound recordings; termination of transfers and licensing; and international copyright.

3-158 **Copyright Handbook.** By Donald F. Johnston. New York: R. R. Bowker, 1978. 309p.

This handbook is the publisher's book-length version of how the law should be interpreted. It covers user rights; copyrightable material; copyright notice; deposit copies; registration; ownership, transfer, and licenses; exclusive rights; infringement remedies; duration; international considerations; fair use; library reproduction; limits on exclusive performance and display rights; and compulsory licenses. The appendixes include the full text of both the 1909 and the 1976 laws, the CONTU guidelines, and examples of many of the copyright forms.

3-159 **Development and Use of Patent Classification Systems.** By the U.S. Patent Office. Washington: GPO, 1966.

An introduction to the use of the patent classification system.

3-160 **Foreign Patents: A Guide to Official Patent Literature.** By Francis J. Kase. Dobbs Ferry, NY: Oceana, 1973. 358p. illus. index.

This latest contribution to the literature is very specific in relationship to foreign patents and is particularly directed to patent attorneys and agents who must consult patent literature. The author notes "it is intended as an introduction to the foreign official patent literature, such as printed patent specifications and patent journals. It is essentially a guide to the meaning of the dates and numbers on patent specifications and abstracts, and should assist the searcher in ascertaining the state of prior knowledge." The material is organized alphabetically by country. Each geographical entry contains: 1) the name and address of the patent office or body that deals with the granting of industrial rights; 2) an English translation of any existing printed specifications; 3) a description of the official patent journals and other relevant publications excluding monographs, and collections of patent laws and regulations; and 4) the duration of the patent. Also of value is F. Newby's *How to Find Out about Patents* (London: Pergamon, 1967. 177p.). This is particularly good for its explanation of British patents.

3-161 **Index to the U.S. Patent Classification.** By the U.S. Patent and Trademark Office. Washington: U.S. Patent and Trademark Office, 1977. 223p.

This is an alphabetical list of subject headings referring to specific classes and subclasses of the patent classification system.

3-162 **Inventor's Patent Handbook.** By Stacy V. Jones. Rev. ed. New York: Dial Press, 1969. 226p.

A simple guide to patents stressing procedures for application. Also pertinent is Terrence W. Fenner's *Inventor's Handbook* (New York: Chemical Publishing Co., 1969. 309p.).

3-163 **Manual of Classification.** By the U.S. Patent and Trademark Office. Washington: Patent and Trademark Office, 1979- . A looseleaf, continually revised service.

Lists the numbers and descriptive titles of all the classes and subclasses. Once a field of interest has been identified by a precise class and subclass number, it becomes relatively easy to monitor the weekly issues of the *Official Gazette*.

3-164 **Official Gazette of the United States Patent and Trademark Office.** Patents. Washington: GPO, 1872- . weekly.

3-164a **Index to Patents Issued from the United States Patent and Trademark Office.** Washington: GPO, 1920- . annual.

The *Official Gazette* is a weekly listing of patents issued with a brief description and sketch for each patent. Preceding the list are notices, decisions, suits, and reissues of former patents. The patents are listed in four groups: general and mechanical, chemical, electrical, and design. In the first three groups, patents are numbered consecutively. The design patents have a separate numbering system. Each issue has an index of patentees, a classified index according to the *Manual*, and a geographical index of inventors. Over 4.1 million patents have been issued to date and are available on 16mm microfilm. The *Index of Patents* lists patentees, reissue patentees, design patentees, plant patentees, defensive publications and disclaimers and dedications.

3-165 **Uniterm Index to U.S. Chemical and Chemically Related Patents.** v. 1- . New York: IFI/Plenum Data, 1950- . bimonthly; with annual cumulations.

Organized on the principle of coordination of unit terms used to describe the elements of chemical patents, making it possible to identify quickly patents containing particular interactions. The uniterm index is called a "Dual Dictionary." In addition, there are separate patentee and assignee indexes.

3-166 **World Patents Index—WPI.** London: Derwent Publications Ltd. Documentation Services, 1974- . weekly.

3-166a **General Patents Index—GPI.** London: Derwent Publications Ltd. Documentation Services, 1970- . weekly.

3-166b **Central Patents Index—CPI.** London: Derwent Publications Ltd. Documentation Services, 1970- . weekly.

Among this publisher's computer-based current awareness services, the most significant is the worldwide patent alerting system, issued in several forms. The data base consists of basic patent specifications from 24 countries, added at the rate of about 12,000 weekly. *WPI* is published in four printed sections: General, Electrical, Mechanical, and Chemical, with six index approaches in each. Three printed abstracts journals—P, General; Q, Mechanical; and R, Electrical—constitute the *GPI* issued also as 19 different sets of cards and as 100 feet of 16mm microfilm, four to a frame, at nine- to ten-week intervals. *CPI* consists of two alerting bulletins, one arranged by country, one by systematic classification, each containing its own Patent Number Index, CPI Class Index, Patentee Index, and Basic Number or Patent Concordance Index.

GOVERNMENT DOCUMENTS AND TECHNICAL REPORTS

3-167 **Correlation Index: Document Series and PB Reports.** By Special Libraries Council of Philadelphia and Vicinity. New York: special Libraries Association, 1953. 271p.

An alphanumeric listing of technical report numbers with the corresponding PB (Publication Board) accession number. The PB number is needed to locate the abstract of the report in the early issues of the *U.S. Government Research and Development Reports* (3-179), then known as the *Bibliography of Technical Reports*.

3-168 **Dictionary of Report Series Codes.** Ed. by Lois E. Godfrey and Helen F. Redman. 2nd ed. New York: Special Libraries Association, 1973. 645p.

This guide to the alphanumeric codes used to identify technical reports includes 25,500 codes designating the various agencies of the Department of Defense, the Atomic Energy Commission, their contractors (including industrial, educational, and professional organizations), agencies of the U.S. government, and similar agencies of foreign governments. The second edition has doubled the number of entries cited in the first and reflects the phenomenal increase in the report literature, which is now estimated to encompass as much as half a million publications a year. Useful data include a glossary, a bibliography of articles on technical codes, and sources of the codes included here. The text proper is divided into three color-coded sections: 1) "Reference Notes—48 explanations of series designations or of the practice of the assigners in expanding on a series designation"; 2) a list of report series codes, arranged by the first letter found in the code, with identification of the related agency; 3) corporate entries with related report series codes.

3-169 **Government Publications: A Guide to Bibliographic Tools.** By Vladimir M. Palic. 4th ed. Washington: Library of Congress, 1975. 441p. index.

The purpose is to outline "bibliographic aids in the field of official publications issued by the United States, foreign countries, and international governmental organizations. It is intended to be a practical guide directing the researcher, the student, and . . . the reference librarian." Catalogs, indexes, bibliographies, checklists and other bibliographic aids that do not have their origins in government agencies are also included. The guide is organized into three parts: part 1, United States of America: Federal, State, and Local; part 2, international Governmental Organizations: General, UN, League of Nations, Other; and part 3, Foreign Countries: General, Western Hemisphere, Europe, Africa, Near East, and Asia. Entries give full bibliographic data plus LC classification number and, generally, a brief descriptive annotation.

3-170 **Government Reference Books: A Biennial Guide to U.S. Government Publications.** 1968/69- . Littleton, CO: Libraries Unlimited, 1970- . biennial.

A comprehensive classified list of major government reference publications for a biennial period. The book is arranged in four main sections: 1) general references, 2) social sciences, 3) science and technology, and 4) humanities. Entries provide complete bibliographic description—corporate body; price; LC card number; S/N, ISBN, or ISSN; and SuDocs classification. The work is indexed by personal author, title, subject, and corporate author.

3-171 **Government Reports Announcement and Index.** v. 75, no. 7- . Washington: National Technical Information Services, April 4, 1975- . biweekly.

Previously called *U.S. Government Research and Development Reports* (3-179) and *Government Reports Announcement.* This is a listing of the available reports from NTIS arranged in 22 subject fields: aeronautics; agriculture; astronomy and astrophysics; atmospheric sciences; behavioral and social sciences; biological and medical sciences; chemistry; earth sciences and oceanography; electronics and electrical engineering; energy conversion; materials; mathematical sciences; mechanical, industrial, civil, and marine engineering; methods and equipment; military sciences; missile technology; navigation, communications, detection, and countermeasures; nuclear science and technology; ordnance; physics; propulsion and fuels; and space technology.

3-172 **Government Wide Index to Federal Research and Development Reports.** Washington: Clearinghouse for Federal Scientific and Technical Information, 1965-1971. 71v.

This was a computer-produced subject, personal author, corporate author, report number, and accession number index to the abstract indexes: *U.S. Government Research and Development Reports* (3-179), *Technical Abstract Bulletin, Scientific and Technical Aerospace Reports* (12-167), and *Nuclear Science Abstracts* (7-11). For each subject, personal author, and corporate author entry the title of the report was given on one line with the accession number, price of hard copy, price of microfiche, and one or more of the above four abstract indexes with the citation as to volume and issue. The report number index gives the same information without the title, while the accession number index substitutes the report number for the accession number and leaves out the title. Because of the variant thesauri of subjects and corporate bodies within the agencies producing the abstract journals, there are inconsistencies in the index. It is continued by the *Government Reports Announcement and Index* (3-171).

3-173 **Introduction to United States Public Documents.** By Joe Morehead. Littleton, CO: Libraries Unlimited, 1975. 289p. index.

Designed for students and librarians, Morehead's text is concerned with federal publications—the relationship between their production and distribution, and their control, access, and management in libraries and information centers. The first four chapters discuss the Government Printing Office, the Superintendent of Documents, the depository library system, and administration of documents collections. The remaining seven chapters cover publications of the federal hierarchy, beginning with general guides. A Name/Subject Index and Title/Series Index provide quick access to information, and the thorough, well-structured table of contents approaches the functions of an index.

3-174 **Monthly Catalog of United States Publications.** By U.S. Superintendent of Documents. Washington: GPO, 1895- . monthly.

The basic current bibliography of the government publications that are printed by the Government Printing Office and distributed by the Superintendent of Documents. Monthly indexes cumulated annually. A very recent aid to subject retrieval is *Cumulative Subject Index to the Monthly Catalog of U.S. Government Publications, 1900-1971*, compiled by William W. Buchanan and Edna M. Kanely (Washington: Carrollton Press, 1973-1975. 15v.). U.S. documents constitute a complex field, and the *Monthly Catalog* is not the ultimate bibliography. It does not include documents that are classified, that are published directly by the issuing agency, or that are contracted for by U.S. government agencies. The *Monthly Catalog* is nonetheless important for the science librarian, for it does include many highly significant scientific documents—those of such agencies as the National Bureau of Standards, the National Oceanic and Atmospheric Administration, Geological Survey, the Bureau of Mines, the National Research Council, etc.—many of which are not duplicatively described in the various bibliographies of technical reports.

3-175 **Monthly Checklist of State Publications.** Washington: Library of Congress, Processing Dept., Exchange and Gifts Division, 1910- . monthly.

Includes only those documents submitted by state issuing agencies to the Library of Congress, so it is manifestly incomplete. Still, it is the best available.

3-176 **State Government Reference Publications: An Annotated Bibliography.** By David W. Parish. Littleton, CO: Libraries Unlimited, 1974. 240p.

This is an excellent annotated bibliographic guide to state publications, covering more than 800 representative documents issued by the offices and agencies of various states and U.S. territories. Except for the *Monthly Checklist* (3-175), there is no current guide to the 20,000 official state publications released each year.

3-177 **UNESCO List of Documents and Publications 1972-1976.** New York: Unesco; distr. New York: Unipub, 1979. 2v. 1131p.

Includes all publications and documents issued by Unesco in 1972-1976, including articles published in periodicals. Divided into five parts: list of acronyms used, annotated list, personal name index, conference index, and subject index.

3-178 **U.S. Government Manual.** Washington: Office of the Federal Register, National Archives and Record Service; distr. Washington: GPO, 1935- . annual.

Formerly entitled *U.S. Government Organization Manual.* Provides up-to-date information about the functions and administrative structure of the agencies in the legislative, judicial, and executive branches of the U.S. government. It gives the names of the principal officers directing each department along with a discussion of the authority and activities of the department. Organizational charts are provided. Indispensable for identifying federal agencies responsible for generating or sponsoring scientific reports.

3-179 **U.S. Government Research and Development Reports.** Washington: Clearinghouse for Federal Scientific and Technical Information, 1946-1971. 71v.

The earlier titles of this publication were: *Bibliography of Scientific and Industrial Reports*; *Bibliography of Technical Reports*; and *U.S. Government Research Reports*. The abstracting service announced to the public the availability of new reports of U.S. government-sponsored research and development that did not appear in *Nuclear Science Abstracts* (7-11) and STAR (12-167). It has been superseded by *Government Reports Announcement and Index* (9-68).

3-180 **U.S. Government Scientific and Technical Periodicals.** Comp. by Philip A. Yannerella and Rao Aluri. Metuchen, NJ: Scarecrow Press, 1976. 263p. bibliog. index.

This directory provides comprehensive treatment of 266 U.S. government scientific and technical periodicals. Complete information is given for each entry including SuDocs class number, date of first issue, starting date of latest title and/or SuDocs class number, latest frequency, issuing agency and its address, depository item number, subscription price, history, and several other informational items.

4

History of Science

Most scientists and engineers fail to understand the importance of the history of science. They may recognize the value of it, but, because they lack training in history, few understand its ramifications on modern research. There are several definitions of the history of science. Jean Lindsay, in the "History of Science," says that it is "an attempt to find out what questions man has asked at various times about the physical universe and what answers he has found satisfying"; it is "also partly a study of problems of practical technique; how men managed to capture gases, or measure the amount of blood pumped through the heart in an hour. It therefore involves some study of scientific instruments and of the technique of scientific experiment. It also involves a study of the interaction of scientific ideas with the other ideas commonly held at the time and expressed, for example, in poems and plays. It is a study of what ordinary educated men thought about the universe and themselves at any particular time, and so it is part of the general history of civilization."[1] George Sarton, in *A Guide to the History of Science*, states that many scientists reject the past — "The past is finished, irremediable, permanent; there is nothing we can do about it, and hence it is better not to worry about it. In the second place, some men of science will admit interest in history and realize its importance and difficulties, but they are not interested in the history of science."[2]

The books listed in this chapter deal with the history of science and technology. They are only selective, for any guide containing sources on the literature of science and engineering can only include a representative selection of books in this area. There are hundreds of items that cover this topic and far too many to include here, but the standard sources are covered in addition to some of the newer titles. Books with a long tradition of editions are not included, but are useful in the study of the development of ideas and concepts. Early textbooks and scientific journals are also excellent sources for the science historian. In fact, almost any text in science and engineering could be used as a sourcebook for the science historian. Large research libraries, recognizing this fact, are attempting to preserve at least one copy of every edition of all books published in the fields of science and engineering.

---

[1]Jean Lindsay, in *Results of the Scientific Revolution, A Symposium* (Glencoe, IL: The Free Press, 1951), pp. 9-10.

[2]George Sarton, *A Guide to the History of Science* (Waltham, MA: Chronica Botanica Co., 1952), p. 44.

GENERAL

4-1   **Asimov's Guide to Science.** By Isaac Asimov. Rev. ed. New York: Basic Books, 1972. 945p. index.

First published in 1960 under the title *Intelligent Man's Guide to Science.* A revised edition was published in 1965, but this revision incorporates recent scientific developments. This popularly written work will appeal primarily to laymen with little formal background in the sciences.

4-2   **The Awakening Interest in Science during the First Century of Printing, 1450-1550: An Annotated Checklist of First Editions Viewed from the Angle of Their Subject Content.** By Margaret Bingham Stillwell. New York: Bibliographical Society of america, 1970. 399p. index.

Arranged in two parts. The first is an author list of 900 selected titles arranged under six subject categories: astronomy, mathematics, medicine, natural science, physics, and technology. Each entry provides full bibliographic description, list of references, and occasional citations to monographs. Part 2 contains an analytical list of editors, places of printing, and names of printers and translators.

4-3   **Catalog of Books in the American Philosophical Society Library, Philadelphia.** Westport, CT: Greenwood Press, 1970. 28v.

Photographically reproduced catalog cards of one of the major collections in the history of science.

4-4   **Catalogue of Scientific Papers, 1800-1900.** By Royal Society of London. London: Clay, 1867-1902; Cambridge: University Press, 1914-1925. 19v.

4-4a   **Subject Index.** Cambridge: University Press, 1908-1914. 3v. in 4. (Reprint editions available. See *Guide to Reprints*).

Virtually the ultimate bibliography of the scientific literature of the nineteenth century. It covers periodical articles and the proceedings and transactions of societies. Entries are arranged alphabetically by author and chronologically under each author. In each entry, the author's name, full title of article, title of periodical, volume, date, and inclusive paging are given. Titles in East European languages are followed by a translation of the title, usually into English, but sometimes into French or German. All other entries are in the language of the paper without translation. Three separately published subject indexes cover only pure mathematics, mechanics, and physics, containing brief title, author's name, periodical, volume, date, and pagination. The indexes were classed according to those in the *International Catalogue of Scientific Literature* (4-15) and originally were to include an index for each of the 17 sections. Continued for a brief 14 years into the twentieth century as the *International Catalogue of Scientific Literature.*

4-5   **Catalogue of the History of Science Collections of the University of Oklahoma Libraries.** By Duane H. D. Roller and Marcia M. Goodman. London: Mansell, 1976. 2v.

Since strong collections in the field of the history of science are few, and printed catalogs even fewer, these two volumes are especially valuable. The collection now amounts to over 40,000 volumes and an additional 10,000 in microform making it one of the larger history of science collections in the country. The catalog reproduces in dictionary arrangement the main entry cards, some analytics and, for works dealing with persons, the subject cards as well.

4-6   **Critical Bibliography of the History of Science and Its Cultural Influence.** no. 1- . Beltsville, MD: History of Science Society, 1913- .
This major serial bibliography, arranged in a closely classified order, has appeared at least once a year in the official quarterly journal of the History of Science Society under the title *ISIS, International Review Devoted to the History of Science and Its Cultural Influences* (Beltsville, MD: ISIS, 1913- . quarterly.).
In 1971, a cumulation of the first 90 critical bibliographies was published as *ISIS Cumulative Bibliography* (4-14).

4-7   **Dictionary of Inventions and Discoveries.** Ed. by E. F. Carter. 2nd ed. rev. New York: Crane, Russak, 1976. 208p.
This edition contains some 3,000 brief entries. A good book to browse through, but not comprehensive for the serious researcher.

4-8   **Dictionary of the History of Ideas: Studies of Selected Pivotal Ideas.** Philip P. Wiener, editor in chief. New York: Scribner's, 1973-1974. 5v. index.
Among the pivotal ideas discussed in this set are those that have been central to the history of science: monism, dualism, matter, life (its uniqueness or its assimilability to matter), atomism, and primary qualities. The separate index does an excellent job of interrelating every aspect and phase of all of these interlocking ideas.

4-9   **Encyclopedia of Philosophy.** Paul Edwards, editor in chief. New York: Macmillan, 1967; repr. New York: Macmillan, 1973. 8v. in 4. index.
This important work is included here because of the major emphasis it places on the pivotal contributions of mathematicians, physicists, biologists, and chemists to the development of modern philosophy. This and the *Dictionary of the History of Ideas* (4-8) are important to any history of science collection.

4-10   **A Guide to the History of Science: A First Guide for the Study of the History of Science with Introductory Essays on Science and Tradition.** By George Sarton. Waltham, MA: Chronica Botanica, 1952. 316p. illus.
A good general literature guide to the history of science written by the foremost authority in the field.

4-11   **Historical Studies in the Physical Sciences.** v. 1- . Philadelphia, PA: University of Pennsylvania Press, 1969- .
An annual devoted to articles on the history of the physical sciences from the eighteenth century to the present. An effort is made to cover new directions and methods of research in modern history of science. It incorporates *Chymia*, history of chemistry annual.

4-12   **A History of Science.** By George Sarton. New York: Norton, 1970. 2v. illus.
This is a classic history that was originally published by Harvard University Press in 1952. Volume 1 covers ancient science through the golden age of Greece; the

second volume covers Hellenistic science and culture in the last three centuries B.C.

4-13 **History of Science.** Ed. by Rene Taton. New York: Basic Books, 1963-1966. Trans. by A. F. Pomeraus. 4v. illus. plates. diagrams.

Original French title was *Histoire generale des sciences* (Paris: Presses Universitaires de France, 1957-1064. 4v.). Excellent survey history for all periods up to about 1960.

4-14 **ISIS Cumulative Bibliography: A Bibliography of the History of Science Formed from ISIS Critical Bibliographies 1-90, 1913-65.** Ed. by Magda Whitrow. London: Mansell, in conjunction with the History of Science Society, 1971. 2v.

George Sarton's bibliographies that appeared in *ISIS* between 1913 and 1965, have been cumulated in these two volumes. Volume 1 and volume 2, part 1, describe reference works, monographs, pamphlets, and articles relating to the work and contributions of individual scientists. Where necessary materials are subdivided by such publication forms as bibliographies, biographies, and collected editions. Volume 2, part 2, is arranged alphabetically by institutions (e.g., The Royal Society of London) with appropriate subdivisions. Both volumes follow LC rules of cataloging, filing, and transliteration. This qualifies as the most significant bibliographic tool ever published in the history of science. For bibliographies after 1965, see *Critical Bibliography of the History of Science and Its Cultural Influence* (4-6).

4-15 **International Catalogue of Scientific Literature, 1st-14th, 1901-1914.** London: published for the International Council by the Royal Society, 1902-1921; repr. New York: Johnson Reprint Corporation, 1975. 238v. in 32.

This was an annual bibliography of papers, monographs in serials, and independent works in the sciences and their application. Citations are arranged in a decimal classification order. Only 14 annual issues were published. Each issue is divided into 17 separate parts, each of which covers one subject, as follows: A, Mathematics; B, Mechanics; C, Physics; D, Chemistry; E, Astronomy; F, Meteorology; G, Mineralogy; H, Geology; J, Geography, Mathematical and Physical; K, Paleontology; L, General Biology; M, Botany; N, Zoology; O, Human Anatomy; P, Anthropology; Q, Physiology; and R, Bacteriology. Each of these sections has its own schedules, indexes in four languages, author catalog, and subject catalog. This is the most important bibliography in the science field for this period. A continuation of the Royal Society's *Catalog* (4-4).

4-16 **Introduction to the History of Science.** By George Sarton. Baltimore, MD: Williams and Wilkins, 1927-1948; repr. Huntington, NY: Robert E. Krieger Publishing Co., 1975. 3v. in 5. illus. (Carnegie Institution of Washington. Publications No. 376).

During his lifetime George Sarton was the undisputed dean of the history of science as a separate scholarly discipline. The work is both a narrative and a bibliographic history. While monumental in its coverage for the period through the early 1900s, it must now be supplemented by *Scientific Thought* (4-23).

4-17 **Lexikon der Geschichte der Naturwissenschaften: Biographien, Sachwörter und Bibliographien.** Mit einer Einfuhrung "Die Zeitalter der Naturforschung" und einer Übersichtstabelle von Josef Mayerhofer. Vienna: Holinek, 1959- . In progress.

Published to date are: Liferungen 1, Introductory Articles and Aachen-Achord; 2/3, Achard-Bewegurg; 4/5, Bewegury-Daniel von Morley; 6, Daniel von Morley-Dodel, Arnold; liferungen 1 through 6 compose band I. Projected to become the most comprehensive bio-bibliographical encyclopedia survey of the history of science through the close of the nineteenth century.

4-18 **A List of Books on the History of Science.** Prepared by A. G. S. Josephson. Chicago: John Crerar Library, 1911. 279p.

4-18a **Supplement.** 1917.

4-18b **2nd Supplement.** 1942-1946.

Covers mathematics, astronomy, chemistry, physics, geology, paleontology, mineralogy, and crystallography. An excellent guide from a major science library. See also the *John Crerar Library Catalog* (3-139)

4-19 **Repertorium Commentationum a Societatibus Litterariis Editarum.** By Jeremias David Reuss. Secondum Disciplinarum Ordinem Digessit. I. D. Reuss. Gottingen: Dieterich, 1801-1821; repr. New York: B. Franklin, 1962. 16v.

Limited to the indexing of Society publications to 1800, this was the precursor of the Royal Society's *Catalogue of Scientific Papers* (4-4). Sixteen volumes were published, as follows: 1) Historia naturalis, generalis et zoologia; 2) Botanica et mineralogia; 3) Chemia et res metallica; 4) Physica; 5) Astronomia; 6) Oeconomia; 7) Mathesis, mechanica, hydrostatica, hydraulica, hydrotechnia, aerostatica, pneumatica technologia, architectura civilis, scientia navalis, scientia militaris; 8) Historia; 9) Philologia, linguae, scriptores graeci, scriptores latini, litterae elegantiores, poesis, rhetorica, ars antiqua, pictura, musica; and 10-16) Scientia et ars medica et chirurgica.

4-20 **Science since 1500: A Short History of Mathematics, Physics, Chemistry, Biology.** By H. T. Pledge. 2nd ed. London: HMSO for the Ministry of Education, Science Museum, 1966. 360p. graphs. plates. maps. index.

A survey history of modern science containing a profusion of footnote references to further readings, it hence serves also as a bibliographic essay.

4-21 **Scientific Books, Libraries, and Collectors: A Study of Bibliography and the Book Trade in Relation to Science.** By John L. Thornton and R. I. J. Tully. 3rd ed. rev. London: Library Association; distr. Detroit, MI: Gale Research Co., 1971. 508p. illus. index.

Thornton's guide to the scientific literature has become the classic bibliographical history of science. Twelve revised and updated chapters survey the production, distribution, and storage of scientific literature from the pre-Gutenburg era through the mid-twentieth century. Six of these chapters discuss scientific authors and their publications; the remainder of the volume provides introductory commentary on scientific societies, the periodical literature, scientific libraries — private and public — and scientific publishing and bookselling. In addition, chapter 9 is a useful survey of guides to the literature, current and retrospective bibliographies, and other control tools. There is an 85-page bibliography.

4-22 **Scientific Societies in the United States.** By Ralph Samuel Bates. 3rd ed. Cambridge, MA: MIT Press, 1965. 326p.

A history of scientific societies in the United States, presented in narrative form with numerous bibliographic references. The separate chapters cover Scientific Societies in 18th Century America; National Growth, 1800-1865; The Triumph of Specialization, 1866-1918; American Scientific Societies and World Science, 1919-1944; The Increase and Diffusion of Knowledge; The Atomic Age, 1945-1955; and Scientific Societies in the Space Age.

4-23 **Scientific Thought, 1900-1960: A Selective Survey.** Ed. by R. Harre. New York: Oxford University Press, 1969. 277p. illus. index.

Establishes and traces the primary presuppositions of modern science as they have developed since 1900. Among the 12 major topics treated are molecular biology, ecological genetics, and cell biophysics. A substantial introduction to the present stage of scientific theory.

4-24 **Short History of Scientific Ideas to 1900.** By Charles Joseph Singer. Oxford: Clarendon Press, 1959. 525p.

A concise elementary treatment of the development of science up to the twentieth century. For the current century, see *Scientific Thought* (4-23).

HISTORY OF MATHEMATICS

4-25 **Bibliography and Research Manual of the History of Mathematics.** By Kenneth O. May. Buffalo, NY: University of Toronto Press, 1973. 818p. illus.

Consists of two parts, the first of which is an excellent but short manual on retrieval, storage, and analysis of information. Although the work was written for research workers in the history of mathematics, part 2 is a bibliography on the history of mathematics "directed toward historians and mathematicians and focuses on utility for information retrieval rather than on matters of antiquarian interest." The more than 30,000 entries are arranged under the following headings: Biography; Mathematical Topics; Epimathematical Topics; Historical Classifications; and Information Retrieval. Each entry gives bibliographic information sufficient for locating the reference, but incomplete in that it does not provide titles of articles or contents of books.

4-26 **Bibliography of Non-Euclidean Geometry.** By D. M. Y. Sommerville. London: The University of St. Andrews, 1911; repr. New York: Chelsea, 1970. 410p. index.

The 4,500 entries are arranged chronologically from the fourth century B.C. to 1911, with 125 items added to this reprint from 1911 to date.

4-27 **The Development of Mathematics.** By Eric Temple Bell. 2nd ed. McGraw-Hill, 1945. 637p.

Directs primary attention to the several mathematical discoveries of the last 6,000 years.

4-28   History of Mathematical Notations. By Florian Cajori. LaSalle, IL: Open Court Publications, 1951-1952. 2v. illus. index.

Describes the first appearance of each symbol and its diffusion among mathematicians around the world. Where symbols have encountered opposition through competition, the results of such conflicts are carefully narrated.

4-29   History of Mathematics. By David Eugene Smith. New York: Dover, 1958. 2v. (Repr. of 1923-1925 ed.).

A good textbook history of elementary mathematics through calculus.

4-30   History of Mathematics, from Antiquity to the Beginning of the 19th Century. By Joseph Frederick Scott. 2nd ed. London: Taylor & Francis, 1960; repr. New York: Barnes and Noble, 1969. 266p. illus.

A chronology of the development of the main mathematical ideas over the past 2,000 years. Contains excellent appended biographical notes. This work presupposes a moderately high level of prior knowledge or at least a highly structured interest in the field.

4-31   Source Book in Mathematics, 1200-1800. Ed. by Dirk Jan Struik. Cambridge, MA: Harvard University Press, 1969. 427p. illus. (Source Books in the History of the Sciences).

Contains 75 excerpts from the writings of Western mathematicians from the thirteenth to the end of the eighteenth century. The selection is confined to pure mathematics or to those fields of applied mathematics that have a direct bearing on the development of pure mathematics.

4-32   The World of Mathematics: A Small Library of the Literature of Mathematics from a'h'mose the Scribe to Albert Einstein. By James Roy Newman. New York: Simon and Schuster, 1956-1960. 4v.

This is a readable account of mathematics for the person who has relatively little mathematical background.

HISTORY OF ASTRONOMY

4-33   Astronomy of the Ancients. Ed. by Kenneth Brechen and Michael Feirtag. Cambridge, MA: MIT Press, 1979. 224p. illus.

Eight articles introducing the reader to ancient astronomers through history, archaeology, technology and mythology. Covers Indian medicine wheels, pictographs and petroglyphs, first scientific instruments, naked-eye astronomy, Sirius enigmas, Stonehenge, Gorgon's eye, and language of archaic astronomy.

4-34   History of Astronomy. By Giogio Abetti. New York: Schumann, 1952. 338p. illus. (The Life of Science Library, No. 24).

A translation from the Italian of *Storia dell'astronomia*. It is divided into three parts: part 1, From Beginning to Copernicus; part 2, Reformation of Astronomy; and part 3, The Modern Era. A thorough, balanced history.

4-35   A History of Nautical Astronomy. By Charles Henry Cotter. New York: American Elsevier, 1968. 387p. illus.

Nautical astronomy or, as it is sometimes designated, celo-navigation, is the determination of the position of ships at sea by reference to celestial objects.

When man first began to venture out on the high seas beyond visual contact with land formations, he was forced to solve at each succeeding stage of adventuring increasingly complex problems of observation, instrumentation, and chartmaking. Cotter's book expertly traces in developmental order the story of problems encountered and solutions reached.

4-36   **Source Book in Astronomy.** By Harlow Shapley. New York: McGraw-Hill, 1929. 412p. illus. (Source Books in the History of the Sciences).

4-36a   **Source Book in Astronomy: 1900-1950.** By Harlow Shapley. Cambridge, MA: Harvard University Press, 1960. 423p. illus. (Source Books in the History of the Sciences).

These two excellent sourcebooks contain excerpts from the classical works in astronomy from its beginnings to 1950.

HISTORY OF PHYSICS

4-37   **Early Physics and Astronomy: A Historical Introduction.** By Olaf Pedersen. New York: American Elsevier, 1974. 413p. (History of Science Library).

Contains a general account of some of the major topics in ancient and mediaeval science. Intended to give only an idea of the complexity of the history of science rather than a detailed history. Includes a bibliography and biographical appendix.

4-38   **Historic Researchers, Chapters in the History of Physical and Chemical Discovery.** By Thomas Wightman Chalmers. New York: Fernhill, 1968. 288p. illus.

A reprint of a series of articles that appeared in *Engineer* on the works and research of earlier scientists.

4-39   **Resources for the History of Physics: I. Guide to Books and Audiovisual Materials; II. Guide to Original Works of Historical Importance and Their Translations into Other Languages.** Ed. by Stephen G. Brush. Hanover, NH: University Press of New England, 1972. 86 + 90p. index.

A survey of historical contributions to the development of physics as a separate discipline.

4-40   **Source Book in Physics.** By William F. Magie. Cambridge, MA: Harvard University Press, 1963. 620p. illus. (Source Books in the History of the Sciences).

Includes excerpts from the contributions made by physicists in the three centuries ending with the year 1900. Its range of subject coverage includes magnetism and electricity, mechanics, properties of matter, sound, heat, and light.

4-41   **Sourcebook on Atomic Energy.** By Samuel Glasstone. 3rd ed. New York: Van Nostrand Reinhold, 1967. 883p.

Describes in simple language, with a minimum of mathematics, the most important developments in those fields of science encompassed by the general term atomic energy; it is, therefore, a history of nuclear science.

HISTORY OF CHEMISTRY

4-42   **Catalog of the Edgar Fahs Smith Memorial Collection in the History of Chemistry, Pennsylvania University Library.** Boston: G. K. Hall, 1960. 524p.

An author and subject catalog of one of the finest history of chemistry collections comprising more than 13,000 books, manuscripts, and prints.

4-43   **Discovery of the Elements.** By Mary Elviva Weeks. 7th ed. Easton, PA: Journal of Chemical Education, 1968. 896p. illus.

A complete and accurate account of the individual discovery of all elements through lawrencium.

4-44   **Evolution of Chemistry: A History of Its Ideas, Methods, and Materials.** By Eduard Farber. 2nd ed. New York: Ronald Press, 1969. 437p. illus.

A reliable survey of the historical development of chemistry from its origins to the present time.

4-45   **History of Chemistry.** By James Riddick Partington. London: Macmillan, 1961-1970. 4v.

The definitive history of chemistry by a noted authority. The text is detailed, well documented, and exhaustive through the late 1960s.

4-46   **Select Bibliography of Chemistry, 1492-1902.** By Henry Carrington Bolton. Washington: Smithsonian Institution, 1893-1904; repr. Millwood, NY: Kraus Reprint, 1973. 4v. (Smithsonian Miscellaneous Collection Vol. 36, 40:7, 41:3, 44:5).

Bolton's bibliography lists about 18,000 chemical books published in Europe and America from the rise of literature, 1492, to the close of the year 1902. The main volume, covering the years 1492 to 1892, is divided into seven sections of differing degrees of thoroughness. The seven sections are: I, Bibliographies; II, Dictionaries; III, History; IV, Biography; V, Chemistry, Pure and Applied; VI, Alchemy; and VII, Periodicals. The supplements each carried the work farther and also picked up many items missed in the original work. An eighth section published later covered the academic dissertations with emphasis on France, Germany, Russia, and the United States.

4-47   **Source Book in Chemistry, 1400-1900.** By Henry Marshall Leicester. Cambridge, MA: Harvard University Press, 1963 (c.1952). 554p. illus. (Source Books in the History of the Sciences).

4-47a   **Source Book in Chemistry, 1900-1950.** By Henry Marshall Leicester. Cambridge, MA: Harvard University Press, 1968. 408p. illus. (Source Books in the History of the Sciences).

Collection of excerpts from the works of noted chemists beginning with Vannoccio Biringuccio (1480-1539) and continuing through 1950. Biographies are not included, but a bibliography of biographies is included.

HISTORY OF BIOLOGICAL SCIENCE AND AGRICULTURE

4-48   **British Natural History Books from the Beginning to 1900: A Handlist.** Comp. by Richard B. Freeman. Boston: G. K. Hall, 1974. 300p.

Arranged by main entry, this handlist includes all books about wild animals and plants of the British Isles, all books about wild animals and plants in general published or printed in the British Isles, as well as those published abroad by British authors.

4-49 **Encyclopedia of American Agricultural History.** By Edward L. Schapsmeier and Frederick H. Schapsmeier. Westport, CT: Greenwood Press, 1975. 467p. index.
Close to 2,500 entries concerning agricultural history; includes specialized and general agricultural/Western terms, names of agencies, individuals and periodicals.

4-50 **Guide to the Literature of Botany: Being a Classified Selection of Botanical Works Including Nearly 6,000 Titles Not Given in Pritzel's "Thesaurus."** By Benjamin Dayden Jackson. New York: Hafner, 1881; repr. New York: Harper and Row, 1964. 626p.
This is an outstanding bibliographic guide that lists over 9,000 works covering the early botanical literature. Of great value for the historical analysis of botanical literature.

4-51 **History of Biology.** By Eldon J. Gardner. 3rd ed. Minneapolis, MN: Burgess, 1972. 464p.
The only general history of the total field that provides adequate coverage of the developments since circa 1930.

4-52 **The History of Biology: A Survey.** By E. Nordenskiold. Trans. from the Swedish by L. B. Eyre. London: Kegan Paul, 1929; repr. New York: Tudor, 1960. 629p. illus.
Despite its age, still the best history of the field.

4-53 **History of Biology to about the Year 1900.** By Charles Joseph Singer. 3rd ed. New York: Abelard-Schuman, 1959. 579p. illus.
A good general history for the period to 1900.

4-54 **A History of Genetics.** By Alfred Henry Sturtevant. New York: Harper and Row, 1965. 165p.
A good readable general history of genetics.

4-55 **A Hundred Years of Biology.** By Benjamin Dawes. London: Duckworth, 1952. 429p.
An outline of the historical background and main trends in the development of biological theory. There is an extensive bibliography at the end of each chapter.

4-56 **North American Forest and Conservation History: A Bibliography.** By Ronald J. Fahl. Santa Barabara, CA: ABC-Clio Press, 1977. 408p. index.
This is a single-alphabet author list of 8,181 briefly annotated entries up to 1975. It covers scholarly and trade journals, monographs, yearbooks, dissertations, and theses on the history of forestry and related sciences.

4-57   **North American Forest History: A Guide to Archives and Manuscripts in the United States and Canada.** Comp. by Richard C. Davis. Santa Barbara, CA: ABC-Clio Books, 1977. 376p. bibliog. index.

This work records 3,830 manuscript groups in 358 repositories with some oral history and photography. An excellent source for the history of North American forestry.

4-58   **Selected References to Literature on Marine Expeditions, 1700-1960, Fisheries-Oceanography, Washington University Library.** Boston: G. K. Hall, 1972. 517p.

This selected list includes over 9,000 citations to oceanographic research, ships and expeditions from the eighteenth to the mid-twentieth centuries, arranged alphabetically by author.

HISTORY OF GEOSCIENCE

4-59   **Birth and Development of the Geological Sciences.** By Frank Dawson Adams. New York; Dover, 1954 (c.1938). 506p. index.

This is probably the best general history on geology available, covering the geological ideas of ancient Greece to those of modern science. Each chapter deals with a specific geological topic, such as origin of mountains, earthquakes, and origin of springs and rivers.

4-60   **Exploring the Ocean World: A History of Oceanography.** By Clarence P. Idyll. Rev. ed. New York: T. Y. Crowell, 1972. 296p. illus. (part col.).

An excellent general history of oceanography, each of whose 11 chapters is by a different author. Topics covered are: the science of the sea, underwater landscape, biology of the sea, physics of the sea, chemistry of the sea, food from the sea, farming the sea, mineral resources and power, underwater archaeology, marine ecology and pollution, and man beneath the sea.

4-61   **Mapmakers of America: From the Age of Discovery to the Space Era.** By S. Carl Hirsch. New York: Viking, 1970. 176p. illus.

A small compact history of map making. Although not comprehensive, it is still useful.

4-62   **Source Book in Geology.** By Kirtley Fletcher Mather. New York: McGraw-Hill, 1939. 702p. illus. (Source Books in the History of the Sciences).

4-62a **Source Book in Geology, 1900-1950.** By Kirtley Fletcher Mather. Cambridge, MA: Harvard University Press, 1967. 435p. illus. (Source Books in the History of the Sciences.

Collections of excerpts from the writings of early geologists.

HISTORY OF TECHNOLOGY

4-63   **Bibliography of the History of Electronics.** By George Shiers. Metuchen, NJ: Scarecrow Press, 1972. 323p. index.

Contains 1,820 descriptive annotations of books, articles, and reports on the history of electronics and telecommunications from the 1860s to the 1950s.

4-64   **The Early Years of Modern Civil Engineering.** By Richard Shelton Kirby. New Haven: Yale University Press, 1932. 324p. illus.

Assembles scattered insights on civil engineering history of the eighteenth and the first part of the nineteenth centuries. Each chapter covers a specific part of civil engineering, such as roads and pavements, railroads, bridges, and sewers. Another chapter offers biographical sketches of practitioners. Select bibliographies at the end of each chapter.

4-65   **History of Technology.** By Charles Joseph Singer. New York: Oxford University Press, 1954-1958. 5v. illus. (part col.).

The definitive history of technology by an outstanding historian of science. The individual volumes cover: v. 1, From Early Times to the Fall of the ancient Empires; v. 2, Mediterranean Civilizations and the Middle Ages; v. 3, From Renaissance to the Industrial Revolution; v. 4, Industrial Revolution; and v. 5, The Late 19th Century. Each chapter was prepared by a specialist. Full bibliographies, tables, maps, and illustrations conclude each volume.

4-66   **The Science of Mechanics in the Middle Ages.** By Marshall Clagett. Madison, WI: University of Wisconsin Press, 1959. 711p. illus.

A specialized historical study book of the criticism and resultant modification of Aristotelian mechanics that took place from the thirteenth through the fifteenth centuries.

4-67   **Scientific Instruments.** By Harriet Wynter and Anthony Turner. New York: Scribner's, 1975. 239p. illus. (part col.). bibliog. index.

This book is for collectors since it concentrates on the smaller or more usually found instruments dating from about 1550 to 1850. Among the instruments described are astrolabes, quadrants, spheres, orreries, compasses, charts and other navigational gear, chronometers, sundials, perpetual calendars, gnomons, surveying rods, theodolites, alidades and levels, spectacles, telescopes, pocket globes, microscopes, and backstaffs.

4-68   **A Short History of Technology from the Earliest Times to A.D. 1900.** By Thomas K. Derry. New York: Oxford University Press, 1970 (c.1960). 783p. illus. (Oxford Paperbacks, 231).

A good general history of technology well suited to small and medium-sized public libraries.

HISTORY OF MEDICINE

4-69   **Bibliography of the History of Medicine.** By U.S. National Library of Medicine. v. 1- . Washington: GPO, 1965- . annual.

This annual bibliography will be cumulated every five years. Covers journal articles, monographs, and analytic entries for symposia, congresses and similar composite publications, as well as historical chapters in general monographs.

4-70   **A Biographical History of Medicine, Excerpts and Essays on the Men and Their Work.** By John H. Talbott. New York: Grune and Stratton, 1970. 1211p. illus. index.

This unique compilation consists of over 550 biographical essays on the men (and a handful of women) who have made significant contributions to the medical sciences from 2250 B.C. (Hammurabi) through the first half of the twentieth century. Each entry, which is illustrated by a photograph or composite illustration of the biographee, includes a description of the subject's life and contributions and excerpts from his or her writings.

4-71   **Garrison and Morton's Medical Bibliography: An Annotated Check-List of Texts Illustrating the History of Medicine.** By Leslie T. Morton. 3rd ed. Philadelphia, PA: Lippincott, 1970. 872p. index.
"This bibliography is an attempt to bring together in convenient form references to the most important contributions to the literature of medicine and ancillary sciences, and, by means of annotations, to show the significance of individual contributions in the history and development of the medical sciences." There are 7,534 entries in this edition, chronologically arranged by medical category and including full name of author, with dates, titles, publisher or journal citation, and in most cases a brief annotation.

4-72   **History of Medical Illustrations from Antiquity to A.D. 1600.** By Robert Herrlinger. London: Pitman Medical and Scientific Publishing Co., 1980. 178p. illus. (part col.). index.
An outstanding book on medical illustrations. It is a translation of *Geschichte der medizinischen Abbildung.*

4-73   **History of Medicine.** By Arturo Castiglioni. 2nd ed. rev. and enlarged. New York: Knopf, 1947; repr. New York: Jason Aronson, 1973. 1192p. illus.
This is a comprehensive, well-written history of medicine.

4-74   **History of Medicine.** By Ralph Herman Major. Springfield, IL: Charles C. Thomas, 1954. 2v. illus.
Provides a "continuous account of the stream of medical history, punctuated with the names of eminent physicians often accompanied by a brief biography." Written for the medical student and medical practitioner. At the end of each section there is a biographical addendum covering those persons who were just mentioned in the text.

4-75   **History of Medicine.** By H. E. Sigerist. New York: Oxford University Press, 1951-1961. 2v. illus. (Yale University. Department of the History of Medicine. Publications No. 27, 38).
The two volumes are devoted to 1) Primitive and Archaic Medicine and 2) Early Greek, Hindu and Persian Medicine. The author's premature death cut short what had been projected in eight volumes. These two periods, however, receive their definitive treatment here. Volume 1 contains exhaustive bibliographies.

4-76   **Source Book of Medical History.** By Logan Clendening. New York: Dover, 1960 (c.1942). 685p.
This is a collection of original contributions in the field of medicine. It is arranged in chronological order beginning with the *Kahun Papyrus* of 1900 B.C. to Wilhelm Conrad Roentgen, *On a New Kind of Rays* (1895).

5

Mathematics

Mathematics explores the forms, arrangement, and relationships of numbers. It uses defined literal, numerical, and operational symbols. In the study of mathematics, one encounters angles, triangles, squares, polygons, algebra, curves, differential equations, calculus, integrals, integration, series, space, topology, statistics, tensors, vectors, trigonometry, theory of probability, and computer science. Mathematics serves as a basis for most of the other science fields; no science is exempt from using mathematics.

The ancient Greeks accepted the science of mathematics as basic to all sciences. Euclid, the father of geometry, developed his theorems in the third century B.C. and published them in *The Elements*. About 250 A.D., Diophantus' produced his *Arithmetica* that was used by Fermat in his development of analytic geometry. It was not until the seventeenth century that the present system of signs for arithmetic calculations, percent, and fractions was introduced and standardized, to facilitate commercial transactions. The Arab, al-Khwarizmi, developed algebra. He was also an astronomer, which shows how the two sciences worked together. Ptolemy, who lived from 90-268 A.D. was an astronomer and geographer and set forth the concepts of trigonometry. In the later 1600s and early 1700s, Isaac Newton and Gottfried Wilhelm von Leibnitz developed calculus while Descartes, during the early 1500s founded the science of analytic geometry.

Since then, several individuals have contributed to the development of mathematics. Leonhard Euler (1707-1783) provided a proof of the binomial theorem; Carl Gauss, Lobachevsky, Bolyai and Riemann worked on non-Euclidean geometry. Bertrand Russell and Alfred North Whitehead produced *Principia Mathematica* (1910-1913) which set in motion the research in symbolic logic. In 1854 George Boole developed what is now known as Boolean algebra, an integral part of programming electronic computers.

Societies have also contributed to the study of mathematics and have produced numerous mathematical publications. The oldest existing mathematical society is the Mathematische Gesellschaft in Hamburg, founded in 1690. Its major publication is its *Mitteilungen*. The Moscow Mathematical Society, founded in 1864, is the oldest national mathematics society and publishes the *Uspekhi Matematicheskikh Nauk*. The London Mathematical Society, with its *Journal* and *Proceedings*, was founded in 1865. The New York Mathematics Society was founded in 1888 and changed to the American Mathematical Society in 1894. It is the oldest mathematical society in the United States. Several important serials and periodicals published by the American Mathematical Society include its *Proceedings, Memoirs, Transactions, Mathematical Reviews, Colloquium Publications, Mathematical Survey, Bulletin*, and *Symposium Proceedings*. Among other international mathematical societies are the Chinese Mathematical Society, Societe Mathematique de France, Deutsche Mathematiker Vereinigung, Edinburgh Mathematical Society, Unione Matematica Italiana,

Mathematical Society of Japan, Mathematical Association of America, Society for Industrial and Applied Mathematics, and Canadian Mathematical Congress.

Mathematics is one of the best bibliographically controlled disciplines in the sciences. Because it is unusual for a math book to become dated, libraries tend to retain all material and to maintain good access records. Mathematicians respect their literature and usually aid with the control of their materials. They strive to have complete collections, especially serial runs.

In the previous edition of this book the mathematical sources were separated into pure mathematics, statistics and computer science. In this edition, all sources are combined into major groupings based on format, rather than on subject.

GUIDES TO THE LITERATURE

5-1   **Computers and Data Processing Information Sources.** By Chester Morrill. Detroit, MI: Gale Research Co., 1969. 275p. (Management Information Guide, 15).
Annotated bibliography of literature on data processing budgeting, finance, organization, personnel, records, and equipment, and of reference books for administrative offices. Appendixes: directories of associations, manufacturers, and services. Still useful despite its age.

5-2   **Current Information Sources in Mathematics: An Annotated Guide to Books and Periodicals, 1960-1972.** By Elie M. Dick. Littleton, CO: Libraries Unlimited, 1973. 281p. index.
Topically arranged in 33 chapters, some 1,600 titles of the most important monographic English-language publications appearing from 1960 to 1972 are listed by full bibliographic citation and annotated. Essential works are coded to reviews in *Mathematical Reviews* (5-24). Also covers reference books, periodicals, professional organizations, and publishers. Supplements Parke's *Guide to the Literature of Mathematics and Physics* (5-6).

5-3   **Guide to Computer Literature.** By Alan Pritchard. 2nd ed. Hamden, CT: Shoe String Press, 1972. 194p.
Originally published in London by Clive Bingley, this guide is arranged by primary, secondary, and tertiary sources. Thus, it begins with research reports, journal articles, patents, proceedings, etc.; moves on to abstracts, bibliographies, review services, and handbooks; and interprets tertiary sources as libraries, directories, and service bureaus. Also includes reports on cost of information and statistical studies of the nature of computing literature.

5-4   **Guide to Reference Sources in the Computer Sciences.** By Ciel Carter. New York: Macmillan Information, 1974. 237p.
Compiled on the principle that sources of information are of prime importance, the first chapters concentrate on societies and "supersocieties" and their symposia and conference literature and on research centers, followed by the more traditional chapter sequence of bibliographies and reference works, keyed back to the first chapters whenever appropriate. Final sections list journals (English language first), publishers, and acronyms of organizations. Annotations are descriptive and critical. Title and association indexes.

5-5   **Guide to Tables in Mathematical Statistics.** By Joseph Arthur Green-
         wood and H. O. Hartley. Princeton, NJ: Princeton University Press,
         1962. 1014p. diagrams.
Sponsored by the Committee on Statistics of the Division of Mathematics of the
National Academy of Sciences—National Research Council. Classified arrange-
ment; appendixes include "Contents of Books of Tables; and "Mathematical
Tables Listed in This Guide and in Fletcher, Etc." Author and subject indexes.
Updated by *Statistical Theory and Methods Abstracts* (5-26).

5-6   **Guide to the Literature of Mathematics and Physics Including Related
         Works on Engineering Science.** By Nathan Grier Parke. 2nd ed. rev.
         New York: Dover, 1958. 436p. index.
Preceded by chapters on research and library use, the main bibliography of some
5,000 titles is arranged under some 120 subject headings with occasional notes.
"Bibliographic aids" is one of the lengthy categories. Definite emphasis on
applied mathematics. For more current materials see Dick's *Current Information
Sources in Mathematics* (5-2).

5-7   **How to Find Out in Mathematics: Guide to Sources of Information.** By
         John E. Pemberton. 2nd ed. rev. Elmsford, NY: Pergamon, 1969. 193p.
         (Commonwealth and International Library).
Excellent, comprehensive manual of sources for librarians and students. Covers
careers for mathematicians; organization of information; dictionaries, encyclopedias, and theses; periodicals and abstracts; societies; mathematical education; sources for mathematical and computer tables, for history, and for
biography; and bibliographies. The last part is devoted to sources for applications of mathematics; probability and statistics; operational research; and actu-
arial science. Special features: evaluation and acquisition of mathematical books;
government and mathematics; and Russian sources.

5-8   **Use of Mathematical Literature.** Ed. by A. R. Dorling. Boston: Butter-
         worths, 1977. 260p. index. (Information Sources for Research and
         Development).
This is a graduate level guide that covers the general character of mathematical
literature—e.g., the structure of the discipline, major organizations and journals,
and reference materials; history of mathematics; combinatorics; rings and
algebras; group theory; measure and probability; topology; and mathematical
programming.

5-9   **Using the Mathematical Literature: A Practical Guide.** By Barbara
         Kirsch Schaefer. New York: Marcel Dekker, 1979. 141p. index. (Books
         in Library and Information Science, v. 25).
This book is intended to provide an insight to the vast and varied amount of
mathematical literature and to act as a guide to its exploitation. It is not a comprehensive bibliography, but rather a starting point for further investigation.

ABSTRACTS, INDEXES, AND BIBLIOGRAPHIES

5-10   **Bibliography of Basic Texts and Monographs on Statistical Methods,
         1945-1960.** By William R. Buckland and Ronald A. Fox. 2nd ed. New
         York: published for the International Statistical Institute by Hafner,
         1963. 297p. index.

A very careful selection of almost 200 English language entries on statistical methods and their applications. Arrangement is by broad subject categories with an author index.

5-11 **Bibliography of Multivariate Statistical Analysis.** By T. W. Anderson, Somesh Das Gupta, and George P. H. Styan. New York: Wiley, 1973. 642p. index.

A comprehensive compilation of just about everything that has been written (in almost any language) on multivariate analysis of continuous variables, up to 1966 for articles and 1970 for books. There are entries for some 6,093 research papers and 213 books, arranged alphabetically by author. The items have also been grouped according to 20 subject-matter areas, briefly described, and subdivided into 94 distinct topics; for each of these one can locate the authors who have written something of relevance. For locating pertinent writings on more narrowly specified topics, there is an index containing roughly 450 entries.

5-12 **Bibliography of Statistical Bibliographies.** By Henry Oliver Lancaster. Edinburgh: Oliver and Boyd, 1968. 103p. index.

Covers the range of subjects that might be taught in a university department of statistics, mathematical statistics, or probability. It is arranged in two classes—personal (contributors to statistical theory) and subject (methodology and applications). Each entry is a complete citation, giving author, year, title, publisher, and number of pages. "Author index" refers to names of authors, editors, compilers, and biographees.

5-13 **Bibliography of Statistical Literature.** By Maurice G. Kendall and Alison G. Doig. Edinburgh: Oliver and Boyd, 1962-1968. 3v.

Comprehensive listing of the significant contributions to statistics from the sixteenth century to 1959. For current information consult *Statistical Theory and Methods Abstracts* (5-26).

5-14 **Bibliography on Time Series and Stochastic Processes.** By Herman Wold. Cambridge, MA: MIT Press, 1965. 516p.

Covers research on time series and stochastic processes up to 1965. For current information on this topic consult *Statistical Theory and Methods Abstracts* (5-26).

5-15 **Computer and Control Abstracts.** v. 1- . London: Institution of Electrical Engineers; New York: Institute of Electrical and Electronics Engineers, 1966- . monthly. (Science Abstracts Series C).

An international abstracting journal covering systems and control theory, control and computer technology and applications, systems and equipment. Over 24,000 items are abstracted each year. Abstracts are short, and the language of the original article is indicated. Monthly and annual indexes: author, bibliography, books, conferences, patents, and subject. The supplementary current awareness publication of the two institutes is *Current Papers on Computers and Control* (1966- .) with approximately 2,000 references per issue.

5-16 **Computer and Information Systems Abstract Journal.** v. 1- . Riverdale, MD: Cambridge Scientific Abstracts, 1962- . monthly.

An international abstracting journal for material on computer software; computer applications; computer mathematics; and computer electronics. Covers periodicals, government reports, conference proceedings, books, dissertations and patents. Indexes: acronyms, subjects, authors, and sources. Started as *Information Processing Journal*, which changed to *Computer and Information Systems* with current title in 1978.

5-17 **Computer Literature Bibliography, 1946-1967.** Ed. by W. W. Youden. New York: Arno Press, 1970 (c.1967). 844p.

Encompasses some 11,500 citations from journals, books, workshop reports, and conference proceedings to achieve exhaustive bibliographic control of electronic data processing to 1967. The work was originally published by the U.S. National Bureau of Standards in two volumes, covering the years 1946-1963 and 1964-1967, respectively, and is now conveniently reprinted in one volume. However, no attempt was made to integrate the contents; the two volumes are printed consecutively. Access is provided by a bibliography section, which includes the full title and author(s) of each article.

5-18 **Computing Reviews.** v. 1- . New York: Association for Computing Machinery, 1960- . monthly.

Provides short signed reviews of computer literature in classified arrangement, emphasizing machine aspects such as hardware and software; however, there are also sections on textbooks, proceedings, the mathematics of computation, functions, and applications to natural sciences and engineering. There are occasional features and multiple reviews and some authors' rebuttals. Lists of reprints available from ACM of their publications and of the original bibliographies on specific topics published at irregular intervals are included along with the reviews. Author index for each volume and cumulative permuted KWIC indexes.

5-19 **Current Mathematical Publications.** v. 1- . Providence, RI: American Mathematical Society, 1969- . biweekly.

Produced in the editorial offices of *Mathematical Reviews* (5-24). Each issue contains a list of items to be reviewed in *Mathematical Reviews*. Items are listed by subject and then alphabetically by author, giving complete bibliographical citation. Formerly called *Contents of Contemporary Mathematical Journals* and *New Publications* until 1975.

5-20 **Guide to Mathematical Tables.** By Aleksandr Vasil'evich Lebedev and R. M. A. Fedorova. New York: Pergamon, 1960. 586p. supplements.

A translation of the Russian *Spravochnik Po Matematicheskikh Tablits*. Similar to Fletcher's *Index to Mathematical Tables* (5-21) but omitting many of the more obscure tables.

5-21 **Index to Mathematical Tables.** By Alan Fletcher. 2nd ed. Reading, MA: Addison-Wesley, 1962. 2v.

An indispensable working index to tables in mathematical works of international coverage, from the sixteenth century to 1961. By "working" index, Fletcher intends to exclude tables of historical interest only. The two volumes are arranged in four parts as follows: volume 1 contains part I, Historical Introduction and Index According to Functions; volume 2 contains part II, Bibliography (of sources of tables); part III, Errors (in published tables); and part IV, Index to

Part I. Information given for each table in part I includes number of decimals or figures, interval and range of argument, facilities for interpolation, authorship, and date. To find a complete citation of a table, one refers to part II, the alphabetical bibliography, using the abbreviated citation—author's name and date—provided in the reference in part I.

5-22 **Information Science Abstracts.** v. 1- . Philadelphia, PA: Documentation Abstracts, 1966- . bimonthly.

First published as *Documentation Abstracts*. Abstracts about 600 articles in each issue from over 350 journals covering material in documentation and related fields.

5-23 **Jahrbuch über die Fortschritte der Mathematik.** Berlin: DeGruyter, 1868-1942. 60v.

This abstracting service was the major one for the period covered. It was written in German, and provided international coverage using a classified arrangement. Essential for the period prior to 1940 when *Mathematical Reviews* (5-24) began publication.

5-24 **Mathematical Reviews.** v. 1- . Providence, RI: American Mathematical Society, 1940- . 2v. per year in 12 numbers.

This monthly abstracting service has international coverage of mathematical works. Most abstracts are in English, but French and German abstracts are also included. Abstracts are numbered consecutively, and each gives the author, title, periodical citation and/or publisher. Each book and monograph has a star preceding the author's name and a price following the citation. Journal issues on a given topic may be reviewed as a whole and supplied with an ad hoc title. Titles of articles in French, German, and English are given in their original language, while those in other languages are translated into English with an indication of the original language. The abstracts themselves are arranged by subject and are signed. Reprints from other science abstracting services are identified by abbreviations. There is an author index and a brief subject index to the categories, including symposia, proceedings, biographical collections, and bibliographies. Important to note is that this service uses its own method of transliterating Russian, which varies considerably from the Library of Congress method; for convenience, comparative transliteration tables from several sources are printed at the end of the issues.

5-25 **Mathematics of Computation.** v. 1- . Providence, RI: American Mathematical Society, 1943- . quarterly.

Volumes 1 through 13 were called *Mathematical Tables and Other Aids to Computation*. This quarterly periodical is devoted to avances in numerical analysis, applications of computational methods, mathematical tables, high-speed calculations, and other aids to computation. An important part of this work is the section called "Reviews and Descriptions of Tables and Books," which also lists unpublished mathematical tables (UMT) kept in the editorial office of the periodical, copies of which may be requested at a nominal cost.

5-26 **Statistical Theory and Method Abstracts.** v. 1- . Edinburgh: published for the International Statistical Institute by Oliver and Boyd, 1959-1970; London: Longman, 1971- . quarterly.

Merged in 1964 with *International Journal of Abstracts: Statistical Theory and Methods, 1954-1963.* Classified arrangement of such topics as probability, distributions, variance analysis, experiment design, sampling, new tables, and stochastic theory and time series analysis. Author index.

5-27 **Zentralblatt für Mathematik und ihre Grenzgebiete.** v. 1- . Berlin: Springer, 1931- .
An international, classified abstracting service with author and subject indexes. Signed abstracts are written in English, French, or German, although the editorial apparatus is in German. "Grenzgebiete" means literally "boundary areas"; hence, the service covers physics, astrophysics, and geophysics.

ENCYCLOPEDIAS

5-28 **Condensed Computer Encyclopedia.** By Philip B. Jordain. New York: McGraw-Hill, 1969. 605p. diagram. bibliog. index.
Intended primarily for the businessman and the student rather than the computer specialist, this work defines technical terms, explains interpretive languages such as LISP, and translates jargon into plain English. Excellent use of illustrative examples to clarify such functions as operation of loops and truth tables. Somewhat dated but still useful.

5-29 **Encyclopedia of Computer Science.** Ed. by Anthony Ralston. New York: Petrocelli/Charter, 1976. 1523p. illus. index.
This is a book intended to serve the layman, the non-specialist in the field, and the specialist needing information outside his area of specialization with over 480 signed articles covering all aspects of computer science. Prior to all the articles, the editor has included a broad taxonomy of computer science, with five main subject areas listed: computer science, data processing, information science, information processing, and symbol manipulation. One appendix covers the full citations for abbreviations and acronyms common in computer science, as well as a rundown on mathematical notation and various units of measure. Another covers three useful numerical tables: power of two, octal-decimal integer conversions, and octal-decimal fraction conversion.

5-30 **Encyclopedia of Computer Science and Technology.** v. 1- . Executive eds. Jack Belzer, Albert G. Holzman, and Allen Kent. New York: Marcel Dekker, 1975- .
A comprehensive encyclopedia that surveys the computer field and its broad spectrum of component subfields.

5-31 **Encyclopedia of Computers and Data Processing.** v. 1- . Detroit, MI: International Electronics Information Services, 1978- . illus.
A comprehensive encyclopedia covering all aspects of computers and data processing. Many of the articles are signed and include bibliographies. When finished, this should be a welcome addition to any reference collection.

5-32 **Encyclopedic Dictionary of Mathematics.** By the Mathematical Society of Japan. Cambridge, MA: MIT Press, 1977. 2v. illus. index.
This is a translation of the Japanese *IWanami Sugaku Ziten (IWanami Mathematical Dictionary).* The bulk of the dictionary consists of well printed and

arranged material on every major mathematical topic known plus biographical articles. There are 23 specialized tables that range from algebraic equations, knot theory, conformal mappings, and interpolation to statistical hypothesis testing. There are numerous indexes and appendixes to supplement the text.

5-33   **Enzyklopädie der mathematischen Wissenschaften, mit Einschluss ihrer Anwendungen.** v. 1- . Hrsg. im Auftrage der Akademie der Wissenschaften zu Berlin. 2. völlig neubearb. Aufl. Leipzig: Teubner, 1950- .

It took from 1898 to 1935 to publish the first edition of six volumes in 23. The second edition is also appearing slowly. As is typical of German scholarly encyclopedias, this work is noted for its comprehensive, accurate, and detailed treatment and is directed to the professional scientist. Volumes focus on main subfields of mathematics and on fields of application.

5-34   **Fundamentals of Mathematics.** Ed. by H. Behnke et al. Cambridge, MA: MIT Press, 1974. 3v.

This major work should serve as an excellent comprehensive reference source; it is a translation from the second German edition of *Grundzuge der Mathematik* (Göttingen: Vandenhoeck and Ruprecht, 1960. 5v. in 6). Contents: v. 1, Foundations of Mathematics: The Real Number System and Algebra; v. 2, Geometry; v. 3, Analysis.

5-35   **Universal Encyclopedia of Mathematics.** Foreword by James R. Newman. New York: Simon and Schuster, 1969. 598p. diagrams. tables.

A general encyclopedia designed for secondary school and college students. Topical coverage is from arithmetic through calculus, followed by separate parts on formulas and on tables of functions with explanations. Based on a German work, Joseph Meyers' *Grossen Rechenduden* (1964).

5-36   **VNR Concise Encyclopedia of Mathematics.** Ed. by W. Gellert and others. New York: Van Nostrand Reinhold, 1977. 1v. (various paging). illus. (part col.). index.

This is an English version of the *Kleine Enzyklopadie der Mathematik.* Part 1 covers traditional areas of elementary mathematics such as plane and solid geometry, algebraic equations, and development of the number system; part 2 goes into the diverse aspects of higher mathematics, such as set theory, calculus, and numerical analysis; and part 3 contains brief surveys of various facets of contemporary mathematics. There are some basic tables, an index of mathematicians, and a detailed subject index.

DICTIONARIES

5-37   **Computer Dictionary.** By Donald D. Spencer. 2nd ed. Ormond Beach, FL: Camelot Publishing, 1979. 192p. illus.

This is a layman's glossary of 2,500 words, phrases, and acronyms, covering people, machines, and organizations important to the computer's historical and present development; programming languages; as well as terms used in personal computing, computer-assisted instruction, management, and society.

5-38 **Computer Dictionary and Handbook.** By Charles J. Sippl and Charles P. Sippl. 2nd ed. Indianapolis, IN: Howard W. Sams, 1972. 778p. illus.
Includes 22,000 definitions that are well written and clear, with extensive explanations. Acronyms and abbreviations are given in the main list of terms but may also be found under special headings in appendixes. A great many cross-references are included, and often a definition is repeated under variant forms of an expression. About one-third of the book consists of textbook data on such subjects as computer system principles and procedures, computer systems personnel, management science, model building techniques, operations research, number systems, flowcharting, and computer languages.

5-39 **Data Communications Dictionary.** By Charles J. Sippl. New York: Van Nostrand Reinhold, 1976. 545p.
Over 14,500 terms, concepts, and abbreviations currently used in the fields of data communications, carrier systems, special data networks, and data systems equipment and components are defined. There is also a list of acronyms and abbreviations at the end of the volume.

5-40 **Diccionario matematico; expañol-ingles, ingles-español. Mathematics Dictionary; Spanish-English, English-Spanish.** By Mariano Garcia Rodriguez. New York: Hobbs, Dorman, 1965. 78p.
Only equivalents of words and phrases in Spanish and English are presented, but this dictionary is an asset to libraries that serve Spanish-speaking communities.

5-41 **Dictionary of Computers.** By Anthony Chandor, John Graham, and Robin Williamson. 2nd ed. New York: Penguin Books, 1977. 440p.
A glossary of some 3,000 words, phrases, and acronyms used in connection with computers. Entries are explained in simple English, with basic concepts explained at length when necessary and advanced words extensively cross-referenced. An introductory article on computing precedes the glossary.

5-42 **Dictionary of Data Processing.** By Jeff Maynard. New York: Crane Russak, 1976 (c.1975). 269p. illus.
This is a small well-done dictionary of data processing covering more than 4,000 terms intended primarily for computer users and computer and data process managers. Includes six appendixes, including a list of common acronyms and abbreviations, and an explanation of the basic flowchart symbols.

5-43 **Dictionary of Logical Terms and Symbols.** By Carol Horn Greenstein. New York: Van Nostrand Reinhold, 1978. 188p. illus. bibliog.
The various notational systems used by logicians, engineers, and computer scientists are interrelated in this dictionary. It includes multicolumn tables that enable translation from one notational system to another, and presents the many ways in which key English expressions can be formalized. The main symbolic forms of Boolean algebra, set theory, quantification theory, two-termed relations, logic gates, and epistemic, doxastic, deontic, and tense logic are presented in alternative systems, such as Peano-Russell, Hilbert, and Polish notation, as well as in English. It also includes program flowchart symbols, Euler and Venn diagrams for categorical propositions and syllogisms, consistency trees, truth-functional and binary truth tables, and 66 circuit diagrams.

5-44 **Dictionary of Mathematical Sciences.** By Leo Joseph Herland. 2nd ed. rev. and enlarged. New York: Ungar, 1965. 2v.
Volume 1, German to English; volume 2, English to German. Only equivalents are given under each entry term.

5-45 **Dictionary of Mathematics.** By T. Alaric Millington and William Millington. New York: Barnes and Noble, 1971 (c.1966). 259p.
Contains brief definitions in all areas of mathematics serving as a general dictionary for beginning students.

5-46 **Dictionary of Statistical Terms.** By Maurice G. Kendall and William R. Buckland. 3rd ed. rev. and enlarged. Published for the International Statistical Institute. New York: Hafner, 1971. 166p.
The standard, professionally prepared dictionary of basic statistical terminology, containing well over 2,500 entries. It provides evaluations of controversial usage and for further clarification refers to entries in the *Bibliography of Statistical Literature* (5-13) and *Statistical Theory and Method Abstracts* (5-26). Essential for all collections.

5-47 **Dictionary of Symbols of Mathematical Logic.** Ed. by R. Feys and F. B. Fitch. Amsterdam: North Holland Publishing Co.; distr. New York: Humanities Press, 1969. 175p. (Studies in Logic and the Foundations of Mathematics).
Explains and interprets the concepts and notations of formalized deductive systems of symbolic logic used in standard mathematical philosophical works.

5-48 **Elseviers's Dictionary of Computers, Automatic Control, and Data Processing.** In Six Languages: English/American, French, Spanish, Italian, Dutch and German. Comp. and arranged on an English alphabetical basis by W. E. Clason. 2nd ed. rev. Amsterdam: Elsevier, 1971. 474p.
Includes close to 4,000 entries in all areas of information processing and computers. Terms are listed in English followed by their equivalents in the other languages.

5-49 **Funk and Wagnalls Dictionary of Data Processing Terms.** By Harold A. Rodgers. New York: Funk and Wagnalls, 1970. 151p.
Although neither comprehensive in coverage nor thorough in its definitions, this work presents material in very readable manner. Appendixes offer invaluable additional information—e.g., mathematical symbols, character codes, and acronyms.

5-50 **Glossary of Computing Terminology.** By C. L. Meek. New York: CCM Information Corporation, 1972. 372p.
Each term included is accompanied by a reference symbol that indicates the source from which the definition originates. In providing access to a wide variety of definitional sources, the work effectively supplements more conventional dictionaries.

5-51 **International Dictionary of Applied Mathematics.** Ed. by W. F. Freiberger. Princeton, NJ: Van Nostrand, 1960. 1173p.

One of the best encyclopedic dictionaries, defining some 8,000 terms and methods with some appropriately lengthy explanations in the application of mathematics for advanced scientists in engineering and the physical sciences. French, German, Spanish, and Russian indexes to the English language main section.

5-52　**Layman's Dictionary of Computer Terminology.** By Norman Sondak and Eileen Sondak. New York: Hawthorn, 1973. 203p.

A simplified, non-technical dictionary that is helpful to the novice. Most of the principal words and acronyms currently used have been included, and the definitions are clear and concise. Multiword terms are listed according to current usage and not inverted. Another advantage to this dictionary over many other such publications is the presence of pronunciation guides. Short biographies of important people in the field are also included.

5-53　**MAST. Minimum Abbreviations of Serial Titles – Mathematics.** By Mary L. Tompkins. North Hollywood, CA: Western Periodicals, 1969. 427p.

This unique index, which looks complicated at first glance, depends on an understanding of the concepts concerning "minimum abbreviations" or MINABBS, each of which "in general consists of the initial letters of a word up to but not including the second vowel." For example, Memoirs would be Mem., Societa would be Soc., and Institute would be Inst. The user takes the titles of a serial as it appears in the citation of a footnote or bibliography, applies the MINABBS concept and checks the index. The index gives the official title in brief form plus an index number that refers to the second section. Here, in numerical order, the user finds the full bibliographical, unabbreviated listing of over 2,000 serials cited in the literature of mathematics with place of publication and date of first issue.

5-54　**Mathematics Dictionary.** By Glen James and Robert C. James. 4th ed. Princeton, NJ: Van Nostrand Reinhold, 1976. 509p. illus. index.

The best dictionary available for general coverage of arithmetic through calculus plus the fields of probability and statistics. It also includes biographic sketches of individuals who have contributed to the development of mathematics. More than 8,000 terms are arranged alphabetically by leading word but many associated and analogous terms are subordinated.

5-55　**Mathematische Wörterbuch mit Einbeziehung der theoretischen Physik.** Ed. by Josef Naas and Hermann Ludwig Schmid. Berlin: Akademie Verlag; New York: Pergamon, 1961. 2v.

A comprehensive German dictionary containing definitions, articles on trends and biographies of some 400 mathematicians. Bibliographical notes follow some articles. Appendixes include a list of symbols and a list of 82 branches and applications of mathematics.

5-56　**Microcomputer Dictionary and Guide.** By Charles J. Sippl and David A. Kidd. Champaign, IL: Matrix, 1976 (c.1975). 516p. illus.

The dictionary offers over 5,000 entries of terms, definitions and products, and explanations of procedures and applications. Seven useful appendixes cover symbols, units and constants of electronics; mathematics definitions; statistics definitions; electronics and computer acronyms and abbreviations; computer language summaries, APL, BASK and FORTRAN; computer number and binary switching systems; and definitions of programmable calculator terms.

5-57 **Russian-English Dictionary of the Mathematical Sciences.** By A. J. Lohwater and S. H. Gould. Providence, RI: American Mathematical Society, 1974 (c.1961). 267p.

Prepared under the joint auspices of the U.S. National Academy of Sciences, the Academy of Science of the USSR, and the American Mathematical Society. Provides equivalents for more than 10,000 terms. Intended for scientists' use at any level of mathematics and theoretical physics.

5-58 **Standard Dictionary of Computers and Information Processing.** By Martin H. Weik. 2nd ed. rev. Rochelle Park, NJ: Hayden Book Co., 1977. 390p. illus. bibliog.

Contains full explanations, practical examples, and supplementary information for over 12,500 hardware and software terms. It includes terminology from such fields as information retrieval, mathematics, electronics, and photomicrography. Includes many practical examples, numerous illustrations, and cross-references to other closely related terms plus some definitions that are close to encyclopedic in length.

5-59 **Technical Dictionary of Data Processing, Computers and Office Machines: English/German/French/Russian.** By Erich Burger with the cooperation of Wolfgang Schuppe. New York: Pergamon, 1970. 1463p.

The vocabulary is presented in four languages, in tabular form, thus eliminating the need for a separate index. Some problems are evident with compound entries, especially those in Russian. As with most multilingual dictionaries, full definitions are not given, only equivalents in other languages.

HANDBOOKS AND TABLES

5-60 **Barlow's Tables of Squares, Cubes, Square Roots, Cube Roots and Reciprocals of All Integers Up to 12,500.** By Peter Barlow. Ed. by L. J. Comrie, 4th ed. London: Spon, 1941; new impression New York: Chemical Publications, 1960. 258p.

Comparable to the *CRC Handbook of Tables for Mathematics* (5-61), it is the preferred handbook for British libraries.

5-61 **CRC Handbook of Tables for Mathematics.** 4th ed. Cleveland, OH: Chemical Rubber Co., 1970. 1120p.

Originally published as a supplement to the *Handbook of Chemistry and Physics* (3-76), it is now the preferred manual used by students and professionals in all sciences because of its comprehensiveness and accuracy for pure and applied mathematics and its selection of probability and finance tables. Supplementing the tables are explanations of their use, basic notes and formulas for algebra, geometry, and analytic geometry, calculus forms, differential equations, Laplace transforms, etc.

5-62 **CRC Handbook of Tables for Probability and Statistics.** By William H. Beyer. 2nd ed. Cleveland, OH: Chemical Rubber Co., 1968. 642p.

A collection of the relatively standard statistical tables. The 13 sections cover probability and statistics; normal distribution; binomial, Poisson, hypergeometric, and negative binomial distribution; student's t-distribution; chi-square distribution; F-distribution; order statistics; range and studentized range; correlation coefficient; non-parametric statistics; quality control; miscellaneous statistical tables; and miscellaneous mathematical tables.

5-63 **Catalog of Special Plane Curves.** By J. Dennis Lawrence. New York: Dover, 1972. 218p. illus. index.
An illustrated book of some special plane algebraic and transcendental curves. The first two chapters give a brief description of properties and types of derived curves. The remaining chapters are devoted to individual curves.

5-64 **Collected Algorithms from CACM.** New York: Association for Computing machinery, 1978. 2v. looseleaf. index.
Contains all algorithms published in the Algorithms Department of Communications of the ACM (CACM) and the Association for Computing Machinery's *Transactions on Mathematical Software (TOMS).* New algorithms, certifications, remarks, and translations are issued quarterly. Each entry lists the identifying title of the algorithm, source, author and his institutional address, certification status of the algorithm, and remarks. Beginning with no. 493, the algorithms are available in machine-readable form.

5-65 **Eight-Place Tables of Trigonometric Functions for Every Second of Arc, with an Appendix on the Computation to Twenty Places.** By Jean Peters. Bronx, NY: Chelsea, 1971 (c.1968). 954p.
Eight place values for sine, cosine, tangent, and cotangent are given for every second of the arc $0°$-$90°$. An appendix covers the calculation of sine and cosine to 21 places. A monumental compilation accepted as authoritative in the scientific community.

5-66 **Engineering Mathematics Handbook: Definitions, Theorems, Formulas, Tables.** By Jan J. Tuma. 2nd ed. rev. and enlarged. New York: McGraw-Hill, 1979. 394p. illus. bibliog. index.
This is "a concise summary of the major tools of engineering mathematics . . . prepared with the intent to serve as a desk-top reference book for engineers, scientists, and architects." Each of the parts of the text is grouped according to the type of mathematics covered: 1—algebra, plane and solid geometry, trigonometry, analytic geometry, and elementary functions; 2—differential calculus, infinite series, integral calculus, vectors, complex variables, Fourier series, and special functions; 3—ordinary and partial differential equations and related topics; 4—numerical methods, probability, statistics, and related tables of numerical coefficients; 5—indefinite integrals.

5-67 **Guide to the Applications of the Laplace and 8-transforms.** By Gustav Doetsch. 2nd ed. New York: Van Nostrand Reinhold, 1971. 240p. index.
This edition has been rewritten to simplify mathematical theory and to stress the practical applications of the Laplace transform for the engineer and physicist.

5-68 **Handbook of Applied Mathematics.** By Edward E. Grazda and Martin Ernest Jansson. 4th ed. New York: Van Nostrand Reinhold, 1966; repr. 1977. 1119p. illus.

A self-instructing ready reference work on the application of mathematics to the practical trades. The first six chapters are a review of the operations of arithmetic, algebra, geometry, trigonometry, and differential calculus. The remaining chapters apply these mathematical skills to such fields as carpentry, electricity, and machine-shop work.

5-69   **Handbook of Mathematical Functions: With Formulas, Graphs and Tables.** Rev. ed. Washington: U.S. National Bureau of Standards; distr. Washington: GPO, 1971. 1046p.

This handbook includes every special function usually needed by scientists and engineers in their research. It contains an increase in the number of functions, more extensive numerical tables, and larger collections of mathematical properties of the tabulated functions over the first edition.

5-70   **Handbook of Mathematical Tables and Formulas.** By Richard S. Burington. 5th ed. New York: McGraw-Hill, 1973. 500p. illus. index.

This book is composed of two parts: the first includes the main formulas and theorems of algebra, geometry, trigonometry, calculus, vector analysis, sets, logic, matrices, linear algebra, numerical analysis, differential equations, some special functions, Fourier and Laplace transforms, complex variables, and statistics; the second includes tables of logarithms, trigonometry, exponential and hyperbolic functions, powers and roots, probability distributions, annuity, etc. Students and researchers will find this book useful — particularly the tabulations.

5-71   **Handbook of Mathematics.** Ed. by L. Kuipers and R. Tinman. Trans. by I. N. Sneddon. New York: Pergamon, 1969. 782p. (International Series of Monographs in Pure and Applied Mathematics, v. 99).

Translated from the Dutch, this is a scholarly compendium of the subject. Contents: history of mathematics, number systems, linear algebra, analytical geometry, analysis, sequences and series, theory of functions, ordinary differential equations, special functions, vector analysis, partial differential equations, numberical analysis, the Laplace transform, probability, and statistics.

5-72   **Handbook of Numerical Matrix Inversions and Solution of Linear Equations.** New York: Wiley, 1968; repr. Huntington, NY: Krieger, 1975. 171p.

Covers direct methods, iterative methods, error measures, and scaling, with emphasis on computer solutions. Special features: a glossary of matrix terminology, a list of theorems on matrix algebra, and some test matrices.

5-73   **Handbook of Probability and Statistics with Tables.** By Richard Stevens Burington and Donald Curtis May, Jr. 2nd ed. rev. New York: McGraw-Hill, 1970. 462p. illus. tables. bibliog.

The first edition was published in 1953. This enlarged edition includes a new chapter on non-parametric methods, and some new material was added in such areas as regression theory, analysis of variance, and sampling techniques. As in the first edition, there are many formulas, definitions, specialized tables, and other material for ready reference use. In general, this handbook provides only basic material in probability theory; more advanced methods are not discussed.

5-74 Mathematical Handbook for Scientists and Engineers: Definitions, Theorems, and Formulas for Reference and Review. By Granino Arthur Korn. 2nd ed. New York: McGraw-Hill, 1968. 1130p. illus. (McGraw-Hill Handbooks).

A well organized handbook covering mathematical definitions, theorems and formulas necessary for the scientist, engineer, or student. The format enables the user to select quickly the information needed without having to read through unnecessary discussions. The important formulas and definitions are enclosed in boxes; the main text of each topic presents a concise review in large print; finally, the more detailed discussions and advanced topics are given in smaller print. Includes a glossary of symbols and notations.

5-75 Mathematics Manual: Methods and Principles of the Various Branches of Mathematics for Reference, Problem Solving, and Review. By Frederick S. Merritt. New York: McGraw-Hill, 1962. 378p.

The manual ranges in scope from simple arithmetic through higher mathematics including matrices, tensors, probabilities, and statistics. It gives the important definitions, principles, theorems, corollaries, relationships, and methods of the most commonly used branches of mathematics, with illustrative examples and references.

5-76 New Table of Indefinite Integrals, Computer Processed. By Melvin Klerer and Fred Grossman. New York: Dover, 1971. 198p.

Conventional tables of integrals were used as the primary sources for this table of over 2,000 indefinite integrals.

5-77 SI Metric Handbook. By John L. Feirer. New York: Scribner's; Peoria, IL: Charles A. Bennett, 1977. 1v. (various paging). illus. index.

The publishers suggest that the volume is designed "to serve as a reference for all individuals interested in metric conversion in basic occupational areas." The book is divided into volume A, SI Measuring Systems and ISO Standards, and volume B, Applied Metrics. Volume A contains a history and definition of the SI measuring system (all basic units, conversion methods, key individuals, a worldwide history); precision measuring tools (micrometers, gauges, "rules"); and material on national and international standards with an explanation of the difference between ISO and SI metric standards. Volume B covers design and drafting; metrics in metalworking; metrics in other areas such as woodworking, graphic arts, and auto mechanics; metrics in office and business practice; and metrics in the domestic sciences — food, clothing, agriculture, and health.

5-78 Selected Tables in Mathematical Statistics, Volume 1. Ed. by H. L. Harter and D. B. Owen. Chicago: Markham, 1970. 405p. (Markham Series in Statistics).

The first of several volumes of collected statistics assembled by the Institute of Mathematical Statistics. Five tables, each checked for accuracy before being entered, are included in this volume: Tables of the Cumulative Non-central Chi-square Distribution; Tables of the Exact Sampling Distribution of the Two-sample Kolmogorov-Smirnov Criterion D_{mn}; Critical Values and Probability Levels for the Wilcoxon Rank Sum Test and the Wilcoxon Signed Rank Test; The Null Distribution of the First Three Product-Moment Statistics for Exponential, Half-Gamma, and Normal Scores; and Tables to Facilitate the Use of Orthogonal Polynomials for Two Types of Error Structures.

5-79 **Simplified Guide to Automatic Data Processing.** By William A. Bocchino. 2nd ed. Englewood Cliffs, NJ: Prentice Hall, 1972. 351p. illus. index.

The text of this guide, which presupposes no expertise, provides a basic glossary and explains the input-processing-output cycle of data processing in non-technical language, with useful illustrations provided by IBM. Systems analysis techniques (e.g., the learning curve, linear programming, Monte Carlo method, queuing theory, simulation, game theory, PERT, and flow charts) are briefly explained and illustrated.

5-80 **Statistical Tables for Science, Engineering, Management and Business Studies.** By J. Murdoch and J. A. Barnes. 2nd ed. New York: Halsted Press (Wiley), 1973 (c.1970). 40p.

Even though some of the tables have been photographically reduced, the selection represents 32 of the commonly used statistical and mathematical tables covering cumulative binomial and Poisson probabilities; percentage points and ordinates of the normal distribution; exponential function; percentage points of t, x2, and F distribution; the correlation coefficient; statistical quality control tables; significance tables for runs; attribute single sampling tables; random number tables; logarithms; antilogarithms; logarithms of factorials; natural logarithms; squares; and square roots.

DIRECTORIES

5-81 **Computer Directory and Buyers' Guide.** v. 23- . Newtonville, MA: Berkeley Enterprises, 1974- . annual as the October issue of *Computers and People.*

Consists of numerous rosters and lists in all areas of the computer industry. Also contains several technical sections. An excellent source for up-to-date information on firms, products and services, facilities, and hardware leasing.

5-82 **Computer Industry Annual.** 1st- . Comp. by International Computaprint Corporation. Concord, MA: Computer Design Publishing Corporation, 1969/70- . illus. index.

Consists of essays and charts analyzing the comparative characteristics of data communication, graphic (cathode ray tube and plotter), memory keyboard, magnetic tape, paper tape, punch card, printing, optical character equipment, and timesharing services. Two directory sections lists services and products by supplier and manufacturer.

5-83 **Computer Yearbook.** Detroit, MI: American Data Processing, 1952- . biennial.

Consists of two sections: series of survey and state-of-the-art articles and a directory and reference section. The directory section includes book publishers, associations, computer manufacturers, bank automation, university computer centers, and companies and private schools offering data processing instruction. A good series of which the older volumes become state-of-the-art volumes useful for historical reference.

5-84 **Directory of Online Databases.** v. 1- . Santa Monica, CA: 1979- . quarterly.

This is a guide to about 400 bibliographic and nonbibliographic data bases and data base families. For each entry in the directory, information is given on type of data base, subject, producer, online service, content, coverage, and updating. The addresses of the producers and online services are given and there are subject, producer, service, and data base name indexes.

5-85   **International Directory of Computer and Information System Services 1974.** 3rd ed. London: Europa Publications; distr. Detroit, MI: Gale Research Co., 1974. 636p. index.

This directory lists the names, addresses, telephone numbers, and principal officers of institutions and companies, plus information (when available) on educational facilities, type of services provided, operating conditions, type of computer installation, etc. Most major countries are included, though the listings for some are highly selective.

5-86   **Makers of Mathematics.** By Alfred Hopper. New York: Random House, 1958. 402p. illus. (Modern Library Paperbacks, P38).

An introduction to the great personalities in mathematics and the concepts that they developed for the foundations of modern mathematics. The author assumes that the reader has no advanced mathematical background.

5-87   **Men of Mathematics.** By Eric Temple Bell. New York: Simon and Schuster, 1965. 591p.

Bell's book is an interesting account of 34 men who were prominent in the development of mathematics from its beginning. Two criteria were used to determine inclusion: first, the known appeal of the man's life and character; second, the importance of the man's works to modern mathematics. The book is arranged in chronological order.

5-88   **Who's Who in Computers and Data Processing 1971: A Biographical Dictionary of Leading Computer Professionals.** 5th ed. Chicago: published jointly by the New York Times Book and Educational Division and Computers and Automation, Berkeley Enterprises; Quagrangle Books, 1971. 3v. index.

This directory has been published for more than 20 years in various formats and under various titles. Although the current edition is dated for up-to-date information it is valuable for historical data. Volume 1, Systems Analysts and Programmers; volume 2, Data Processing Managers and Directors; volume 3, Other Computer Professionals. The work provides some 15,000 capsule biographies of the who's who type.

5-89   **World Directory of Mathematicians.** 5th ed. Uppsala, Sweden: International Mathematical Union, 1974. 779p.

An alphabetical list of mathematicians, giving name and address only; also includes a geographical list that gives just the name.

6

Astronomy

Astronomy is the study of the sun, moon, planets, satellites of the planets, comets, stars, and nebulae, including their pasts, futures, distances, dimensions, movements, physical characteristics, and the laws that govern them. Early astronomy was limited to astronometry: positional classification, measurement, and observation of the heavenly bodies. Astronomers like Flamsteed, Bradley and Bessel compiled detailed astronometric star catalogs and atlases that are invaluable to modern investigations of stellar proper motions. Astronometry remains an important subfield in astronomy due to current applications in navigation, surveying, mapping, and calculating time. Modern theory and instrumentation have broadened the scope of astronomical observation. Inventions, such as the Schmidt camera with wide-angle lens and fast focus, have sighted and imprinted images of celestial objects hitherto unknown; and computer science has vastly increased the speed and accuracy of recording and relating data.

Theoretical astronomy deals with analytic solutions of the past, present and future positions and motion of celestial bodies. Advances in theory are due to advanced mathematical precepts and physical principles. Theoretical astronomy began with Copernicus (1473-1543) who published his theories in *On the Revolutions of Celestial Bodies* near the end of his life. Galileo, relying on telescopic observations, supporting Copernicus' presentation and brought it to popular attention. Kepler (1571-1630) produced his *Three Principles of Elliptic Planetary Motion*, which were later verified by Newton's exposition of the *Law of Universal Gravitation.*

Celestial mechanics, a branch of theoretical astronomy, is the study of the gravitational interaction of heavenly bodies. Astrophysics, another category of astronomy, analyzes the physical composition of bodies in space. Advances in astrophysics are due to improved refractive, reflective, and radio telescopes, as well as auxiliary aids such as photoelectric cells and spectroscopes. These instruments, measuring sensitivity beyond the unaided human eye, register images of the outer ranges of light waves, such as ultraviolet and infrared, which are transcribed through photographic techniques into pictures observable by the eye. Simultaneous analysis is used to determine composition and motion of light intercepted from celestial bodies. The resulting photos and analyses are part of the literature the librarian locates through retrospective bibliographies of observatory libraries.

Another important category of theory is cosmogony, which is concerned with the creation of the universe — an almost unverifiable study, yet supported by scientific evidence and principles. Its source stems from a work by Immanuel Kant, entitled, *Universal Natural History and the Theory of the Heavens* (1755), in which he speculates that nebulae are clusters of stars located at greater distances than previously believed. This insight is the basis of his nebular hypothesis on the origin of the planetary system. A bolder hypothesis of the twentieth century, presented in 1948 by T. Gold and H. Bondi, is continuous creation which links continuous creation with F. Hoyle's relativity theory.

In this century, radio engineering advanced the radio telescope, a concept introduced in primitive form by K. G. Jansky and Grote Reber. The radio telescope was refined by application of the radar principle, when the radio astronomer, Hey, observed an obstruction in radar detection in England during World War II. He determined that the origin of the obstruction was radio activity of immense solar flares not encountered by soundings of the quiet sun. In this instance, a problem presented in a technological advance simultaneously revealed a phenomenon that furthered man's understanding of his universe. In like manner, this century has brought forth another development, space exploration, resulting in yet another challenge for the librarian in controlling the explosion of technical and popular literature.

It is apparent that astronomy relies on its older observational literature for the advancement of modern theory. For this reason, retrospective bibliographies of astronomy assume major importance for the science librarian, and such works are carefully noted in this chapter.

GUIDES TO THE LITERATURE

6-1 **Astronomical Atlases, Maps & Charts: An Historical and General Guide. . . .** By Basil J. W. Brown. London: Search Publishing Co., 1932; repr. London: Dawsons, 1968. 197p. illus. facsims.
Topics covered include stellar atlases, charts, and catalogs; solar and spectrographic charts; planetary charts; celestial globes; zodiacs and planispheres; and maps and charts of other celestial objects.

6-2 **Astronomy and Astrophysics: A Bibliographical Guide.** By D. Kemp. London: Macdonald Technical & Scientific; New York: Archon, 1970. 584p. glossary. index.
Selective guide to major books, articles, technical and non-technical reports of theory and data arranged in 75 subject categories. The 3,642 entries are listed chronologically when suitable; entries are usually accompanied by short critical annotations. Sections include: Reference Media, Star Catalogs, Ephemerides, Glossary of Abbreviations, author and subject indexes.

6-3 **Astronomy of Star Positions.** By Heinrich Eichhorn. New York: Frederick Ungar Publishing Co., 1974. 357p. index. bibliog.
This is a guide to the wealth of information that is contained in star catalogs. The subtitle of the work is "A Critical Investigation of Star Catalogues, the Methods of Their Construction, and Their Purpose."

6-4 **Guide to the Literature of Astronomy.** Robert A. Seal. Littleton, CO: Libraries Unlimited, 1977. 306p. bibliog. index.
This is a guide that presents a selection of basic sources for the beginner and for the non-specialist. It is arranged in four sections by topic and type of materials: Reference Sources in Astronomy—guides, book reviews, abstracts and indexes, dictionaries, etc.; General Materials—general works, amateur astronomy, history, etc.; Descriptive Astronomy—solar system, stars, etc.; and Special Types—instrumentation, space science, radio astronomy, etc.

ABSTRACTS, INDEXES, AND BIBLIOGRAPHIES

6-5   **Astronomischer Jahresbericht.** Berlin: G. Reimer, 1899-1968. 68v.
An exhaustive classified international bibliography of publications in astronomy
and astrophysics, radio astronomy, space exploration, and related fields. Specific
publication forms include articles, technical reports, books, conference proceedings, translations, and observatory reports. Only scattered abstracts are provided; however, there are citations to abstracts in *Mathematical Reviews* (5-24)
and *Physics Abstracts* (7-12) and to book reviews. This series is continued by
Astronomy and Astrophysics Abstracts (6-6).

6-6   **Astronomy and Astrophysics Abstracts.** v. 1- . Berlin: Springer, 1969- .
semiannual.
This is a successor to *Astronomischer Jahresbericht* (6-5) and retains all the
former's scholarly features. Coverage remains international though the editorial
apparatus is in English. Time lag in reporting is up to eight months. Titles of entries are given in the language of the author, with English translations whenever
possible; titles of Russian papers are always given in English. Abstracts are written in English, French, or German; authors' abstracts are used by preference. The
arrangement is by a classification scheme of 108 categories, including sections for
periodicals, proceedings, books, and activities in addition to the specific subjects.
In the serial numbering of the abstracts the first three numerals identify the
entry's location in the classification. Six categories are subdivided by specific
names: observatories, institutes; societies, organizations; solar eclipses; comets;
novae; and supernovae. Entries with limited relevance to astronomy or
astrophysics are listed without abstracts. Includes author and subject indexes.

6-7   **Bibliographie générale de l'astronomie.** By J. C. Houzeau and A. Lan-
        caster. Bruxelles: 1882-1889. 2v.; repr. London: Holland Press, 1964. 2v.
Covers books and manuscripts in the first volume; society publications and
periodical articles in the second. Systematically arranged. Special features include
lengthy historical introduction; manuscript locations. Author index.

6-8   **Bibliography of Astronomy, 1881-1898.** Buckinghamshire, England:
        University Microfilms Limited, 1970. On microfilm.
Prepared from standard slips of some 52,000 items; recorded at the Observatoire
Royal de Belgique. A printed "Guide to the Bibliography of Astronomy,
1881-1898," prepared by J. B. Sykes in 1970, is also available on microfilm.

6-9   **Bibliography of Non-Commerical Publications of Observatories and
        Astronomical Societies.** Rev. ed. Utrecht, The Netherlands: "Sonnen-
        borgh" Observatory, 1973. 1v. (various paging). index.
Lists 202 observatories and astronomical institutes on the Sonenborgh exchange
list and their publications. Each entry in this mimeographed volume contains the
address of the institution; current and discontinued series including titles,
numbers, and dates; and availability of documents.

6-10   **Bibliography on Meteorites.** By Harrison Scott Brown. Chicago: Univer-
        sity of Chicago Press, 1953. 686p. (An International Catalogue of
        Meteorites).

Chronologically arranged bibliography of the world's literature on meteorites and related subjects published from 1491 to 1950. Lists more than 8,650 items consisting of books, pamphlets, technical reports, and the contents of 1,068 technical journals. Unless marked by an asterisk, all items were checked in the original sources, to assure correct, complete identification. Following each citation is a list of the abstracts, translations, and reviews of the entry. Author index.

6-11   **Catalog of the Naval Observatory Library, Washington, D.C.** Boston: G. K. Hall, 1976. 6v.

Covers about 75,000 volumes of the library; strong in history of astronomy, observatory publications, astrometry, journals, almanacs, ephemerides, and navigation.

ENCYCLOPEDIAS

6-12   **All about Telescopes.** By Sam Brown. 2nd ed. Barrington, NJ: Edmund Scientific Co., 1975. 192p. illus. index.

A non-technical book that has everything anyone would want to help in building a telescope. Included are chapters on getting acquainted with the telescope, observing the sky show, photography with your telescope, mirror grinding and testing, telescopes you can build, telescope mounts, collimation and adjustments, and telescope optics.

6-13   **Amateur Astronomer.** By Patrick Moore. 7th ed. London: Butterworths, 1971. 356p.

A good book for casual reading that combines a readable text with charts and tables. Includes discussions on telescopes, the solar system, the sun, moon, aurorae, planets, comets, meteors, stars, nebulae, and galaxies.

6-14   **Astrophysical Quantities.** By C. W. Allen. 3rd ed. London: The Athlone Press, University of London; distr. Atlantic Highlands, NJ: Humanities Press, 1974 (c.1973). 310p. index. references.

Includes all of the astronomical constants and numerical quantities that a professional astronomer and student would need. Information is arranged in the following sections: Introduction; General Constants and Units; Atoms; Spectra; Radiation; Earth; Planets and Satellites; Interplanetary Matter; Sun; Normal Stars; Stars with Special Characteristics; Star Populations and the Solar Neighbourhood; Nebulae, Sources and Interplanetary Space; Clusters and Galaxies; and Incidental Tables.

6-15   **Cambridge Encyclopaedia of Astronomy.** Simon Mitton, editor in chief, New York: Crown, 1977. 481p. illus. (part col.). index.

This is a standard source for both beginners and advanced amateurs containing superb illustrations. After an introduction to the study of the universe, the chapters cover stars and the nature of cosmic matter; our sun, the solar system, and its members; intergalactic space and the galaxies, including radio galaxies; cosmology and life in the universe; and the state of astronomy in 1977.

6-16   **Concise Encyclopedia of Astronomy.** By A. Weigert and H. Zimmermann. 2nd English ed. London: Adam Hilger; distr. New York: Crane, Russak, 1976 (c.1975). 532p. illus.

Contains over 1,500 entries on all aspects of astronomy as well as on such related topics as astrophysics and space travel. Entries include the names of noted astronomers, names of celestial bodies, astronomical instruments, and topics such as the history of astronomy, cosmogony, and cosmology. The encyclopedia is addressed to interested laypersons; consequently, mathematical formulas were kept to a minimum, and illustrations, photographs, and tables are extensively used throughout the encyclopedia.

6-17 **Encyclopedia of Astronomy.** By Gildert E. Satterthwaite. New York: St. Martin's, 1971. 527p.

This generally good one-stop source provides 2,200 entries covering all important aspects of the subject: historical events; planetary, stellar, and galactic systems; equipment; and some biographical sketches of prominent astronomers. Definitions are concise and clear, accompanied by illustrations and approximately 200 diagrams. Recent discoveries made through the space program are not represented.

6-18 **Flammarion Book of Astronomy.** By Gabrielle Camille Flammarion. Trans. from French by Annabel Bernard Pagel. London: Allen and Unwin; distr. New York: Simon and Schuster, 1964. 670p. illus. plates. diagrams. tables. planispheres.

First published as *Astronomie populaire* (Paris: Flammarion, 1881. 839p.). This book exhibits flair and enthusiasm in its exposition of the universe in which we live. Many of the first sections (on the solar system and heavenly bodies) have been unevenly updated; however, instrumentation and space exploration were new to this edition. The excellent photogravure illustrations (917, some in color) lend a handsome touch.

6-19 **Illustrated Encyclopedia of Astronomy and Space.** Ed. by Ian Ridpath. New York: T. Y. Crowell, 1976. 240p. illus. (part col.). index.

This encyclopedia is an excellent, up-to-date combination of astronomical and astronautical knowledge that will serve the needs of the general reader, the space buff, and amateur astronomer. Coverage of the United States and Soviet Union space programs is unusually detailed with 1976 as the cutoff date for information on a given topic.

6-20 **Larousse Guide to Astronomy.** By David Baker. New York: Larousse, 1978. 288p. illus. (col.). index.

This is an excellent general introduction to the world of astronomy. Starting with a general introduction to amateur telescopes, the volume continues with a discussion of stellar dynamics, properties of the solar system, and galactic theory, and concludes with a brief history of celestial observation by man. It does have a British slant.

6-21 **Messier's Nebulae and Star Clusters.** By Kenneth Glyn Jones. New York: American Elsevier, 1969 (c.1968). 480p. illus. bibliog. indexes.

Describes all the nebulous objects found in Messier's eighteenth century catalog. Provides historical and biographical background about the nebulae and clusters and about the astronomers themselves. Well illustrated with photos, sketches, and diagrams, including coordinates, star charts, and classification.

6-22   **New Guide to the Planets.** By Patrick Moore. 3rd ed. New York: Norton, 1972. 224p. illus. index.

The first chapters of this work cover the wandering stars, birth and movement of the planets, and rocket exploration. Each remaining chapter covers a particular planet, incorporating recent research, with two final chapters on space exploration and planetary life. A good general survey, but not a handbook in the strictest sense.

6-23   **New Space Encyclopaedia: A Guide to Astronomy and Space Exploration.** New, rev. ed. New York: Dutton, 1973. 326p. illus.

Prepared for the general reader, this handbook contains some 800 entries of varying length, from two-line identifications of technical terms to multi-page descriptions of space medicine, missiles, satellites, the Apollo Project, etc. The text is quite readable, with many cross-references and illustrations. Updating since the 1969 edition has been mainly through the appendix.

6-24   **1001 Questions Answered about Astronomy.** By James S. Pickering. Rev. ed. New York: Dodd, Mead & Co., 1975. 420p. illus. index.

A good general book that covers the sun, earth, moon, planets, stars, comets, galaxies and other astronomical phenomena.

6-25   **Solar System.** Ed. by G. P. Kuiper and B. M. Middlehurst. Chicago: University of Chicago Press, 1953-1963. 4v. illus. diagrams. maps. charts. tables.

A standard work of reference in four volumes: volume 1, The Sun; volume 2, The Earth as a Planet; volume 3, Planets and Satellites; volume 4, The Moon, Meteorites and Comets. Bibliographic references for chapters. Analytic index of subjects and definitions in each volume.

6-26   **Stars and Stellar Systems: Compendium of Astronomy and Astrophysics.** Ed. by G. P. Kuiper and B. M. Middlehurst. Chicago: University of Chicago Press, 1960-1976. 9v. illus. tables. maps.

Partially sponsored by the National Science Foundation. Presentation of stellar astronomy as an empirical science, "co-ordinated and illustrated by the application of theory." Contents: volume 1, Telescopes; volume 2, Astronomical Techniques; volume 3, Basic Astronomical Data; volume 4, Clusters and Binaries; volume 5, Galactic Structures; volume 6, Stellar Atmosphere; volume 7, Nebular and Interstellar Matter; volume 8, Stellar Structure; volume 9, Galaxies and the Universe. Extensive bibliographic references. Subject indexes.

6-27   **Yearbook of Astronomy.** New York: Norton, 1962- . annual.

Directed much more to the amateur astronomer, it reports astronomical events for the year with appropriate data and discussion articles plus a directory of astronomical societies.

DICTIONARIES

6-28   **A-Z of Astronomy.** By Patrick Moore. New York: Scribner's, 1976. 192p. illus.

This book is designed primarily for the novice astronomer covering astronomical terms, instruments, events, places, phenomena, observatories, and principles with brief biographical essays about the more important astronomers.

6-29   **Amateur Astronomer's Glossary.** By Patrick Moore. New York: Norton, 1967. 162p. illus. (The Amateur Astronomer's Library).
A reliable dictionary of the more common terms that the amateur astronomer encounters.

6-30   **Astronomical Dictionary: In Six Languages.** By Josip Kleczek. Praha: Nakladatelstvi' Ceskoslovenske academié Ved; distr. New York: Academic, 1961. 972p.
Multilingual glossary keyed to the initial English list, which has equivalent words in all the other languages. Russian, German, French, Italian, and Czech lists refer by accession number to the English equivalent.

6-31   **Dictionary of Astronomical Terms.** By Ake Wallenquist. Ed. and trans. from the Swedish by Sune Engelbrektson. Garden City, NY: Natural History Press, 1966. 265p.
Directed to college students and the novice astronomer, this dictionary was compiled by a noted astronomer, assisted by astronomers at the American Museum — Hayden Planetarium. Unusually good for quick reference.

6-32   **Dictionary of Astronomy, Space, and Atmospheric Phenomena.** By David F. Tver. New York: Van Nostrand Reinhold, 1979. 281p. illus.
This reference work for the layperson and amateur astronomer provides up-to-date definitions on a number of topics besides those listed in the title, including physics and mathematics. The entries are brief and there are numerous line drawings. Provides a minimum of mathematics definitions.

6-33   **Facts on File Dictionary of Astronomy.** Ed. by Valerie Illingworth. New York: Facts on File, 1979. 378p. illus.
This book for students and specialists contains information not found in other astronomy dictionaries. The entries are brief but the information is current.

6-34   **Glossary of Astronomy and Astrophysics.** By Jeanne Hopkins. Chicago: University of Chicago Press, 1976. 169p.
A comprehensive dictionary of astronomical and astrophysical terms.

6-35   **Navigation Dictionary.** 2nd ed. Washington: U.S. Navy Department, Oceanographic Office; distr. Washington: GPO, 1969. 292p. (H. O. Publ. 220).
Terms in current usage for the navigator of any type of craft. Terms have been borrowed from many fields (such as astronomy, cartography, mathematics, etc.), but the definitions used are concerned with navigation.

HANDBOOKS AND FIELD GUIDES

6-36   **Air Almanac.** 1933- . Washington: U.S. Nautical Almanac Office; distr. Washington: GPO, 1941- . 3 times per year.
Issued jointly by Her Majesty's Nautical Almanac Office, this convenient compendium contains atronomical data required for air navigation. Charts and tables set forth information on rising, setting, and twilight of the sun and moon. Also included are a navigational star chart, a table of symbols and abbreviations, a standard time table, and sky diagrams. This work was formed by the union of the *American Air Almanac* and the *British Air Almanac*.

6-37 **Amateur Astronomer's Handbook.** By James Muirden. Rev. ed. New
York: T. Y. Crowell, 1974. 404p. illus. index.
A practical guide particularly well suited for the amateur who wishes to establish
an observatory with a small to medium-sized telescope. There is instruction on
techniques of viewing the various celestial bodies and on astronomical
photography. Features include glossary, reading lists, tables of recurring
astronomical phenomena, such as eclipses, comets and meteor showers.

6-38 **Amateur Astronomer's Handbook.** By John Benson Sidgwick. 3rd ed.,
prepared by R. O. Gamble. London: Faber and Faber, 1971. 577p. illus.
A very good reference book with advanced technical information on the instrumental and theoretical background of practical astronomy. An extensive
bibliography divided into 32 categories is included. His companion volume,
directed to practical application of the theory, is *Observational Astronomy for
Amateurs* (6-48).

6-39 **American Ephemeris and Nautical Amanac.** Washington: U.S. Nautical
Almanac Office; distr. Washington: GPO, 1852- . annual.
An ephemeris is a list or table of the computed positions that a celestial body
occupied or will occupy on certain dates. The almanac covers phenomena such as
eclipses; universal and sidereal time; the sun; the moon; heliocentric ephemerides
of major planets; elements of the moon, the sun, and major planets; geocentric
ephemerides of major and minor planets; day numbers; mean places of 1,078
stars; eclipses of the sun and moon; ephemerides for physical observation;
satellites; tables of risings and settings; and tables of miscellany. Also published
in Great Britain by Her Majesty's Nautical Almanac Office under the title *The
Astronomical Ephemeris.*

6-40 **Apparent Places of Fundamental Stars.** Heidelberg, West Germany:
Astronomisches Rechen-Institut, 1940- . annual.
Contains the mean and apparent places for over 1,500 stars giving R. A. and
Dec., mean place, magnitude, spectral class, etc.

6-41 **Astronomy: A Handbook.** Ed. by G. D. Roth. Rev. ed. by Arthur Beer.
New York: Springer-Verlag, 1975. 567p. illus. index.
This is a translation of a German edition containing in-depth articles on all major
aspects of astronomy. For the advanced amateur.

6-42 **Astronomy Data Book.** By J. Hedley Robinson. New York: Halstead
Press, Wiley, 1972. 271p. illus.
A useful handbook with information on many aspects of astronomy such as
eclipses, comets, planets, stars, meteors, and other stellar and non-stellar objects.
A good introduction to astronomy for the student.

6-43 **Astrophysical Formulae: A Compendium for the Physicist and
Astrophysicist.** By Kenenth R. Lang. Berlin; New York: Springer-
Verlag, 1974. 735p. references. index.
The material is presented in textual form rather than in long lists of tables and
charts. The lists of references indicate where one can find the derivations of the
long and complicated formulas. The five sections cover: Continuum Radiation;
Monochromatic (Line) Radiation; Gas Processes; Nuclear Astrophysics and High
Energy Particles; and Astrometry and Cosmology.

6-44   **Burnham's Celestial Handbook: An Observer's Guide to the Universe beyond the Solar System.** By Robert Brunham, Jr. Rev. and enlarged ed. New York: Dover, 1978. 3v. illus. index.

This comprehensive guide includes descriptions of more than 7,000 celestial objects, with detailed information on several hundred. The work is arranged alphabetically by constellation and includes photographs, diagrams, charts, and detailed descriptive information. The double star section for each constellation has data on the names of the pair, separation, position angle, approximate magnitudes, right ascension and declination, and notes on spectral type, binary characteristics, and motion.

6-45   **Complete Nautical Astronomer.** By Charles H. Cotter. New York: American Elsevier, 1969. 336p. index.

All of the essential materials on theory, practical application, and instrumentation have been brought together in this up-to-date text on nautical astronomy. Appendixes cover trigonometry and calculus as they relate to nautical astronomy. This volume is a companion to *The Complete Coastal Navigator* (London: Hollis & Carter, 1964. 320p.).

6-46   **Field Book of the Skies.** By William Tyler Olcott. 4th ed. New York: Putnam, 1954. 482p. illus. (Putnam's Nature Field Books).

This field book is considered one of the best for the more experienced amateur astronomer. Arranged in outline form, it includes information on mythology and identification of the constellations and individual stars.

6-47   **Field Guide to the Stars and Planets, Including the Moon, Satellites, Comets, and Other Features of the Universe.** By Donald Howard Menzel. Boston: Houghton Mifflin, 1964. 397p. illus. (The Perterson Field Guide Series, 15).

This is one of the very popular guides and ranks with Olcott in its usefulness to the amateur astronomer. Monthly star charts and a photographic atlas of the sky are prominent features. There is also a chapter on telescopes and one on photography. There is a glossary, bibliography, index.

6-48   **Observational Astronomy for Amateurs.** By John Benson Sidgwick, Rev. ed. London: Faber and Faber, 1971. 376p. index.

A book for the serious amateur with detailed explanations. Concentrates on the various heavenly bodies and how to observe and identify them.

6-49   **Pictorial Guide to the Moon.** By Dinsmore Alter. 3rd ed. New York: T. Y. Crowell, 1973. 216p. illus. index.

A good general guide to the moon with well-written text and good illustrations.

6-50   **Pictorial Guide to the Planets.** By Joseph H. Jackson. 2nd ed. New York: T. Y. Crowell, 1973. 248p. illus. index.

Provides a good treatment of the solar system. The planets, including earth, are analyzed in unusual detail. The pictures are current with the publication date, including some from the latest Mariner flights. Thirteen useful tables that follow the chapters cover data on celestial bodies in the solar planetary system and space exploration therein, with special emphasis on the earth and moon.

6-51   **Pictorial Guide to the Stars.** By Henry C. King. New York: T. Y. Crowell, 1967. 167p. illus. index. bibliog.
A well-illustrated book covering the stars, nebulae, galaxies, the Milky Way, and the universe as a whole. Has good text and excellent illustrations for the student and layperson.

6-52   **Standard Handbook for Telescope Making.** By N. E. Howard. New York: T. Y. Crowell, 1959. 326p. illus. index.
Presents in clear language all aspects of building your own telescope. In particular it describes how to build an 8-inch, f/7 Newtonian reflector.

6-53   **Tables of Minor Planets.** By Frederick Pilcher and Jean Meeus. Jacksonville, IL: F. Pilcher, Illinois College, 1973. 104p. references.
This book lists over 1,800 asteroids with statistical tables that describe all possible aspects of each. Gives orbital elements, magnitudes, and a list of discoverers plus several other special tables.

6-54   **Telescope Handbook and Star Atlas.** By N. E. Howard. 2nd ed. New York: T. Y. Crowell, 1975. 226p. illus. index. bibliog.
This is a good introduction to telescopes and practical astronomy for the amateur. Covers an explanation of telescopes and the types of celestial objects that may be viewed, followed by star maps, a gazetteer, and a Messier catalog.

6-55   **Whitney's Star Finder: A Field Guide to the Heavens.** By Charles A. Whitney. New York: Knopf, 1974. 97p.
A good starfinder as well as a practical guide for observing sunspots, rainbows, halos, comets, meteors, the moon and planets. There are discussions of twilight (or dawn) colors, of telling time, and date by the sun.

ATLASES AND STAR CATALOGS

6-56   **Atlas of Representative Stellar Spectra.** By Yasumasa Yamashita, Kyoji Nariai, and Yuji Norimoto. New York: Wiley, 1978. 129p. illus. (A Halsted Press Book).
A good atlas that adds to previously published atlases of stellar spectra, providing a relatively large number of spectra with different dispersions from those of spectra found in earlier atlases. It consists of 45 plates containing spectra of 197 MK standard stars and 19 plates of 77 peculiar stars. Two tables list standard stars and peculiar stars.

6-57   **Atlas of the Universe.** Ed. by Patrick Moore. Foreword by Sir Bernard Lovell. Chicago: Rand McNally, 1970. 272p. illus. (mainly col.) index.
This beautifully executed atlas contains at least 1,500 photographs, maps, and other illustrations. Observations on exploration of space are followed by atlas sections: the earth from space; the moon; the solar system; and the stars. The appendix contains a catalog of stellar objects, glossary, and the "Beginner's Guide to the Heavens."

6-58   **Color Star Atlas.** By Patrick Moore. New York: Crown, 1973. 112p. index.

Designed to meet the needs of the beginning viewer. Star maps and annotated constellation charts are excellent. There are chapters on introductory theory, stellar types, and galaxies.

6-59   **Concise Atlas of the Universe.** By Patrick Moore. New York: Rand McNally, 1974. 192p.

A less expensive atlas for the layman, in which lunar exploration has been reduced to essentials. Includes material on the Russian moon lander; the comet, Kahoutek; and Pioneer 10 photos of Jupiter.

6-60   **New Photographic Atlas of the Moon.** By Zdenek Kopal. New York: Taplinger, 1971. 311p. illus.

This superb atlas is an updated version of a 1965 lunar atlas by the same author. Included are excellent recent telescopic pictures made with a large telescope at Pic-du-Midi in France, and the best recent space photographs. The text starts with the earliest telescopic discoveries of Galileo in 1610 and ends with the most recent data from instruments left on the moon's surface by the Apollo astronauts. There are 200 pictures with explanatory captions; those of particular interest are the well-known craters Copernicus and Plato and the complete photographic coverage of the moon's far side.

6-61   **Palomar Observatory Sky Atlas, July 16, 1954.** By National Geographic Society. Pasadena, CA: printed by the Graphic Arts Facilities of the California Institute of Technology, 1954-1958. 10p., and atlas of 1,618 plates in 9v. and Supplement (Southern Extension).

"Consists of photographic reproductions of red- and blue-sensitive photographs of 879 fields, 6.6° square (sq), and covers the entire sky north of -27° declination. This Atlas . . . has the most complete coverage and the best quality of any in existence and contains an enormous wealth of information that is just beginning to be explored."

6-62   **Revised New General Catalogue of Nonstellar Astronomical Objects.** By Jack W. Sulentic and William G. Tifft. Tucson, AZ: University of Arizona Press, 1973. 384p.

The information contained in the *New General Catalogue of Nebulae and Clusters of Stars* (London: Royal Astronomical Society, 1888-1895. 2v.), is herein presented with errors and duplications corrected. This revision is closely tied to the *Palomar Observatory Sky Atlas* (6-61). The authors point out that before their completion of this work it was difficult to identify a particular object in the *Palomar Observatory Sky Atlas* because of the multitude of catalog or literature sources; also, the converse problem—that of identifying the object unambiguously on the *Atlas* when given the specific object—posed difficulties. For the 7,840 objects bearing an *NGC* number, these difficulties are now eliminated; rectangular coordinates have been provided for each *NGC* object, directly measured from the *Atlas* prints. In the final form, this revised *NGC* provides a working collection of basic data for a large fraction of all brighter objects (other than stars) of interest to observers.

6-63   **Second Reference Catalogue of Bright Galaxies: Containing Information on 4,364 Galaxies with References to Papers Published between 1964 and 1975.** By Gerard de Vaucouleurs, Antoinette de Vaucouleurs, and

Harold G. Corwin, Jr. Austin, TX: University of Texas Press, 1976. 396p. bibliog. (University of Texas Monographs, no. 2).

This is a supplement to the *Reference Catalogue of Bright Galaxies* (Austin, TX: University of Texas Press, 1964. 268p.) compiled by Mr. de Vaucouleurs. It consists of 4,364 of the brightest galaxies that have been the object of some individual study. The first part of the catalog consists of coordinates, classification, diameter, magnitude, color, radial velocity, and radio data on individual galaxies. The second part contains references to the published literature of specific galaxies from January 1, 1964 to July 1, 1975.

6-64   **Star Atlas and Reference Book (Epoch 1950) for Students and Amateurs . . . ; The Reference Handbook.** By J. Gall Inglis and A. P. Norton; ed. by R. M. G. Inglis. 15th ed. Edinburgh and London: Gall and Inglis, 1964. 57p. with 18 maps. illus. diagrams. tables.

Recommended as first choice for amateurs. The subtitle indicates that it shows "over 9,000 stars, nebulae and clusters, with descriptive lists of objects mostly suitable for small telescopes; notes on planets, star nomenclature, etc." Stars down to the seventh magnitude are included. Special features: a list of star catalogs and a bibliography of standard works of reference and of astronomical catalogs.

6-65   **Star Atlas of Reference Stars and Nonstellar Objects.** By the Smithsonian Institution. Cambridge, MA: MIT Press, 1969. 13p. illus. 152 charts in box.

Prepared for use with the Laboratory's *Star Catalog* (6-66). Chart size is 38x32-cm. A transparent location grid is provided. Short bibliography.

6-66   **Star Catalog.** By the Smithsonian Institution Astrophysical Observatory. Washington: Smithsonian Institution, 1966. 4v.

A comprehensive catalog for the epoch and equinox of 1950.0 giving positions and proper motions of 258,997 stars; it combines several earlier catalogs for the epoch 1950: *Third and Fourth Fundamental Catalogs* of the "Berliner Jahrbuch" (Heidelberg: published by the Astronomisches Rechen-Institute for the International Astronomical Union); Astronomische Gesellschaft, Leipzig. *Erste und zweiter Katolog fur das Aguinoktium 1950*; and B. Boss, *General Catalogue of 33,342 Stars for the Epoch 1950* (Washington: Carnegie Institution of Washington, 1937. 5v.). In the Smithsonian catalog the following information is given for each star: right ascension and declination for equator, equinox, and epoch 1950.0; standard deviation of the position epoch 1950.0; right ascension and declination from equator and equinox 1950.0; standard deviations; mean epochs; annual proper motions; visual magnitudes; photographic magnitudes; spectral type; Durchmusterung number; source catalog; star number for source catalog; and explanatory notes. This is also available on magnetic computer tape.

6-67   **Times Atlas of the Moon.** Ed. by H. A. G. Lewis. London: Times Newspapers Ltd.; distr. New York: New York Times, 1969. 110p. maps. illus. (part col.). index.

The 60 carefully drawn and attractively colored maps of the near side lunar surface are based on U.S. Air Force 1:1,000,000 lunar charts, supplemented by information from the U.S. Army Topographic Command, NASA, and the U.S. Geological Survey. On the maps, contour lines are shown where data are available, and spot elevations are indicated in meters. All major and many minor topographical features, with the exception of crater-contained rilles (clefts), are clearly delineated and labeled; and a geographic coordinate grid is superimposed on each map. The text describes techniques of lunar flight, lunar landscape and geology, and mapping methodology.

6-68   **True Visual Magnitude Photographic Star Atlas.** By C. Papadopoulos. New York: Pergamon, 1979. 3v.

This atlas presents the stars in their "apparent magnitude," exactly as the eye sees them through a telescope.

DIRECTORIES

6-69   **Directory of Physics and Astronomy Staff Members.** v. 1- . New York: American Institute of Physics. 1959/60- . annual.

Lists approximately 18,000 physics and astronomy personnel at 2,300 institutions. Each entry gives the address and phone number of the department of physics and/or astronomy at a particular institution, the list of faculty and staff, etc. There is no biographical data.

6-70   **Graduate Programs in Physics, Astronomy, and Related Fields.** New York: American Institute of Physics, 1971- . annual.

An authoritative guide listing information on over 200 doctoral and 300 master's degree programs in the United States and Canada. Arranged geographically giving information on the graduate faculty, the research specialties of the staff, course requirements, financial aid, number of students, etc.

6-71   **List of Radio and Radar Astronomy Observatories.** Washington: National Academy of Sciences, National Academy of Engineering, Committee on Radio Frequencies, 1970- . irregular.

Divided into U.S. and foreign observatories, each entry gives information about radio telescopes such as location, information officer, sponsors, type, size, height, sky coverage, collecting area, polarization, and other data.

6-72   **Observatories of the World.** By Thornton Page. Cambridge, MA: Smithsonian Astrophysical Observatory, 1967. 41p.

This is a directory of optical and radio astronomical observatories, excluding other types, such as meteorological and seismic. Arrangement is by country, subdivided by the two types. Information in each entry includes observatory; its city and founding date; list of telescopes and the size of mirrors and/or observing programs. Introductory material explains various observatory equipment and observing programs and offers a brief history of observatories. Still useful in spite of its age.

6-73   **U.S. Observatories: A Directory and Travel Guide.** By H. T. Kirby-Smith. New York: Van Nostrand Reinhold, 1976. 173p. illus. index.

The first part of this excellent book covers the 15 major U.S. observatories in detail: Harvard/Smithsonian, U.S. Naval, Leander McCormick, Green Bank, Allegheny, Yerkes, McDonald, Sacramento Peak, Kitt Peak, Lowell, Flagstaff, Hale (Wilson and Palomar), and Lick, giving a description of the setting, details of construction, important optical/mechanical features, and the major astronomical research use of each telescope. The second part lists 300 observatories, planetariums, and astronomical museums, arranged by state.

7

Physics

Physics is the science of energy, transformations of energy, structure of matter, and interactions of the elementary constituents of nature. The father of physics is said to be Galileo, who in 1638 published his *Discourses Concerning Two New Sciences*, in which he described the sciences as "coherence and resistance to fracture" and "uniform, accelerated, and violent or projectile motions." The beginnings of classical physics is attributed to Sir Isaac Newton. In 1687 he postulated in *Philosphiae Naturalis Principia Mathematica*, three laws of motion governing classical dynamics. From these two respected scientists developed the disciplines of physics.

Classical physics consists of several specialized areas. One of these areas is mechanics, the description of the behavior of material when acted on by a force. The material can be in the form of particles, rigid bodies, liquids, or gases. Mechanics is further subdivided into statics, the study of bodies at rest; kinematics, the study of bodies in motion; and dynamics, the study of motion in relation to forces causing the motion. Acoustics and thermodynamics are also part of classical physics. Acoustics describes the laws and theories of waves, vibrations, and audible and inaudible sounds, and thermodynamics covers all the laws that are related to heat. Electricity and magnetism are basic in any course of classical physics. Electricity is a phenomenon that was known early in history. For example, the Greeks, knew that when amber was rubbed it would attract small pieces of paper. Their word for amber was elektron, the derivation of our word electricity. Electricity is largely mathematical, as a result of Coulomb's experiments. Closely associated with electricity is magnetism. Its basic law, also developed by Coulomb, states that the force between two magnetic poles is proportional to the product of the strengths of the poles and inversely proportional to the square of the distance between them. Finally, there is optics, the study of light. Light results from chemical reactions in combustion or from electrical energy. When studying optics, one investigates many facets of light, such as its speed, calculated at 186,000 miles per second; and reflection and refraction properties. Modern physics is concerned with the structure of matter: molecules, atoms, nuclei, and fundamental particles. Quantum theory, theory of relativity, nuclear physics, atomic physics, and solid-state physics are specialized branches of modern physics. Studies in modern physics focus on the development of nuclear structure by reasons of electronic structure. Most of the development in modern physics occurred between 1897, with the discovery of the electron, and 1942, when large-scale releases of nuclear energy were possible. Important concepts were also developed in the first 30 years of the twentieth century, such a Planck's quantum theory of radiation, 1905; Bohr's theory of the hydrogen atom, 1913; the discovery of the positron and neutron, 1932; and Bohr's theory of nuclear structure, 1936. It was also during this century that Einstein developed his relativity theory, which produced numerous corrections for the Newtonian laws pertaining to velocities approaching the speed of light and the interconvertibility of mass and energy.

The science librarian must be aware of the early resources in physics, including theories published in obscure journals in the early 1800s or late 1700s. Until about 1870, physicists thought of their discipline as the sovereign center of natural philosophy embracing chemistry, astronomy, metallurgy, meteorology, and geology. These fields have, until recently, been autonomous. Today, physics is reasserting its historic claim to this all-inclusive meaning through the specialized fields of physical chemistry, meteorological physics, physical astronomy, astrophysics, biophysics, and geophysics, accompanied by mathematics, the basic language in which all physical concepts are written. Arthur Beiser states:

> It's almost impossible to draw definite boundaries separating the physical sciences from one another. The central theme of physics is the nature of matter and fundamental interactions – gravitational, electromagnetic, nuclear, and so on – it exhibits; that of chemistry is the combination of different species of matter into composite substances; that of geology and the other earth sciences is our planet, from core to outer atmosphere; and that of astronomy is the structure and evolution of the universe. These themes are different, but their pursuit leads to considerable overlapping. Thus astrophysics, geophysics, geochemistry, physical chemistry, and chemical physics are all established disciplines. The principles and concepts of physics, however, underlie the other disciplines of physical science, and as the latter grow more sophisticated, they draw that much more upon physics for the tools of discovery.[1]

GUIDES TO THE LITERATURE

7-1   **How to Find Out about Physics: A Guide to Sources of Information Arranged by the Decimal Classification.** By Bryan Yates. New York: Pergamon, 1965. 175p. (The Commonwealth and International Library. Libraries and Technical Information Division).

Covers sources of information for physics up to 1963. It contains the usual types of information on indexes, abstracts, and other reference sources. Some of the entries are annotated. Dated, but still useful.

7-2   **Physics Literature: A Reference Manual.** By Robert H. Whitford. 2nd ed. New York: Scarecrow Press, 1968. 272p.

A survey of physics literature at the college level. Arranged by approach – bibliographical, historical, biographical, experimental, mathematical, educational, terminological, and topical. Some items are briefly annotated. Dated but still useful for some sections.

7-3   **Use of Physics Literature.** Ed. by Herbert Coblans. Woburn, MA: Butterworths, 1975. 290p. index. (Information Sources for Research and Development).

A book intended for scientists, engineers, students, and librarians. The first five chapters are related to libraries, the structure and control of physics literature, and the basic general reference tools in physics. The remaining chapter covers

---

[1]Arthur Beiser, *The World of Physics* (New York: McGraw-Hill, 1960), p. 2.

specific subject literature related to subfields of physics, such as history of physics, theoretical physics, nuclear and atomic physics, instrumentation, etc.

ABSTRACTS, INDEXES, AND BIBLIOGRAPHIES

7-4   **Acoustics Abstracts.** v. 1- . Brentwood, Essex: Multi-Science Publishing Co., 1967- . bimonthly.
About 1,200 abstracts a year are included. Covers vibration, shock, audio frequencies, ultrasonics, noise, physiological, psychological aspects of sound and bioacoustics. Annual indexes by author and subject.

7-5   **CINDA 76/77: An Index to the Literature on Microscopic Neutron Data.** Vienna: International Atomic Energy Agency; distr. New York: Unipub, 1976. 2v. CINDA 78 (Suppl. 4 to CINDA 76/77), Suppl 5- . 1978- .
This compilation contains bibliographic references to measurements, calculations, and evaluations of neutron cross-sections and other microscopic data. It contains over 150,000 bibliographic extracts compiled from scientific journals, report series, conference proceedings, and unpublished materials. The material is entered first by element and mass number, then by cross section or other quantity. For each of these the following is given: type of work (experimental, theoretical, etc.); type of reference (journal, report, etc.); energy range of incident neutrons; reference citations; laboratory; comments; and numerical data. This compilation is the result of worldwide cooperation through the U.S. Energy Research and Development Administration; USSR Nuclear Data Centre; NEA Neutron Data Compilation Centre at Soclay, France; and IAEA Nuclear Data Section at Vienna.

7-6   **Current Papers in Physics.** v. 1- . London: Institution of Electrical Engineers, 1966- . bimonthly.
Contains about 65,000 titles of research articles from more than 900 of the world's physics journals.

7-7   **Current Physics Index.** v. 1- . New York: American Institute of Physics, 1975- . quarterly.
Provides timely information about physics research currently being published in a core set of 44 physics journals. Arranged by subject, each entry gives title, authors and affiliations, bibliographic reference and DPM number.

7-8   **Current Physics Microform.** New York: American Institute of Physics. monthly.
A monthly microfilm version of the full text of all articles in the primary journals published by AIP and its member societies during the preceding month. Member societies include American Geophysical Union, American Institute of Aeronautics and Astronautics, American Society for Metals, Division of Physical Chemistry of the American Chemical Society, Electron Microscopy Society of America, Geological Society of America, Instrument Society of America, Nuclear and Plasma Sciences Society of IEEE, Philosophical Society of Washington, Society for Applied Spectroscopy, and Physics Clubs of Chicago, Milwaukee, New York, Philadelphia, and Richmond.

7-9   **Dictionary Catalog of the Princeton University Plasma Physics Laboratory Library.** Boston: G. K. Hall, 1970. 4v.

7-9a   **Supplement.** 1- . 1973- .
This catalog covers a comprehensive collection of technical journals and reports plus some monographs. It is in alphabetical format so that each report can be located under authors, originating organization, report number, and subject headings.

7-10   **INIS Atomindex: An International Abstracting Service.** v. 1- . Vienna: International Atomic Energy Agency; distr. New York: Unipub, 1970- .
All fields of nuclear science are covered in this international abstracting service, including not only science and technology but also economic, legal, and social aspects. In addition to the more conventional sources it includes patents, technical reports, and standards. *Nuclear Science Abstracts* (7-11) was discontinued since this publication contained essentially the same information.

7-11   **Nuclear Science Abstracts.** Washington: Energy Research and Development Administration, 1948-1976. 33v.
Prior to February 15, 1975, the issuing agency was the Atomic Energy Commission. The *Abstracts of Declassified Documents*, which appeared from July 1947 to June 1948, is considered the beginning of the *Nuclear Science Abstracts*. In the cumulative report number index, this is considered as volume 2, even though there is another volume 1 that begins with July 1948. *NSA* covered the international literature on nuclear science and technology that appeared in the technical reports of the Energy Research and Development Administration and its contractors; of other U.S. government agencies as well as other governments; and of foreign and domestic universities and research laboratories. Patents, books, conference papers, periodical articles, dissertations, and translations issued throughout the world are also abstracted. Each entry gives the issuing agency, title, author, date, pagination, report number, language if not English, and abstract. There are corporate, personal author, subject, and report number indexes in each issue. There are also cumulative indexes. The report number index also gives the availability of the report — whether it is automatically assigned to a depository library, available at cost from NTIS, or published in some source, which is indicated. It was discontinued because *INIS Atomindex* (7-10) contains essentially the same information.

7-12   **Physics Abstracts.** v. 1- . London: Institution of Electrical Engineers, 1898- . bimonthly.
Physics Abstracts is Section A of *Science Abstracts*, of which the other two sections are: B, *Electrical and Electronics Abstracts* (12-44) and C, *Computer and Control Abstracts* (5-15). *Physics Abstracts* is the only abstracting service that systematically covers the complete domain of modern physics. The documents abstracted are all the primary sources: journals, reports, books, dissertations, patents, and conference papers published in all languages of the world. More than 85,000 items per year are abstracted. Abstracts are short and descriptive. There is a subject and author index in each issue which cumulates every six months. An added feature not present in many abstracting journals is the indication of the author's affiliation. Arrangement of abstracts is in accordance with the following ten decimally divided subjects: 0) General; 1) Physics of elementary

particles and fields; 2) Nuclear physics; 3) Atomic and molecular physics; 4) Classical areas of phenomenology; 5) Fluids, plasmas, and electrical discharges; 6) Condensed matter: structure, thermal and mechanical properties; 7) Condensed matter: electronic structures, electrical, magnetic and optical properties; 8) Cross-disciplinary physics and related areas of science and technology; 9) Geophysics, astronomy and astrophysics. Includes author, bibliography, books, corporate author, and conference indexes.

7-13 **Physics Briefs/Physikalische Berichte.** v. 1- . New York: American Institute of Physics, 1979- . semimonthly.
An abstracting journal that covers all the fields of physics and related areas. Information is taken from over 2,400 series and nonserial periodicals published in all countries and in all languages including books, patents, reports, theses and conference papers. Core journals in physics are abstracted completely. Issued in cooperation with Deutsche Physikalische Gesellschaft and the Fachinformationszentrum Energie, Physik, Mathematik. Supersedes *Physikalische Berichte* (7-14).

7-14 **Physikalische Berichte.** New York: Verlag Chemie International, 1920-1978. 57v.
An abstracting journal that covered all fields of physics. Superseded by *Physics Briefs/Physikalische Berichte* (7-13).

7-15 **Rheology Abstracts: A Survey of World Literature.** v. 1- . New York: Pergamon, for the British Society of Rheology, 1958- . quarterly.
An abstracting service covering more than 100 journals. Annual author and subject indexes.

7-16 **Science Research Abstracts Journal.** v. 1- . Riverdale, MD: Cambridge Scientific Abstracts, 1973- . ten issues per year for each part.
part A: Superconductivity; MHD (Magnetohydrodynamics), Plasmas, Theoretical Physics. Part B: Laser and Electrooptic Reviews—Quantum Electronics. Indexing all primary publication forms, the two parts together yield 32,000 abstracts a year. Each issue of each part contains subject and author indexes. Subject access is further facilitated by many cross-references. Cumulated annual author-subject indexes are also available, as is a special "Particle Index" for part A.

7-17 **Solid State Abstracts.** v. 1- . Cambridge, MA: Cambridge Communications Corporation, 1957- . ten issues per year.
A comprehensive abstracting index to published materials on physics, metallurgy, crystallography, chemistry, and device technology of solids. Sources include all primary documents as well as books. Obviously designed for libraries and practitioners with an exclusive interest in this subfield of physics.

ENCYCLOPEDIAS

7-18 **Encyclopaedic Dictionary of Physics: General, Nuclear, Solid State, Molecular Chemical, Metal and Vacuum Physics, Astronomy, Geophysics, Biophysics, and Related Subjects.** Ed. by James Thewlis. New York: Pergamon, 1961-1964. 9v.

7-18a **Supplement.** 1966- .
This important physics reference set is known equally well as *Thewlis's Encyclopaedic Dictionary of Physics*. It covers all aspects of physics plus related areas in mathematics, astronomy, aerodynamics, hydraulics, geophysics, meteorology, physical metallurgy, radiation chemistry, physical chemistry, structural chemistry, crystallography, medical physics, biophysics, and photography. The various entries are written at graduate level and each is signed. The index volume lists in detail all items not treated as a separate entry in the main volumes. Volume 9 is a multilingual dictionary of English, French, German, Spanish, Russian, and Japanese terms. This is the best multivolume physics encyclopedia available, in part because it is being updated with regular supplements. A second, revised and enlarged edition in seven volumes is planned in 1981 under the title *Encyclopaedia of Physics and Its Applications*.

7-19 **Encyclopedia of Physics.** Ed. by Robert M. Besancon. 2nd ed. New York: Van Nostrand Reinhold, 1974. 1067p. illus. index.
Designed to give uniform coverage, the encyclopedia presents 344 articles written on three different levels: those on the main divisions of physics are intended for readers with little background in the subject; those on the subdivisions are aimed at readers with more knowledge; and those on the more finely divided areas are geared toward readers with fairly sound backgrounds in both physics and mathematics.

7-20 **Handbuch der Physik. Encyclopedia of Physics.** Ed. by S. Flugge, v. 1- . (2. Aufl.). Berlin: Springer, 1955- .
First published in 1926-1929 in 24 volumes and index, the second edition is to be completed in 54 volumes, of which the following are the topics covered:

1-2.	Mathematical methods.
3.	Principles of classical mechanics and field theory.
4.	Principles of thermodynamics and relativity.
5.	Principles of quantum theory.
6.	Elasticity and plasticity.
7.	Crystal physics.
8.	Fluid dynamics.
9.	Fluid mechanics.
11.	Acoustics.
12.	Thermodynamics of gases.
13.	Thermodynamics of liquids and solids.
14-15.	Low-temperature physics.
16.	Electric fields and waves.
17-18.	Dielectrics.
19-20.	Electrical conductivity.
21-22.	Electron emission. Gas discharges.
23.	Electrical instruments.
24.	Fundamentals of optics.
25.	Crystal optics. Diffraction.
26.	Light and matter.
27-28	Spectroscopy.
29.	Optical instruments.
30.	X-rays.
32.	Structural research.

33.   Corpuscular optics.
34.   Corpuscles and radiation in matter.
35-36. Atoms.
37(1). Atoms (pt. 3). Molecules (pt. 1).
37(2). Molecules (pt. 2).
38(1). External properties of atomic nuclei.
38(2)   Neutron and related gamma ray problems.
39.   Structure of atomic nuclei.
40-42. Nuclear reactions.
44-45. Nuclear instrumentation.
46.   Cosmic rays.
47-49. Geophysics.
50-54. Astrophysics.

This is the only systematic encyclopedic treatise covering the entire domain of physics. This is a trilingual work — German, English, and French. Many volumes are entirely in English; many others have English contributions; all have added title pages in English. The subject indexes in each volume serve as informal multilingual lexicons.

7-21   **Thermophysical Properties of Matter.** By Y. S. Touloukian and C. Y. Ho. New York: Plenum, 1970-1977. 13v. (The TPRC Data Series).

This work is the culmination of 12 years of effort in the generation of tables of numerical data for science and technology. It is expected that each volume will be revised, updated, and reissued in a new edition every fifth year. Each volume contains essentially three sections — text, numerical data with source references, and an index. Volumes 1-3 cover thermal conductivity; volumes 4-6, specific heat; volumes 7-9, thermal radiative properties; volume 10, thermal diffusivity; volume 11, viscosity; and volumes 12-13, thermal expansion. Most of the material is in tabular or graphic format, with explanatory text accompanying each bit of data.

7-22   **Zahlenwerte und Funktionen aus Physik, Chemie, Astronomie, Geophysik und Technik.** By Hans Heinrich Landolt and R. Börnstein. 6. Aufl. Berlin; New York: Springer-Verlag, 1950- .

7-22a **Zahlenwerte und Funktionen aus Naturwissenschaften und Technik — Neue Serie. (Numerical Data and Functional Relationships in Science and Technology — New Series).** Berlin; New York: Springer-Verlag, 1961- .

This comprehensive reference work contains the world's most complete collection of tables and other tabulated forms of information concerning the fundamental properties of physics, chemistry, astronomy, geophysics, and technology. The first edition was published in 1883 under the title *Physikalisch-Chemische Tabellen*. Most of the volumes are in German, but some of the more recently published ones have the introduction and some data in both German and English. At this point, only volume 2, part 9, has an index, but there are detailed tables of contents for each of the other volumes. An index for the entire set is in preparation. The contents schedule of the sixth edition and the new series is as follows:

Vol. 1 — Atom- und Molekularphysik.
Vol. 2 — Eigenschaften der Materie in ihren Aggregatzustanden.
Vol. 3 — Astronomie und Geophysik, 1952.

Vol. 4 — Technik.

New Series

Group I — Kernphysik und Kerntechnik — Nuclear Physics and Technology.

Group II — Atom- und Molekularphysik — Atomic and Molecular Physics.

Group III — Kristall- und Festkorperphysik — Crystal and Solid State Physics.

Group IV — Makroskopische und technische Eigenschaften der Materie — Macroscopic and Technical Properties of Matter.

Group V — Geophysik und Weltraumforschung — Geophysics and Space Research.

Group VI — Astronomie, Astrophysik und Weltraumforschung — Astronomy, Astrophysics and Space Research.

DICTIONARIES

7-23 **Concise Dictionary of Physics and Related Subjects.** By J. Thewlis. 2nd ed. rev. and enlarged. Elmsford, NY: Pergamon, 1979. 370p.

The well-known editor of the *Encyclopaedic Dictionary of Physics* (7-18) has produced a dictionary of more than 7,200 entries, based in part on the older work. Intended primarily for students and non-specialists, the *Concise Dictionary* should also be useful for the researcher interested in aspects outside his or her immediate specialty. The briefly defined terms come not only from physics but also from meteorology, mathematics, photography, astronomy, crystallography, and many other subject areas.

7-24 **Dictionary of Physics and Allied Sciences.** Ed. by Charles J. Hyman and Ralph Idlin. New York: Frederick Ungar Publishing, 1978. 2v.

This is a German-English/English-German language dictionary covering physics, mathematics, astronomy, chemistry, meteorology, mineralogy, geology, and geophysics.

7-25 **Dictionary of Spectroscopy.** By R. C. Denney. New York: Wiley, 1973. 161p. illus. (A Halsted Press Book).

Includes both modern and traditional types of spectroscopy in standard dictionary format. Literature references provide a means of following through on given topics. Directed to general spectroscopists and undergraduate students, but not to specialists.

7-26 **Elsevier's Dictionary of Nuclear Science and Technology in Six Languages: English/American-French-Spanish-Italian-Dutch and German.** Comp. by W. E. Clason. 2nd ed. rev. New York: American Elsevier, 1970. 787p.

Arranged by English word followed in tabular form by subject, definition, and equivalent word in French, Spanish, Italian, Dutch, and German. There is a separate index for each of the languages referring to the number of the English word equivalent.

7-27 **English-Russian Physics Dictionary.** Ed. by D. M. Tolstoi. New York: Pergamon, 1978. 848p.

Contains about 60,000 terms from all basic areas of physics, and gives the Russian equivalent of each term.

7-28 International Dictionary of Physics and Electronics. 2nd ed. Princeton, NJ: Van Nostrand, 1961. 1355p. illus.

Even though this book is over 18 years old, it is still an excellent encyclopedic dictionary of physics and related areas. Included in the introduction is a short history of physics. There are also indexes in French, German, Spanish, and Russian.

7-29 McGraw-Hill Dictionary of Physics and Mathematics. Ed. by Daniel Lapedes. New York: McGraw-Hill, 1978. 1v. (various paging).

This dictionary contains 20,000 definitions of terms in physics, mathematics, and other related fields, such as fluid mechanics and mineralogy.

7-30 Penguin Dictionary of Physics. Ed. by Valerie H. Pitt. New York: Penguin Books, 1977. 428p. illus.

This is an abridgment of the *New Dictionary of Physics* (London: Longman, 1975. 619p.). It is concerned with terminology of contemporary physics and omits all biographies and entries concerned with experimental determinations of constants.

HANDBOOKS

7-31 American Institute of Physics Handbook. Ed. by Dwight E. Gray. 3rd ed. New York: McGraw-Hill, 1972. 1v. (various paging). illus. index.

This internationally accepted physics handbook supplies authoritative reference materials, including tables of data, graphs, and bibliographies, all of which are described with a minimum of narration by leaders in physical methods for research. Each of the nine sections covers a field of physics, each presented in logical sequence, from definitions and concepts through the many subtopics of that particular field. An excellent index helps the user locate specific data needed.

7-32 Handbook of Optics. Ed. by Walter G. Driscoll. William Vaughan, assoc. ed. New York: McGraw-Hill, 1978. 1v. (various paging).

A comprehensive work representing a major contribution to the literature of applied physics. Includes much graphic and tabular information and extensive lists of references. The book is intended for the optical systems designer and engineer.

7-33 Tables of Physical and Chemical Constants and Some Mathematical Functions. By George William Clarkson Kaye and T. H. Laby. 14th ed. rev. New York: Longman, 1973. 386p.

This edition has been thoroughly revised, updated and augmented. Divided into sections, each of which begins with a clear introduction. Throughout there are footnote references to the literature; thus, the work also serves as a bibliography of tables of constants. All tabulated values are in SI units.

DIRECTORIES

7-34 Physics. Nobel Lectures Including Presentation Speeches and Laureates' Biographies; 1901-1921, 1922-1941, 1942-1962. By Nobelstiftelsen, Stockholm. New York: Elsevier, 1964. 3v. illus.

A compilation of the complete Nobel lectures including presentation speeches and laureates' biographies. No biographical portraits.

7-35 **Who's Who in Atoms: An International Reference Book.** 6th ed. London: George G. Harrap; distr. New York: International Publications Service, 1974. 1748p. in 2v.

Contains more than 20,000 entries for nuclear scientists working in 76 countries, giving name, address, qualifications, birthdate, education, appointments, published works, society memberships, and statement of nuclear interests.

8

Chemistry

Chemistry is the science of the properties, composition, and structure of matter. It is concerned with the study of changes that take place when one substance is converted into another substance. The founder of modern chemistry is Antoine Laurent Lavoisier (1743-1794), who in 1789, published his *Elementary Treatise on Chemistry*, in which he proposed that complex compounds are the combined weights of their parts. Other prominent chemists include John Dalton (1766-1844), Sir Humphry Davy (1778-1829), and Justus von Liebig (1803-1873). In 1869, Dmitri Ivanovich Mendelief founded what is now known as the periodic law, which states that properties of the elements are periodic functions of their atomic weights.

Physical chemistry relates chemistry to the entire domain of science by studying the fundamental laws and theories of chemistry. The physical chemist works in many specialized areas, such as, thermochemistry; chemical kinetics; chemical equilibrium; structure of gases, liquids, and solids; atomic and nuclear structure; solutions; electrochemistry; photochemistry; colloids; and quantum theory. Analytical chemistry deals with detection and determination techniques needed to measure quantities of elements and groups of elements. Qualitative analysis measures the relative proportion of elements; quantitative analysis detects and measures individual components present in compounds or mixtures. Qualitative and quantitative analysis provide the basis for analyzing chemical compounds. The field has greatly expanded with the development of analytical instrument techniques, such as chromatography, infrared, ultraviolet, and other types of spectroscopy. These developments have improved the accuracy and sensitivity of determinations and have greatly sped up the procedures. The application of chemistry to biological systems is called biochemistry, which is a broad field embracing many subdivisions. Among these are the chemistry of animal life known as physiological chemistry; the chemistry of plants; nutrition; metabolism; the chemical basis of enzymes; and the action of pharmaceuticals and poisons. Organic chemistry is the study of compounds found in living organisms. Invariably these compounds contain carbon, which in turn divide into two classes: aliphatic hydrocarbons or chain structures, such as alcohols, ethers, aldehydes, ketones, acids, esters and amines; and aromatic hydrocarbons or ring structures, of which benzene, toluene, and dinitrobenzoic acid are examples. Because organic chemistry studies the structure of biological compounds, there is some overlap with biochemistry. The study of salts, acids, and bases, and, in general, all non-carbon compounds, is called inorganic chemistry. The chemical elements in their pure and semipure states fall into this area. Since many of these elements are metals, the fields of interest of inorganic chemistry and metallurgy are closely allied. Other fields of chemistry include agricultural chemistry, colloid chemistry, electrochemistry, forensic chemistry, industrial chemistry, photochemistry, and stereochemistry.

Chemical literature is probably the best indexed and abstracted of all the disciplines, primarily as a result of the efforts of Chemical Abstracts Services. There are numerous treatises, advanced textbooks and comprehensive encyclopedias that aid the librarian in locating what is needed for the researcher.

GUIDES TO THE LITERATURE

8-1 **Chemical Publications: Their Nature and Use.** By Melvin Guy Mellon. 4th ed. New York: McGraw-Hill, 1965. 324p. illus.
Gives comprehensive descriptions of many of the standard reference tools and explains how they are to be used.

8-2 **Guide to Basic Information Sources in Chemistry.** By Arthur Antony. New York: Wiley, 1979. 219p. bibliog. index. (Information Resources Series; A Halsted Press Book).
This is an up-to-date first resource for the chemistry student or librarian needing a guide to literature in the field of pure chemistry. It focuses on the user's approach to the field and includes chapters on bibliographic searching by computers, nomenclature, general compilations of data, specialized data compilations, guides to techniques, safety manuals and guides, style manuals and guides for authors, biographies and directories of people, and product, service, and company directories.

8-3 **Guide to the Literature of Chemistry.** By E. J. Crane and others. 2nd ed. New York: Wiley, 1957. 397p. index.
Lists and describes standard reference books, periodicals, and organizations in chemistry. Includes a general subject index, but no title index. A bit dated but still useful if one keeps in mind that there may be new editions of some of the standard works listed.

8-4 **How to Find Chemical Information: A Guide for Practicing Chemists, Teachers, and Students.** By Robert E. Maizell. New York: Wiley-Interscience, 1979. 261p. bibliog. index. (A Wiley-Interscience Book).
A comprehensive and systematic guide to the literature of chemistry. Following introductory chapters on the organization of chemical information the book covers such topics as current awareness programs, access to primary sources, the publications and other programs of the Chemical Abstracts Service, other abstracting and indexing services, computer-based systems, reviews, encyclopedias, patents, safety and related topics, locating information on physical properties of materials, and chemical marketing and business information.

8-5 **How to Find Out in Chemistry.** By Charles Burman. 2nd ed. Oxford, New York: Pergamon, 1966. 226p. illus.
Discusses libraries, guides, periodicals, abstracts, general and physical chemistry, analytical chemistry, inorganic chemistry, organic chemistry, chemical technology, societies, and official publications. Stresses British publications. Dated but still useful.

8-6 **Use of Chemical Literature.** Ed. by R. T. Bottle. 3rd ed. London: Butterworths, 1979. 306p.
A guide to the use of the chemical literature including periodical literature, abstracting services, translations, handbooks, reference works, etc.

ABSTRACTS, INDEXES, AND BIBLIOGRAPHIES

8-7 **Analytical Abstracts.** v. 1- . Cambridge, England: Society for Analytical Chemistry, 1954- . monthly.

Covers all subfields of analytical chemistry: inorganic, organic, biochemistry, pharmaceutical, food, agriculture, air and water effluents, techniques, and apparatuses. Continues section C of the lapsed *British Abstracts* (8-9).

8-8 **Bibliography of Paper and Thin-Layer Chromatography and Survey of Applications.** 1961- . By Karel Macek and others. New York: Elsevier Scientific Publishing, 1968- . index. (Supplementary volumes to the *Journal of Chromatography*).

This series of bibliographies published as supplements to the *Journal of Chromatography* contains entries for literature pertaining to chromatography, which is the most widely used modern procedure in analytical chemistry today. Future bibliographies in book form will be limited only to the branches of chromatographic separations or to expanding fields of applications considered to be of special importance.

8-9 **British Abstracts.** London: Bureau of Abstracts. 1924-1953.

This was an important service for the years 1924-1953, with good abstracts but poor indexes. *British Abstracts* began as the *Abstracts of Chemical Papers* and the *British Chemical Abstracts* in two sections: A, Pure Chemistry and B, Applied Chemistry. In 1937, section A was divided into Ai, Pure Chemistry (general, physiological, and inorganic); Aii, Pure Chemistry (organic chemistry); and Aiii, Pure Chemistry (biochemistry). Each of the subsections was paged separately and had its own index. In 1938, the name was changed to *British Chemical and Physiological Abstracts*, and in 1941 section B was divided into Bi, General and Inorganic Industrial Chemistry; Bii, Industrial Organic Chemistry; and Biii, Agriculture, Foods, and Sanitation. Section C, Analysis and Apparatus, was added in 1944 and the name was changed to *British Abstracts* in 1945. It had good coverage of the FIAT and BIOS reports on wartime German chemical industry. After the war, however, not being able to compete with *Chemical Abstracts* (8-10) and *Biological Abstract* (9-4), *British Abstracts* was discontinued. Much of the material covered in the various sections of *BA* was then transferred to various British periodicals: Ai and Aii to *Current Chemical Papers* (8-16); Aiii to *British Abstracts of Medical Sciences*, which later became *International Abstracts of Biological Sciences* (9-12); Bi and Bii to *Journal of Applied Chemistry—Abstract Section*; Biii to *Journal of the Science of Food and Agriculture*; and C to *Analytical Abstracts* (8-7).

8-10 **Chemical Abstracts: Key to the World's Chemical Literature.** v. 1- . Easton, PA: American Chemical Society, 1907- . weekly.

Undoubtedly the most comprehensive scientific abstracting service in English. The descriptive abstracts are arranged in 80 sections grouped under the following five broad headings: biochemistry; organic chemistry; macromolecular chemistry; applied chemistry and chemical engineering; and physical and analytical chemistry. Every other week abstracts treat biochemistry and organic chemistry; alternate weeks deal with the remaining three groups. *CA* selectively abstracts this impressive array of source materials. Over 14,000 scientific and engineering journals, patents from 26 countries, conference proceedings, reports and monographs are monitored. Over 55% of this material is in English, with the rest in any one of 66 other languages. Indexes to *CA* are supplied for each weekly issue, each volume (six months), and for fixed collective periods (five or ten years). Each weekly issue contains four indexes: author, keyword subject, numerical patent, and patent concordance. Access in the volume indexes is by author (1907-), subject (1907-1971), general subject (1972-), chemical substance (1972-), ring system (1916-), molecular formula (1920-), numerical patent (1935-), patent concordance (1963-), hetero-atom-in context (1967-1971), and index guide (1967-). The collective indexes combine the volume indexes into single listings. From 1907 to 1956 ten-year collective indexes were produced; since 1956, collective indexes have been published every five years. Computer-readable versions of selected portions of *CA* are available. They include *Chemical-Biological Activities, Ecology and Environment, Energy, Food and Agricultural Chemistry, Materials*, and *Polymer Science and Technology*. The weekly *CA* issues, volume indexes, index guide, and patent concordance are also available in computer-readable versions.

8-11   **Chemical Abstracts Service Source Index.** 1907-1974 Cumulative. Columbus, OH: Chemical Abstracts Service, 1975. 2058p. quarterly supplements.

This cumulative index provides in one list complete identification of over 33,000 publications in chemistry, which are customarily cited by their abbreviated titles. Also lists 400 libraries at which the searcher may gain access to the original documents. For each entry complete title, CODEN, former titles, language, history, frequency, price, publisher, ALA entry, successive title, and various notes are given. Also available in computer-readable format.

8-12   **Chemical Titles.** v. 1- . Easton, PA: American Chemical Society, 1960- . biweekly.

A computer-produced KWIC index to the tables of contents of over 700 international chemical periodicals. It usually tries to list titles of articles during the month in which the journal carrying the article appears. The first part is the KWIC index, the second part a bibliography of all the periodicals indexed in the issue arranged by the title of the periodical. Author index. *Chemical Titles* is thus a current awareness service; abstracts of the same articles and papers usually follow their announcement here in *Chemical Abstracts* within several months. Also available in computer-readable format.

8-13   **Chemisches Zentralblatt.** Berlin: German Chemical Society, 1830-1969. Begun in 1830 as *Pharmaceutisches Central-Blatt*, in 1850 it became *Chemisches-pharmaceutisches Central-Blatt*, and in 1856, *Chemisches Zentral-Blatt*. The German Chemical Society took over publication rights in 1897 and changed the name of the journal to *Chemisches Zentralblatt* in 1907. In weekly issues it covered about 3,000 periodicals in pure and applied chemistry, biochemistry, mineralogy, and pharmacy. Extremely valuable for the East European and Russian literature prior to World War II. The volumes published after World War II deteriorated markedly. The service was discontinued at the end of 1969. This service, like *British Abstracts* (8-9), succumbed before the dominance of *Chemical Abstracts* (8-10).

8-14   **Colour Index.** By Society of Dyers and Colourists. 2nd ed. Yorkshire, England: the society; Lowell, MA: American Association of Textile Chemists and Colourists, 1956-1958. 4v.

8-14a   **Supplement.** 1963. 1124p.
This most comprehensive index to all known commercially produced dyes of the world is divided into three parts: the first, which covers all known commercial names under which a dye is sold, is arranged by usage categories, with the dyes in each group arranged according to the spectral hue. Each dye is assigned a C.I. usage number (e.g., C.I. — Acid Violet #7). Also included are sections on the methods of applying the coloring matter, their usage, the more important fastness properties, and certain other basic data. The second part is arranged by C.I. number. For each entry the structural formula is given as well as methods of manufacture and, when known, the inventor and citations to literature and patents. This part is indexed by compound and chemical formula. Part III contains the indexes and such general information as dye and pigment makers, details of fastness tests, and conversion tables for dyes assigned a different C.I. in the first edition. Patent and commercial names indexes.

8-15   **Current Abstracts of Chemistry and Index Chemicus.** v. 1- . Philadelphia, PA: Institute for Scientific Information, 1960- . weekly.
The first 35 volumes were called *Index Chemicus.* This computer-produced abstracting service "contains abstracts of articles reporting the synthesis, isolation, and identification of new compounds, and of articles reporting new chemical reactions or syntheses." The index portion is weekly, cumulating quarterly and annually. It contains molecular formulas and author, subject, biological activity, labeled compounds, and journal indexes.

8-16   **Current Chemical Papers.** v. 1- . London: Chemical Society, 1954- . monthly.
Continues section Ai and Aii of the lapsed *British Abstracts* (8-9). Like *Chemical Titles* (8-12), *Current Abstracts of Chemistry* (8-15) and *Index Chemicus* (8-15), this monthly reporting service is designed to inform chemists of new work faster than abstracting services possibly can.

8-17   **Electroanalytical Abstracts.** v. 1- . Basel: Birkhauser Verlag, 1963- . bimonthly.
This service, which continues the abstract section of the *Journal of Electroanalytical Chemistry*, specializes in all aspects of fundamental, physico-chemical, and analytical electrochemistry. Abstracts are short and descriptive and may lag behind the articles abstracted by as much as two years.

8-18   **Guide to Gas Chromatography Literature.** By Austin V. Signeur. New York: Plenum, 1964-1974. 3v. index.
Cites references to studies in the theory, methodology, and application of gas chromatography. It is arranged alphabetically by author, with subject indexes.

8-19   **Thin-Layer Chromatography Abstracts, 1971-1973.** By Ronald M. Scott. Ann Arbor, MI: Ann Arbor Science Publishers, 1973. 589p. index.
This book, which is devoted to TLC analysis, is a continuation of the author's *Thin-Layer Chromatography Abstracts, 1968-1971* (Ann Arbor, MI: Ann Arbor

Science Publishers, 1972. 395p.). There are over 1,000 abstracts in this volume, divided into 23 groups of compounds that can be analyzed by this technique. The abstracts are grouped by the journal titles and chronologically by year. The annotations are clear and are generally directed to the practical needs of the user. The first volume in the series was *Thin-Layer Chromatography: An Annotated Bibliography, 1964-1968* (Ann Arbor, MI: Ann Arbor Science Publishers, 1965. 284p.).

ENCYCLOPEDIAS

8-20   **Encyclopedia of Chemical Reactions.** By Carl Alfred Jacobson and C. A. Hampel. New York: Van Nostrand Reinhold, 1946-1959. 8v.
Chemical reactants are arranged alphabetically, first by the formulas of the reactant and then by the reagents. The chemical equations are also listed with a brief statement of the reaction conditions. Bibliographic references are provided.

8-21   **Encyclopedia of Chemistry.** By George Lindenberg Clark. 2nd ed. New York: Van Nostrand Reinhold, 1966. 1144p. illus. index.
Covers broad topics, such as, bonding, coordination theory, resonance and fluorescence phenomena, analysis and molecular orbitals. Contains numerous cross-references.

8-22   **Encyclopedia of Chemistry.** Ed. by Clifford A. Hampel and Gessner G. Hawley. 3rd ed. New York: Van Nostrand Reinhold, 1973. 1198p. illus. index.
A respected one-volume reference book on chemistry. The third edition includes new topics in environmental chemistry and the chemistry of life processes. Bibliographic references to additional literature on specific topics are included.

8-23   **Encyclopedia of Electrochemistry of the Elements.** v. 1- . Ed. by James Plambeck. New York: Marcel Dekker, 1973- .
This excellent encyclopedia is divided into two sections. The first section is covered by volumes 1-10 and deals with inorganic electrochemistry; the second section will cover organic electrochemistry beginning with volume 11. The organic compounds have been classified according to their electroactive groups, and these have been ordered for the most part in the manner used in Beilstein's *Handbuch der Organischen Chemie* (8-55). Includes extensive bibliographies.

8-24   **Encyclopedia of Polymer Science and Technology.** New York: Interscience, 1964-1972. 16v. illus.
Contains articles dealing with chemical substances; polymer properties, methods, and processes; and uses of polymers. Each entry is written by a specialist.

8-25   **Encyclopedia of the Alkaloids.** By John S. Glasby. New York: Plenum Press, 1975-1977. 3v.
This is a reference work for all chemists and biochemists with the compounds arranged in alphabetical sequence by substance, providing their empirical and structural formulas, melting points, and individual commentaries on their occurrence, characterization, special properties and reactions, and relation to their alkaloids. It covers the literature to the end of October 1976.

8-26　**Guide for Safety in the Chemical Laboratory.** By the Manufacturing Chemists Association. 2nd ed. New York: Van Nostrand Reinhold, 1972. 505p. illus. index.

This guide covers all aspects of laboratory safety from buildings and equipment, personal protective equipment, and waste disposal through planning for emergency operations. Appendixes include explosive reactions and hazardous properties of over 1,100 chemicals. The book is comprehensive, authoritative, and practical.

8-27　**Kingzett's Chemical Encyclopedia: A Digest of Chemistry and Its Industrial Applications.** By Charles Thomas Kingzett. 9th ed. New York: Van Nostrand Reinhold, 1966. 1092p. illus. index.

An elementary encyclopedia of general chemistry and its industrial applications. Entries are short; common and trade names are included.

8-28　**Nomenclature of Organic Chemistry.** By the International Union of Pure and Applied Chemistry. 2nd ed. (Section C); 3rd ed. (Sections A and B). New York: Plenum, 1971. 337p.

These rules for organic chemistry nomenclature were developed by the IUPAC to satisfy the need for uniformity in terminological usage. It is divided into three sections: A, hydrocarbons; B, fundamental heterocyclic compounds; and C, characteristic groups containing carbon, hydrogen, oxygen, nitrogen, halogen, sulfur, selenium, or tellurium. A chemical compound index is provided.

8-29　**Organic Chemical Process Encyclopedia.** By Marshall Sittig. 2nd ed. Park Ridge, NJ: Noyes, 1969. 712p. illus.

A collection providing the overviews of synthetic processes for industrial chemicals. All of the information is in the form of flow charts. Emphasis is on petrochemicals.

8-30　**Reagents for Organic Synthesis.** By Louis Frederick Fieser and Mary Fieser. New York: Wiley, 1967-1975. 5v.

A comprehensive encyclopedia of chemical reagents. For each reagent, structural formula, molecular weight, physical constants, preferred methods of preparation or purification, suppliers, and examples of significant uses are given. Moreover, every entry is documented with references and with flow sheets showing at a glance the molar ratio of the reactants, the reaction conditions, and the yield.

DICTIONARIES

8-31　**The Condensed Chemical Dictionary.** Revised by Gessner G. Hawley. 9th ed. New York: Van Nostrand Reinhold, 1977. 957p.

Intended to be a quick reference aid. Three distinct types of information are presented: 1) technical descriptions of chemicals, raw materials, and processes; 2) expanded definitions of chemical entities, phenomena, and terminology; and 3) description or identification of a wide range of trademarked products used in the chemical industries. For each chemical, name, synonym, formula, properties, source or occurrence, derivation, grades, containers, hazards, uses, and shipping regulations are given. Trademarked materials are keyed to a list of manufacturers.

8-32   **Dictionary of Alloys.** By Eric N. Simons. New York: Hart, 1970 (c.1969). 191p.
An alphabetical listing of alloys giving a brief decription of their compositions and physical properties. Proprietary names are included.

8-33   **Dictionary of Biochemistry.** By J. Stenesh. New York: Wiley-Interscience, 1975. 344p.
This dictionary updates the *Encyclopedia of Biochemistry* (9-47), edited by R. J. Williams and E. M. Lansford (New York: Van Nostrand Reinhold, 1967. 876p.). It contains some 12,000 entries covering general and specific topics in chemistry, physics, metabolism, biophysics, genetics, biochemistry, and molecular biology.

8-34   **Dictionary of Chromatography.** By Ronald C. Denney. New York: Wiley, 1976. 191p. illus. bibliog. (A Halsted Press Book).
Written for the "student, laboratory technician or general scientist seeking rapid information on chromatography." References to more detailed studies are provided. It includes the main equations covering theoretical chromatography and lists the equipment available for carrying out chromatography in its many forms.

8-35   **Dictionary of Electrochemistry.** By C. W. Davies and A. M. James. New York: Wiley, 1976. 246p. illus. (A Halsted Press Book).
A companion to A. M. James' *A Dictionary of Thermodynamics* (8-38). Definitions tend to be long, and numerous references are made to significant reference and textbooks by code, with full citations listed in a separate section.

8-36   **Dictionary of Named Effects and Laws in Chemistry, Physics, and Mathematics.** By Denis William George Ballentyne and D. R. Lovett. 3rd ed. New York: Barnes and Noble, 1971. 335p. illus.
Includes some 1,200 entries identifying chemical reactions, procedures, theories, laws, etc., that commemorate their discoverers.

8-37   **Dictionary of Organic Compounds: The Constitution and Physical, Chemical, and Other Properties of the Principal Carbon Compounds and Their Derivatives, Together with the Relative Literature Referenced.** 4th ed. completely rev., enlarged and re-set. New York: Oxford University Press, 1965. 5v.

8-37a   **Supplements.** 5th- . 1969- . annual.
An alphabetically arranged list of organic compounds with references to literature through 1963. The annual supplements add information published for the previous year. Supplements 1 through 4 are included in Supplement 5. Each entry gives common name, structural and molecular formulas, molecular weight, sources, physical properties, chemical properties, and, most important of all, the references to the publication of original preparations of the compound. The early editions were edited by I. M. Heilbron.

8-38   **Dictionary of Thermodynamics.** By A. M. James. New York: Wiley, 1976. 262p. illus. (A Halsted Press Book).
This is a companion to C. W. Davies and A. M. James, *A Dictionary of Electrochemistry* (8-35). Contains lengthy definitions of terms in thermodynamics. There are citations to reference books and textbooks.

8-39 **Glossary of Chemical Terms.** By Clifford A. Hampel and Gessner G. Hawley. New York: Van Nostrand Reinhold, 1976. 281p. illus.

This glossary contains 2,000 informative entries, emphasizing major chemical classifications, important functional terms, basic phenomena and processes, the chemical elements, major compounds, general and miscellaneous terminology, and biographies of outstanding contributors to the science of chemistry.

8-40 **Hackh's Chemical Dictionary. American and British Usage, Containing the Words Generally Used in Chemistry, and Many of the Terms Used in the Related Sciences of Physics, Astrophysics, Mineralogy, Pharmacy, Agriculture, Biology, Medicine, Engineering, etc. Based on Recent Chemical Literature.** Ed. by Julius Grant. 4th ed. New York: McGraw-Hill, 1969. 738p. illus.

An authoritative dictionary covering more than 55,000 words and phrases in chemistry and related fields. Trademarks are identified where possible. Presents both U.S. and British usage.

8-41 **Technical Dictionary of Chromatography.** By Hans-Peter Angele. New York: Pergamon, 1970. 119p.

A list of about 3,000 terms and expressions used in the field of chromatographic analysis. The first section is a parallel listing in English, German, French, and Russian terms. The second part is an index by language.

8-42 **Technical Dictionary of Highpolymers; Hochpolymere; Hautspolymeres.** By Vladimir N. Davydov and Horst Howorka. (Title proper in Russian). Oxford; New York: Pergamon, 1969. 959p.

Contains more than 10,000 terms in English, French, German, and Russian. Entries encompass scientific terms for high polymers as well as terms used in applications. Included is colloquial terminology used by engineers and scientists.

8-43 **Technical Dictionary of Spectroscopy and Spectral Analysis: English, French, German, Russian, with a Supplement in Spanish.** Ed. by Heinrich Moritz and Tibor Torok. New York: Pergamon, 1971. 188p. index.

A polyglot dictionary containing more than 3,500 spectroscopy and spectral analysis terms that have appeared in international works, leading technical journals, and papers delivered at international conferences. Arranged alphabetically by the English term, with German, French, Russian and Spanish indexes.

HANDBOOKS AND TABLES

8-44 **Chemical Tables.** By Bela A. Nemeth. New York: Halsted Press, a division of John Wiley, 1976. 476p.

A handy reference source of chemical tables arranged alphabetically within each major section.

8-45 **The Chemist's Companion: A Handbook of Practical Data, Techniques and References.** By Arnold J. Gordon and Richard A. Ford. New York: Wiley, 1972. 537p. index.

Divided into nine chapters, this work covers properties of atoms and molecules, spectroscopy, photochemistry, chromatography, kinetics and thermodynamics, various experimental techniques, and mathematical and numerical information, including the definitions, values, and usage rules of the newly adopted International System of Units (SI Units). The chapter on spectroscopy is especially well done. Also included is a variety of hard to classify but frequently sought information, such as names and addresses of microanalysis companies and chemistry publishers, descriptions and commercial sources of atomic and molecular models, and safety data for hazardous chemicals. More than 500 references are given.

8-46   **Crystal Structures.** By Ralph W. G. Wyckoff. 2nd ed. New York: Wiley-Interscience, 1963-1971. 6v.

For each crystalline compound treated, a detailed description of crystalline structure is given. Includes diagrams, models, formulas, and x-ray measurements. Comprehensive.

8-47   **Gmelin's Handbuch der Anorganischen Chemie.** By Leopold Gmelin. 8 vollig neue beart. aufl. Berlin: Verlag Chemie, 1924- .

The most comprehensive work in inorganic chemistry in the world. The first edition was issued in 1817-1819 under the name, *Handbuch der Theoretischen Chemie.* The seventh edition was completed in 1927. The eighth edition is being compiled by the Deutsche Chemische Gesellschaft and will cover all the literature on inorganic chemistry to January 1, 1950. The work is arranged into 71 systems according to the Gmelin Institute's classification scheme of non-metals to metals. Each entry covers history, occurrence, properties, scientific methods of preparation and commercial process of manufacture. The entire work is on cards in the Gmelin Institute for the literature after January 1, 1950, coded for machine searching in the Institute's archives. Supplements to the entire handbook are now being published.

8-48   **Handbook for Chemical Technicians.** By Howard J. Strauss. New York: McGraw-Hill, 1976. 1v. (various paging). illus. index.

A handbook that presents fundamental data and a short exposition of the theory on each subject and examples relevant to the use of particular tables or graphs. Sections include: weights and measurements; thermal, electrical, and mechanical units; a two-part review of chemical fundamentals; thermochemistry and kinetics; organic chemistry; metals and alloys; fluid mechanics; engineering operations; and safety practices.

8-49   **Handbook of Chemistry and Physics: A Ready-Reference Book of Chemical and Physical Data.** Cleveland, OH: Chemical Rubber Co., 1913- . annual.

An invaluable mine of tabular data in physics and chemistry. It includes sections on the elements, atomic weights, organic compounds, and physical constants. Revised annually.

8-50   **Handbook of Chromatography.** Ed. by Gunter Zweig and Joseph Sherma. Cleveland, OH: Chemical Rubber Company Press, 1972. 2v. illus. bibliog. index.

Volume I of the *Handbook* contains over 500 tables of chromatographic data and is divided into four sections: gas chromatography, liquid chromatography, paper chromatography, and thin-layer chromatography. There is an alphabetical list of tables for each section and an alphabetical compound index; there is no molecular formula index. Volume II is divided into two parts. Part 1, "Principles and Techniques," presents basic explanation of chromatographic methods. Part 2, "Practical Applications" contains information on detection reagents used in paper and thin-layer chromatography, description and graphic materials with sources of further information, and an extensive list of books on chromatography arranged by publisher. This is a major reference work on chromatography.

8-51 **Handbook of Metal Ligand Heats and Related Thermodynamic Quantities.** By James J. Christensen and Reed M. Izatt. 2nd ed. rev. New York: Marcel Dekker, 1975. index. (Contribution no. 13 from the Center for Thermochemical Studies, Brigham Young University, Provo, Utah).

This is a compilation of the published literature dealing primarily with the heats of metal-ligand interactions in solution. The data are recorded in tabular form, and include author, empirical formula, element, synonym, and reference indexes. Each ligand is given a letter-number designation, followed by empirical and structural formulas and synonyms. Under each ligand are listed the metals with delta-H data, Log K, delta-S values, temperature of delta-H data, method of determining delta-H, experimental conditions, references, and remarks.

8-52 **Handbook of Naturally Occurring Compounds.** v. 1- . By T. K. Devon and A. I. Scott. New York: Academic Press, 1972- . illus. index.

This handbook contains most naturally occurring compounds to which structures have been assigned. Polymeric compounds, synthetic derivatives, and degradation products are excluded. Each structure is shown with its name, molecular formula, molecular weight, optical rotation, melting, literature reference (usually the latest), and classification number. Access is provided by alphabetic, molecular weight and molecular formula indexes. This work is projected in three volumes with volume 3 yet to be published.

8-53 **Handbook of Spectroscopy.** v. 1- . Ed. by J. W. Robinson. Cleveland, OH: Chemical Rubber Company Press, 1974- . illus. bibliog. index.

This handbook provides information on the major fields of spectroscopy. Specifically, those fields include NMR, IR, Raman, UV (absorption and fluorescence), mass-spectrometry, atomic absorption, flame photometry, emission spectrography, and flame spectroscopy. There is one combined author and subject index at the end of each volume. The work is projected for three volumes with volume 3 yet to be published.

8-54 **Handbook of the Atomic Elements.** By R. A. Williams. New York: Philosophical Library, 1970. 125p.

An alphabetical listing of the 103 atomic elements. Gives, in tabular form, all of their physical values based on Carbon-12 and values released by the International Union of Pure and Applied Chemistry. A very useful tool.

8-55 **Handbuch der Organischen Chemie.** By Friedrich Konrad Beilstein. 4. Aufl. Berlin: Springer, 1918-1940. 27v.

8-55a **Erganzungswerk I.** Die Literatur von 1910-1919 umfassend. 1938-1948. 27v.

8-55b   **Erganzungswerk II.** Die Literatur von 1920-1929 umfassend. 1941-1955. 27v.

8-55c   **Erganzungswerk III.** Die Literatur von 1930-1949 umfassend. 1956-1974. 27v.

8-55d   **Erganzungswerk III/IV.** Die Literatur von 1930-1959 umfassend. 1974- .

8-55e   **Erganzungswerk IV.** Die Literatur von 1950-1959 umfassend. 1972- .
This is the world's largest compilation of physical data on organic chemistry. For each compound described, the following data are provided: constitution and configuration; natural occurrence and isolation from natural products; preparation, formation, and purification; structural and energy parameters; physical properties; chemical properties; characterization and analysis; and salt and addition compounds. All data are checked for accuracy and complete references are cited. Written in German, it is divided into a basic series and four supplementary series. Each series comprises 27 volumes (or groups of volumes) in which individual compounds are arranged by the Beilstein classification system. The volumes of supplementary series III and IV, covering the heterocyclic compounds (v. 17-27), are combined into a joint issue, *Erganzungswerk III/IV* (Supplementary Series III/IV).

8-56   **Infrared Band Handbook.** By Herman A. Szymanski and Ronald E. Erickson. 2nd ed. rev. and enlarged. New York: IFI/Plenum, 1970. 2v.
A comprehensive treatment of most structural groups and their environments. Arranged by wave number followed by physical state, special information, structural group, and the reference number.

8-57   **Lange's Handbook of Chemistry.** Ed. by John A. Dean. 12th ed. New York: McGraw-Hill, 1979. 1v. (various paging). illus. index.
A handy reference tool providing ready access to chemical and physical data used in the laboratory and manufacturing. It is divided into the following sections: mathematics, general information and conversion tables, atomic and molecular structures, inorganic chemistry, analytical chemistry, electrochemistry, organic chemistry, spectroscopy, thermodynamic properties, and physical properties.

8-58   **Methoden der Organischen Chemie.** By Josef Houben and T. Weyl. 4. vollig neue gestaltete Aufl. Stuttgart: G. Thieme, 1952- . (to be complete in 16v.).
Written in German, this set is considered the authority on laboratory methods in organic chemistry. It gives detailed coverage of the methods of preparation of organic compounds with full discussion of underlying theories and documentation. Complements Beilstein's *Handbuch der Organischen Chemie* (8-55).

8-59   **A New Handbook of Chemistry.** By Philip S. Chen. Camarillo, CA: Chemical Elements, 1975. 212p. index.
A handbook written in a more popular style. Its purpose is to select the most useful information for beginning students, and it offers sections on five-place logarithms, tables on organic and inorganic compounds, a glossary of chemical terms, etc.

8-60 **Parent Compound Handbook.** Columbus, OH: Chemical Abstracts Service, 1976; with bimonthly supplements.

The Parent Compound File (part 1) is an excellent tool listing chemical structure, CAS Registry Number, current *CA* index name, molecular formula, and Wisewessor Line Notation for nearly 50,000 ring parents indexed by CAS. The Index of Parent Compounds (part 2) allows one to locate entries in part 1 by means of *CA* index names, ring analysis, ring substructure, Wisewessor Line Notation, molecular formula, and CAS Registry Number. This information provides easy entry for ring compounds into the *CA* volume and collective indexes.

8-61 **Registry Handbook.** Columbus, OH: Chemical Abstracts Service, 1974. annual supplements.

This book lists, in ascending CAS Registry Number order, molecular formula and *CA* chemical names for over two million chemical substances registered in the *CA* system from 1965 through 1971 with annual supplements adding to the list. This information may be used as access points into the *CA* volume and collective indexes.

8-62 **The Ring Index: A List of Ring Systems Used in Organic Chemistry.** By Austin McDowell Patterson. 2nd ed. Washington: American Chemical Society, 1960. 1452p.

8-62a **Supplements.** 1963-1965. 3v.

The second edition includes some 7,700 ring systems prepared by the indexing staff of Chemical Abstracts Service. The supplements make a combined coverage of 14,265 ring systems. Current ring systems are indexed in the *Parent Compound Handbook* (8-60).

8-63 **Statistical Manual for Chemists.** By Edward L. Bauer. 2nd ed. New York: Academic Press, 1971. 193p. index.

Designed as a manual for the working chemist, this handbook describes, in detail, basic techniques used in statistical methods, including experimental design and the analysis of variance, analysis of variance by range, control charts, correlated variables, sampling, and control of routine analysis. References to additional readings are appended to most chapters, and the book is well indexed.

8-64 **Tables of Spectral Lines.** By Aleksandr Natanovich Azidel' and others. New York: IFI/Plenum, 1970. 782p.

This main part of this handbook is the section containing the tables of spectral lines by elements. The other, briefer section contains tables of spectral lines in order of decreasing wavelength.

TREATISES

8-65 **Advanced Treatise on Physical Chemistry.** By James Riddick Partington. New York: Wiley, 1949-1962. 5v.

Vol. 1 — Fundamental Principles: The Properties of Gases; vol. 2 — The Properties of Liquids; vol. 3 — The Properties of Solids; vol. 4 — Physico-Chemical Optics; vol. 5 — Molecular Spectra and Structures; Dielectrics and Dipole Moments. A mathematical treatment of physical chemistry.

8-66   **Comprehensive Chemical Kinetics.** Ed. by C. H. Bamford and C. H. M. Tippen. Amsterdam: Elsevier, 1969- .
Covers the practice and theory of kinetics of organic, inorganic, and polymerization reactions. The series is divided into broad subject sections consisting of one or more volumes.

8-67   **Comprehensive Inorganic Chemistry.** Ed. by J. C. Bailar and others. Elmsford, NY: Pergamon, 1973. 5v. illus. bibliog. index.
An excellent, concise work that is easy to use. Elements and groups of elements are surveyed from their discovery to the present. Physical, chemical, and biological properties are examined. The master indexes provide easy subject access.

8-68   **Comprehensive Organic Chemistry.** Ed. by Sir Derek Barton and others. Elmsford, NY: Pergamon, 1979. 6v. illus. bibliog. index.
Focuses on the properties and reactions of the major classes of organic compounds including some synthetic compounds and natural products. Original literature references are provided throughout the text. The indexes are excellent and provide molecular formula, subject, author, reaction, and reagent access.

8-69   **Comprehensive Treatise on Inorganic and Theoretical Chemistry.** By Joseph William Mellor. London: Longman, 1922-1937. 16v.

8-69a   **Supplements.** 1956- .
Covers inorganic chemistry comprehensively. The work is arranged according to the periodic table.

8-70   **Rodd's Chemistry of Carbon Compounds; A Modern Comprehensive Treatise.** By E. H. Rodd. 2nd ed. rev. New York: Elsevier, 1964- .
One of the most comprehensive treatises available on carbon compounds, it is divided into four parts: introduction and aliphatic compounds; alicyclic compounds; aromatic compounds; and heterocyclic compounds. References to the original literature are provided. Each volume has its own index; a master index to the entire set is planned.

8-71   **Treatise on Analytical Chemistry.** By Izaak Maurits Kolthoff and P. J. Elving. New York: Interscience, 1959- .
This treatise is divided into three parts: theory and practice; the elements and inorganic and organic compounds; and analytical chemistry in industry. Each part has several volumes. The treatise is "to be a guide to the advanced and experienced chemist."

DIRECTORIES

8-72   **American Chemists and Chemical Engineers.** Ed. by Wyndham D. Miles. Washington: American Chemical Society, 1976. 544p. index.
This directory provides information on 517 prominent men and women in chemistry and covers a span of 300 years.

8-73   **Chem BUY Direct: International Chemical Buyers Directory.** Ed. by Friedrich W. Derz. Berlin: De Gruyter, 1974- .

An international, comprehensive guide to chemicals. The Chem PRODUCT Index lists over 300,000 chemicals; the Chem SUPPLIERS Directory lists over 23,000 firms; and the Chem ADDRESS provides full name and address of the supplier.

8-74 **Chem Sources—U.S.A.** Flemington, NJ: Directories Publishing Co., 1958- . annual.

8-74a **Chem Sources—Europe.** Flemington, NJ: Directories Publishing Co., 1973- . annual.
These two directories provide a comprehensive listing of chemical products in the United States and Europe listing them by chemical name and indexed by product.

8-75 **Chemical Guide to Europe.** 1st ed.- . Pearl River, NY: Noyes, 1961- . annual.
Lists the chemical firms of Western Europe giving name and address, ownership, plant locations, products, local subsidiaries and affiliates, foreign subsidiaries and affiliates, principal executives, latest annual sales, and number of employees.

8-76 **Chemical Guide to the United States.** 1st ed.- . Pearl River, NY: Noyes, 1962- . annual.
Lists the largest U.S. firms, giving name and address, ownership, annual sales, number of employees, principal executives and titles, plant locations, products, domestic subsidiaries and affiliates, and foreign subsidiaries and affiliates.

8-77 **Directory of Graduate Research.** v. 1- . By the Committee on Professional Training of the American Chemical Society. Washington: American Chemical Society, 1953- . biennial.
A directory of the faculties, publications, and doctoral theses in departments or divisions of chemistry, chemical engineering, biochemistry, pharmaceutical and/or medicinal chemistry at universities in the United States and Canada. Only includes programs that lead to the PhD degree.

8-78 **Great Chemists.** By Eduard Farber. New York: Interscience, 1961. 1642p. illus.
Contains biographical sketches of over 100 chemists from the earliest times to the twentieth century. Includes a portrait for each entry.

8-79 **Nobel Prize Winners in Chemistry, 1901-1961.** By Eduard Farber. Rev. ed. London: Abelard-Schuman, 1963. 341p. illus. (The Life of Science Library no. 41).
A popular treatment of each Nobel Prize winner in chemistry from 1901 to 1961. For each entry there is a portrait, a biographical sketch, an analysis of the prize-winning work, and a discussion of its consequences for theory and practice.

9

Biological Sciences

Biology is the science of living organisms, their nature, functions, reproduction, and relations with the environment. It uses the principles of physics, chemistry, and mathematics; and interacts with anthropology, psychology, sociology, agriculture, medicine, and industry. Carl Linnaeus (1707-1778) is one of the first recognized biologists. He established the binomial system of classifying and identifying organisms by assigning a species name to the genus name, the next higher grouping of similar but not interbreeding organisms. The first Linnaean classification, *Systema Naturae*, was published in 1735. In the tenth edition of *Systema Naturae*, some 4,236 animal species were listed; today almost one million species are known. Control of animal specificity is currently maintained by the International Commission on Zoological Nomenclature; the botanical counterpart is the International Bureau of Plant Taxonomy and Nomenclature.

During the seventeenth century, Leeuwenhoek (1632-1723) introduced microbiology to the scientific world by using a crude microscope to observe organisms invisible to the naked eye. Redi (1627-1697), one of the first biologists to use the scientific method, proved by a controlled experiment that, in decaying meat, maggots developed from exposure to flies and not from spontaneous generation. The outstanding work of Pasteur in the nineteenth century is the precursor to the development of biochemistry. Antoine Lavoisier (1743-1794) and Joseph Priestley (1733-1804) discovered that oxygen is vital in animal life and is produced by plants. In 1838 and 1839 M. J. Schleiden and Teodor Schwann established the cell theory in biology. In the publication of *On the Origin of Species by Means of Natural Selection* (1859), Charles Darwin presented observations to support the gradual evolution of life-forms, by expressing both unity in origin and diversity of form. In the twentieth century, one of the most outstanding research contributions has been the chemical analysis of DNA, due to the efforts of Maurice H. F. Wilkins. Francis H. C. Crick and James D. Watson later supplemented Wilkin's work by forming a molecular model of the DNA molecule.

The literature of biology is extensive. There are numerous fields of specialized study, including virology — viruses; bacteriology — bacteria; mycology — fungi; phycology — seaweeds; bryology — liverworts and mosses; pteridology — ferns ("botany" is the term used to describe the study of all plants); entomology — insects; ichthyology — fishes; herpetology — amphibia and reptiles; ornithology — birds; mammalogy — mammals; biochemistry — chemistry of living organisms; biogeography — patterns of distribution of living organisms; biometrics — mathematics used in biological interpretations; biophysics — physics of living organisms; cytology — formation, structure, and functions of cells; ecology — interrelationships of organisms in the environment; embryology — formation, early growth, and development of organisms; endocrinology — ductless glands and their secretions in animals; evolution — origin and change of plant and animal species; genetics — hereditary transmission of characteristics; histology — microscopic structure of plant and animal tissues; paleontology — fossil organisms of the geologic past; parasitology — organisms living at the expense of larger forms; pathology—manifestations of disease, causes, and processes; physiology—functions and processes sustaining life; and taxonomy—classification of plants and animals.

Because there is so much biological literature, this chapter is divided into four sections. The first section covers general biology; the second, cell biology; the third botany/agriculture; and the fourth, zoology.

GENERAL BIOLOGY

Guides to the Literature

9-1   **Guide to the Literature of the Life Sciences.** By Roger C. Smith and W. Malcolm Reid. 8th ed. Minneapolis, MN: Burgess, 1972. 166p. index.

A revision of Smith's *Guide to the Literature of the Zoological Sciences* (Minneapolis, MN: Burgess, 1967. 238p.), it is a library guide rather than a detailed handbook on the literature of life sciences. The first chapter on literature problems discusses science texts, special literature, research fund sources, and biology education materials. There are chapters on library organization, library research methodology, and preparation of scientific papers. Other chapters cover the major reference sources, emphasizing abstract journals, primary research journals, and taxonomic works. A related work is Pieter Smit's *History of the Life Sciences: An Annotated Bibliography* (New York: Hafner, 1974. 1071p.), which is designed to serve historians of science and librarians. International in coverage, it lists and briefly annotates over 4,000 works from the earliest writings through mid-1971.

9-2   **Library Research Guide to Biology: Illustrated Search Strategy and Sources.** By Thomas G. Kirk, Jr. Ann Arbor, MI: Pierian Press, 1978. 84p. illus. bibliog. index. (Library Research Guides Series, no. 2).

This is a step-by-step guide to the basic tools of information gathering for biology. The chapters cover Refining Your Topic, Reviews, *Science Citation Index* and Author Approach, *Biological Abstracts* and Other Subject Indexes, The Last Six Months, Using Guides in Biology, and Using Other Libraries. Appendixes include lists of review serials, guides to the literature, and basic reference sources for biology courses, as well as a section on how to use *Chemical Abstracts* and *Zoological Record.*

9-3   **Use of Biological Literature.** Ed. by R. T. Bottle and H. V. Wyatt. 2nd ed. Woburn, MA: Butterworths, 1972. 379p. index.

A comprehensive guide for graduate student use citing major research and reference works. The chapters on library use and research methods are straightforward; there is a good discussion of specific libraries important for biological research. Includes discussions of the major abstracting services and patent literature accessibility. The book is arranged by biological discipline.

Abstracts, Indexes, and Bibliographies

9-4   **Biological Abstracts.** v. 1- . Philadelphia, PA: Biosciences Information Service, 1926- . semimonthly.

The most comprehensive abstracting service available in biology and biomedicine. Annually, *BA* contains 149,000 abstracts of original research from over 100 countries. The abstracts, written in English, are arranged by some 85 subject categories, and include complete bibliographic identification plus the address of the main author and language of the original article. There is a list of new books and periodicals received in each issue. Indexes provide five types of access: author (personal or corporate names); biosystematic (broad taxonomic categories); generic (genus-species names); concept (broad subject categories); and subject (specific words). Cumulative indexes are provided semiannually. *Biological Abstracts* together with *Biological Abstracts/RRM* (9-5) constitute the BIOSIS (Biosciences Information Service) data base. Reprint publications of BIOSIS in selected specialities are issued as monthly serials with annual index cumulations. They include *Abstracts of Entomology* (9-162), *Abstracts of Mycology* (9-70), *Abstracts on Health Effects of Environmental Pollutants* (11-7), and *BioResearch Today*. Other important serial publications of BIOSIS are: *BIOSIS List of Serials, Guide to the Indexes*, and *A Guide to the Vocabulary of Biological Literature*.

9-5   **Biological Abstracts/RRM.** v. 17- . Philadelphia, PA: Biosciences Information Service, 1980- . monthly.
Biological Abstracts/RRM (Reprints, Reviews, Meetings) is the successor to *BioResearch Index* (9-7). It lists full citations, including language of the article and author address, of some 125,000 symposia, review articles, and research reports not found in *Biological Abstracts* (9-4). Descriptive words and phrases are added by BIOSIS staff to enrich titles and provide additional access through the indexes. Indexes provide access by author (personal or corporate names); biosystematic (broad taxonomic categories); generic (genus-species names); concept (broad subject categories); and subject (specific words).

9-6   **Biological and Agricultural Index: A Cumulative Subject Index to Periodicals in the Fields of Biology, Agriculture, and Related Sciences.** v. 1- . New York: Wilson, 1916/18- . monthly.
An elementary alphabetical subject index for students and laypersons. The analytic indexing is comprehensive for all articles in some 150 journals, which are selected by librarians as the most frequently cited sources for general use. Until 1964, it was titled *Agricultural Index*. There are quarterly and annual cumulations.

9-7   **BioResearch Index.** Philadelphia, PA: Biosciences Information Service, 1967-1979. 16v.
This was a preliminary publication of bibliographic citations for biomedical research taken from annual institutional reports, bibliographies, letters, notes, preliminary reports, reviews, selected government reports, semipopular journals, symposia, and trade journals. It is continued by *Biological Abstracts/RRM* (9-5).

9-8   **Catalog of the Library of the Marine Biological Laboratory and the Woods Hole Oceanographic Institution, Woods Hole, Mass.** Boston: G. K. Hall, 1971. 12v.

9-8a   **Journal Catalog.** 1971. 418p.

Covers the last 300 years of scientific literature related to marine biology and oceanography, including biology, zoology, botany, physiology, microbiology, medicine, physics, chemistry, mathematics, geology, meteorology, geophysics, fisheries and oceanography. Includes 4,000 separate titles of scientific research journals, 250,000 reprints, 12,000 books, 138 expeditions, and nearly 300,000 journal articles.

9-9 **Catalogs of the Scripps Institution of Oceanography Library, University of California at San Diego.** Boston: G. K. Hall, 1970. 12v. The 12 volumes consist of *Author/Title Catalog*, 7v.; *Subject Catalog*, 2v.; *Shelf List*, 2v.; *Shelf List of Documents, Reports and Translations Collection*, 1v.

9-9a **Supplement.** 1- . 1973- .
This comprehensive catalog covers 80,000 bound volumes and 13,000 reports in oceanography, marine biology, and marine technology from 1633 to the present. It specializes in atmospheric sciences, fisheries, geology, geophysics, and zoology, plus a major collection in oceanographic expedition literature.

9-10 **Catalogues of the Library of the Freshwater Biological Association, Cumbria, England.** Boston: G. K. Hall, 1979. 6v.
Contains some 126,000 cards representing one of the world's finest collections of limnological literature. The library's collection of over 4,000 books, 1,100 journals, serials, and reports and 50,000 reprints from the mid-nineteenth century to the present covers biology, physics, and chemistry of inland waters throughout the world with emphasis on freshwater algae, planktonic and benthic invertebrates, fishes, and lake sediments.

9-11 **Index to Illustrations of the Natural World: Where to Find Pictures of the Living Things of North America.** Comp. by John W. Thompson. Syracuse, NY: Gaylord, 1977. 265p. bibliog. index.
All species of plants and animals are listed alphabetically with the books' three-letter symbols, page numbers, etc. that indicate where the illustrations can be found. Separate indexes to scientific names, book titles, and book code letters. A unique book.

9-12 **International Abstracts of Biological Sciences.** v. 1- . London: Pergamon, 1954- . monthly.
From 1954 to 1956 this service was called *British Abstracts of Medical Sciences*. It is arranged by broad subject categories, and entries are numbered consecutively. Titles of articles are translated into English; translations from *Referativnyi Zhurnal's* (3-17) biology section are included. The strong areas are the physiological and biochemical aspects of anatomy, animal behavior, biochemistry, endocrinology, experimental medicine, microbiology, pathology, and pharmacology. Abstracts are signed. There are quarterly author and subject indexes.

9-13 **Marine Science Contents Tables.** v. 1- . Rome: Fisher Resources Division, FAO (Food and Agriculture Organization); distr. New York: Unipub, 1966- . monthly.
A table of contents index to 100 periodicals in all languages. Also provides information on future conferences on marine science.

9-14 **Natural History Index-Guide.** Comp. by Brent Altsheler. 2nd ed. New York: Wilson, 1940. 583p.

Classified subject guide to 3,365 noteworthy books and periodicals, including field guides, published from 1800 to 1940, selected by an international panel of scientists and explorers. In addition to publications in botanical and zoological categories, there are related works in geology, geography, anthropology, and astronomy. The source bibliography provides full citations and the subjects covered in the cited work. Includes a supplementary bibliography on the prominent natural history authorities of the period covered.

9-15 **Oceanic Abstracts.** v. 1- . Louisville, KY: Data Courier Inc., 1964- . six per year.

Formerly *Oceanic Index* and *Oceanic Citation Journal*, this abstracting journal covers biology, fisheries; geology, meteorology, oceanography; acoustics, optics, positioning, remote sensing; desalination, pollution; engineering materials; diving, offshore, deep sea; ships, submersibles, buoys; and government, legal. It scans books, journals, papers, pamphlets, government publications, and symposia. The abstracts are informative.

9-16 **Oceanographic Index.** Comp. by Mary Sears at the Woods Hole Oceanographic Institution. Boston: G. K. Hall, 1972-1974. 15v.

9-16a **Author Cumulation 1946-1970.** 3v.

9-16b **Author Cumulation 1971-1974.** 1v.

9-16c **Regional Cumulation 1946-1970.** 1v.

9-16d **Regional Cumulation 1971-1974.** 1v.

9-16e **Subject Cumulation 1946-1970.** 4v.

9-16f **Subject Cumulation 1971-1974.** 2v.

9-16g **Marine Organisms, Chiefly Planktonic Cumulation 1946-1973.** 3v.

These are working guides to journal articles, monographs and other literature on the marine sciences held in the Library of the Marine Biological Laboratory and of the Woods Hole Oceanographic Institution. The scope is international with emphasis on biological and physical oceanography and marine chemistry, geology and meteorology. Many related topics, however, have been excluded: limnology, terrestrial geology, basic chemistry, fisheries biology, malacology, algology, etc.

9-17 **Wildlife Abstracts: A Bibliography and Index of the Publications Abstracted in Wildlife Reviews.** Washington: U.S. Fish and Wildlife Service; distr. Washington: GPO, 1935- .

The abstracts are divided into two parts: part 1 is a bibliography of titles arranged by subject headings (the *Wildlife Review* issue and page numbers follow each entry); part 2 is an author and subject index. The subject index covers species and groups of animals in detail.

Encyclopedias

9-18 **Encyclopedia of the Biological Sciences.** Ed. by Peter Gray. 2nd ed. New York: Van Nostrand Reinhold, 1970. 1027p. illus. map. tables. bibliog. index.

Provides clear and accurate information for biologists in those fields in which they are not experts. Definitions are unsigned, but the 800 longer articles are signed and represent the work of nearly 500 contributors. Coverage is comprehensive for developmental biology, ecology, physiology, genetics, anatomy, and taxonomy. Includes about 100 biographical entries. Terms for applied biology, such as medicine and agriculture, are excluded.

9-19   **Grzimek's Encyclopedia of Ethology.** Bernhard Grzimek, ed. in chief. New York: Van Nostrand Reinhold, 1977. 705p. illus. (part col.). bibliog. index.

A good survey of ethology with a European slant. Covers history, goals, organization, and methods of ethology; the structure and function of the nervous system and sense organs; the operation of the major senses; biological clocks; migration and orientation; displacement activities; sexuality and courtship; social behavior; and the development and evaluation of behavior. There are two complimentary volumes: *Grzimek's Encyclopedia of Evolution* (9-20) and *Grzimek's Encyclopedia of Ecology* (11-34).

9-20   **Grzimek's Encyclopedia of Evolution.** Bernhard Grzimek, ed. in chief. New York: Van Nostrand Reinhold, 1977 (c.1976). 560p. illus. (part col.). index.

Describes current theories and research findings on a whole range of evolutionary topics ranging from the origin of life to separate chapters on various geological eras and periods. Complements *Grzimek's Encyclopedia of Ethology* (9-19) and *Grzimek's Encyclopedia of Ecology* (11-34).

9-21   **Wildlife in Danger.** By James Fisher and others. New York: Viking, 1969. 368p. illus. (part col.). index.

Offers encyclopedic coverage of plants and animals, particularly bird and mammal species, in danger of extinction. Provides a detailed description of each species including history and present status, protection measures, habits and geographical distribution, and a photo or sketch for some species. The entries are divided into mammals, birds, reptiles, amphibians, fishes, and plants. Most of the information is taken from the *Red Data Book* published by the Survival Service Commission of the International Union of Nature and Natural Resources, which also sponsored this publication.

Dictionaries

9-22   **Dictionary of Biology.** By Michael Abercrombie, C. J. Hickman, and M. L. Johnson. 6th ed. Baltimore, MD: Penguin Books, 1973. 309p.

Lists about 2,000 definitions and is intended for the layperson and beginning student. Definitions range to three-fourths of a page in length. Also, includes terms from paleontology and archaeological anthropology. Cross-references are included. A longer student dictionary is Edwin B. Steen's *Dictionary of Biology* (New York: Barnes and Noble, 1973. 630p.). It provides simple definitions for about 12,000 terms.

9-23   **Dictionary of Biology: English/German/French/Spanish.** By Günther Haensch and Gisela Haberkamp de Anton. New York: Elsevier Scientific, 1976. 483p. index.

Includes 9,795 terms chosen from all areas of biology. Arranged by English term followed by German, French, and Spanish equivalents.

9-24 **Dictionary of Comparative Pathology and Experimental Biology.** By Robert W. Leader and Isabel Leader. Philadelphia, PA: W. B. Saunders, 1971. 238p. bibliog.

Intended for specialists, the expanded definitions cover biological and behavioral characteristics reappearing throughout the animal kingdom in multiple forms. Emphasis is given to terminology relating genetics to disease. The textual definitions refer to tabulations which are useful presentations of the comparative data. Unique features include a glossary for experimental behaviorists, and information on common laboratory animals.

9-25 **Dictionary of Life Sciences.** Ed. by E. A. Martin. New York: Pica Press; distr. New York: Universe Books, 1977. 374p. illus.

A general dictionary for the student and layperson. The treatment of each term is more encyclopedic and descriptive than most dictionaries.

9-26 **Dictionary of the Biological Sciences.** By Peter Gray. New York: Van Nostrand Reinhold, 1967. 602p.

This dictionary lists 40,000 terms. Words derived from the same root prefix or suffix are assembled together, with selected cross-references from prefaced words. Excellent for college level; complementary to Henderson. Dr. Gray also compiled an abridged version of 8,000 terms, *Student Dictionary of Biology* (New York: Van Nostrand Reinhold, 1973. 194p.).

9-27 **McGraw-Hill Dictionary of the Life Sciences.** Daniel N. Lapedes, ed. in chief. New York: McGraw-Hill, 1976. 907p. illus.

Contains basic, thumbnail definitions of more than 20,000 terms. Useful appendixes include: U.S. customary system and the metric system, the International System (SI), conversion factors for the measurement systems, symbols and atomic numbers for the chemical elements, fundamental constants, spectrum of antibiotic activity, normal clinical chemistry and cytology values, plus animal, bacterial, and plant taxonomy.

9-28 **Source-Book of Biological Names and Terms.** By Edmund Carroll Jaeger. 3rd ed. Springfield, IL: Thomas, 1955. 317p. illus.

An alphabetical list of nearly 12,000 elements from which scientific biological names and terms are derived. Gives Greek, Latin, or other origins, plus concise root meanings and examples of derivatives. The third edition has been updated by the supplement of about 1,000 terms. A special feature includes 300 short biographies of persons commemorated in biological terms. For an instructive presentation of the use of Greek and Latin prefixes and suffixes to formulate modern scientific terms, see Donald M. Ayers, *Bioscientific Terminology: Words from Latin and Greek Stems* (3-56); an extensive index enables the work to serve also as a dictionary to Greek and Latin roots.

Handbooks and Field Guides

9-29 **Biological Research Method: A Practical Guide.** By H. H. Holman. New York: Hafner, 1969. 280p. illus.

Intended for researchers who work with animals or animal products. Chapters cover data collection, records and their interpretation, mathematical and statistical methods, and the organization and preparation of scientific papers.

9-30 **Biology Data Book.** Comp. and ed. by Philip L. Altman and Dorothy S. Dittmer. 2nd ed. Bethesda, MD: Federation of American Societies for Experimental Biology, 1972-1974. 3v. index.

A handbook of basic data for biomedical science serving the research needs in the areas of genetics, cytology, morphology, animal processes, tissues, regulators, toxins, biochemistry, and biophysics. Additional tables outline experimental materials and methods, laboratory diets, culture media, and chemical indicators.

9-31 **Enzyme Handbook.** By Thomas E. Barman. New York: Springer-Verlag, 1969. 2v. suppl. 1974. 517p.

Provides molecular data on some 800 enzymes. For an enzyme to be included in the handbook, its molecular weight, specific gravity, specificity, or kinetic properties must be described in the literature. Arranged according to the Enzyme Commission list of enzymes. Enzyme index excludes systematic names.

9-32 **Manual on Radiation Sterilization of Medical and Biological Materials.** Vienna: International Atomic Energy Agency; distr. New York: Unipub, 1973. 328p. illus. index. (Technical Reports Series, no. 149).

Intended for scientific and technical personnel using ionizing radiation for sterilization and preservation of medical and biological products. Topics include physical characteristics of ionizing radiation; effects on pathological organisms; required facilities; safety; dosimetry; packaging for sterilization; sterility control; legal problems; and practical information on sterilization of specific instruments, pharmaceuticals, and tissues. Bibliographic references are provided.

9-33 **Peterson Field Guide Series.** Sponsored by the National Audubon Society and the National Wildlife Federation. Boston: Houghton Mifflin, 1947- .

These excellent standard field guides for the amateur and professional cover many disciplines in natural history. The following titles are pertinent to the biological sciences:

1) *Field Guide to the Birds.* Rev. and ed. by Roger T. Peterson. 1947.
2) *Field Guide to Western Birds.* By Roger T. Peterson. 1972.
3) *Field Guide to Atlantic Coast Shells.* Ed. by Percy A. Morris. 3rd ed. 1973.
4) *Field Guide to the Butterflies.* By Alexander B. Klots; ed. by Roger T. Peterson. 1951.
5) *Field Guide to the Mammals: Field Marks of All North American Species Found North of the Mexican Boundary.* By William H. Burt. 3rd ed. 1976.
6) *Field Guide to Pacific Coast Shells.* By Percy A. Morris. 1974.
8) *Field Guide to the Birds of Britain & Europe.* Ed. by Roger T. Peterson. 3rd ed. 1975.
9) *Field Guide to Animal Tracks.* By Olaus J. Murie. 2nd ed. 1975.
10) *Field Guide to the Ferns and Their Related Families.* By Boughton Cobb. 1975.
11) *Field Guide to Trees and Shrubs.* By George A. Petrides. 1973.
12) *Field Guide to Reptiles and Amphibians of Eastern and Central North America.* By Roger Conant. 2nd ed. 1975.
13) *Field Guide to the Birds of Texas and Adjacent States.* By Roger T. Peterson. 1963.
14) *Field Guide to Rocky Mountain Wildflowers.* By John J. Craighead. 1974.
16) *Field Guide to Western Reptiles and Amphibians.* By Robert C. Stebbins. 1975.

17) *Field Guide to Wildflowers of Northeastern and North-Central North America.* By Roger T. Peterson and Margaret McKenny. 1968.
18) *Field Guide to the Mammals of Britain and Europe.* By F. H. Van Den Brink. 1968.
19) *Field Guide to the Insects of America North of Mexico.* By Donald J. Borror and Richard E. White. 1970.
20) *Field Guide to Mexican Birds.* By Roger T. Peterson and Edward L. Chalif. 1973.
21) *Field Guide to Birds Nests Found East of the Mississippi River.* By Hal H. Harrison. 1975.
22) *Field Guide to Pacific States Wildflowers.* By Theodore F. Niehaus and Charles L. Ripper. 1976.
23) *Field Guide to Eastern Edible Wild Plants.* By Lee Peterson. 1978.
24) *Field Guide to the Atlantic Seashore: Invertebrates and Seaweeds of the Atlantic Coast from the Bay of Fundy to Cape Hatteras.* By Kenneth L. Gosner. 1979 (c.1978).
25) *Field Guide to Western Birds' Nests of 520 Species Found Breeding in the United States West of the Mississippi River.* By Hal H. Harrison. 1979.

9-34 **Pictured-Key Nature Series.** Dubuque, IA: William C. Brown, 1946- . This is a good series of introductory texts in biology. Each volume is fully illustrated and covers a specific topic. The following titles are included in the series:
How to Know Pollen and Spores. By Ronald O. Kapp. New ed. 1970.
How to Know the Aquatic Plants. By George W. Prescott. 1969.
How to Know the Butterflies. By Paul R. Ehrlich. 1961.
How to Know the Cacti. By E. Yale Dawson. 1963.
How to Know the Eastern Land Snails. By John B. Burch. 1962.
How to Know the Economic Plants. By Harry E. Jaques. 1958.
How to Know the Fall Flowers. By Mabel J. Cuthbert. 1948.
How to Know the Freshwater Algae. By George W. Prescott. 3rd ed. 1970.
How to Know the Freshwater Fishes. By Samuel Eddy. 2nd ed. 1969.
How to Know the Grasses. Ed. by Richard W. Pohl. 2nd ed. 1968.
How to Know the Grasshoppers, Cockroaches and Their Allies. By Jacques R. Helfer. 1963.
How to Know the Immature Insects. By Chu H. F. 1949.
How to Know the Insects. By Harry E. Jaques. 1947.
How to Know the Land Birds. By Harry E. Jaques and Mabel J. Cuthbert. 1947.
How to Know the Lichens. By Mason E. Hale. 1969.
How to Know the Mammals. By Ernest S. Booth. 3rd ed. 1971.
How to Know the Marine Isopod Crustaceans. By George A. Schultz. 1970.
How to Know the Mosses and Liverworts. By Henry S. Conrad. 1956.
How to Know the Protozoa. By L. Jahn. 1949.
How to Know the Seaweeds. By E. Yale Dawson. 1956.
How to Know the Spiders. By B. J. Kaston. 2nd ed. 1972.
How to Know the Spring Flowers. By Mabel J. Cuthbert. 1949.
How to Know the Tapeworms. By Gerald D. Schmidt. 1970.
How to Know the Trees. By Howard A. Miller and H. E. Jaques. 2nd ed. 1972.

How to Know the Trees: An Illustrated Key to the Most Common Species of Trees Found East of the Rocky Mountains. By Harry E. Jaques. 1946.

How to Know the Trematodes. By Stewart C. Schell. 1970.

How to Know the True Bugs: (Hemiptura: Heteroptera). By James A. Slater and Richard M. Baranowski. 1978.

How to Know the Water Bird. By Harry E. Jaques. 1960.

How to Know the Weeds. By R. E. Wilkinson and Harry E. Jaques. 2nd ed. 1972.

How to Know the Western Trees. By Harry Baerg. 2nd ed. 1973.

Living Things, How to Know Them. By Harry E. Jaques. 1946.

Plant Families, How to Know Them. By Harry E. Jaques. 1948.

9-35 **Putnam's Nature Field Books.** New York: Putnam, 1915- .
These are handy little books that introduce the reader to special topics in biology. Titles that pertain to the biological sciences include:

Birds of the Ocean. By Wilfrid B. Alexander. 1963.

Field Book of American Trees and Shrubs. By F. Schuyler Mathews. 1915.

Field Book of Common Mushrooms. By William S. Thomas. 1948.

Field Book of Eastern Birds. By Leon A. Hansman. 1946.

Field Book of Insects of the U.S. and Canada. By Frank E. Lutz. 1948.

Field Book of Marine Fishes of the Atlantic Coast. By Charles M. Breder, Jr. 1948.

Field Book of Ponds and Streams. By Ann Morgan. 1930.

Field Book of Seashore Life. By Roy W. Miner. 1950.

Field Book of Snakes of U.S. and Canada. By Karl P. Schmidt and D. Dwight Davis. 1941.

New Field Book of American Wild Flowers. By Harold W. Rickett. 1963.

New Field Book of Freshwater Life. By Elsie B. Klots. 1966.

New Field Book of Nature Activities and Hobbies. By William Hillcourt. 1970.

New Field Book of Reptiles and Amphibians. By Doris M. Cochran and Coleman J. Gain. 1970.

9-36 **Statistical Tables.** By F. James Rohlf and Robert R. Sokal. San Francisco, CA: Freeman, 1969. 353p. illus.
A convenient compilation for preparing and interpreting statistics in the biological, social, and earth sciences. Each table is accompanied by an explanation of its nature, instructions for finding values in the tables, and source of data. Mathematical tables are omitted except where necessary for statistical work.

9-37 **Statistical Tables for Biological, Agricultural and Medical Research.** By Ronald Aylmer Fisher and Frank Yates. 6th ed. rev. New York: Hafner, 1974. 146p.
A comprehensive collection of statistical tables useful in the biological disciplines. A brief explanation of use accompanies each of the 34 tables. A bibliography of 82 books, monographs, and papers on statistical methods is included.

9-38 **UFAW Handbook on the Care and Management of Laboratory Animals.** By the Universities Federation for Animal Welfare. 5th ed. London: Churchill Livingstone, 1976. 635p. index.

Written by experts in the various fields of laboratory-animal science, this handbook describes procedures for the humane treatment of laboratory animals. Provides references to the original literature.

Directories

9-39   **AIBS Directory of Bioscience Departments and Faculties in the United States and Canada.** By the American Institute of Biological Sciences. Ed. by Peter Gray. 2nd ed. Stroudsburg, PA: Dowden Hutchinson & Ross, 1975. 660p.

Excellent compilation of information on 2,529 bioscience departments in 1,231 colleges and universities. The material is arranged by state or province, and subdivided by institution. Includes institutional and faculty indexes.

9-40   **Naturalists' Directory International.** 1- . South Orange, NJ: PCL Publications, 1878- . biennial.

A geographically arranged list of amateur and professional naturalists. Information provided includes name, address, and subject areas of interest. Emphasis is given to North American naturalists. Two final sections list natural history museums and societies, and natural history periodicals. Occasional supplements are issued.

9-41   **World Directory of Collections of Cultures of Microorganisms.** By S. M. Martin and others. New York: Wiley-Interscience, 1972. 560p. index.

A directory of 352 international collections of microorganisms sponsored by WHO and Unesco. Information is reported in English, French, German, Japanese, Russian, and Spanish. The microorganisms listed include algae; bacteria; fungi; lichens; protozoa; tissue cultures; animal, bacterial, insect, and plant viruses; and yeasts. The indexes provide access by geographic areas (city and country) and name of microorganism (genus-species). Cross-references from the common names are provided.

CELL BIOLOGY

Abstracts, Indexes, and Bibliographies

9-42   **Abstracts of Microbiological Methods.** Ed. by V. B. D. Skerman. New York: Wiley-Interscience, 1969. 883p. (Techniques in Pure and Applied Microbiology Series).

A collection of abstracts on the microbiological methods used by researchers over the past 25 years. The abstracts are arranged in alphabetical order by the name of the method followed by the author, identification of the bacteria under study, citation to the original article, the abstract, and in some cases a short bibliography.

9-43   **Current Advances in Genetics.** v. 1- . New York: Pergamon, 1976- . monthly.

International current awareness service in genetics, molecular biology and the medical sciences. Subdivided into 54 different sections with cross-references by full titles. Each issue contains a journal list and new book notification. Author index.

9-44   **Microbiology Abstracts.** v. 1- . London: Information Retrieval, 1965- . monthly.
Contains some 20,000 abstracts a year and is issued in three sections: A, Industrial and Applied Microbiology; B, Bacteriology; and C, Algology, Mycology and Protozoology. Each section is subdivided into about 20 categories. Book notices and notifications of proceedings are featured in each section. Author index in monthly issues. Quarterly subject indexes. Annual list of periodicals and patents abstracted.

9-45   **Virology Abstracts.** v. 1- . London: Information Retrieval Ltd., 1967- . monthly.
Abstracting service covering all areas of virology research including isolation and identification, phage studies, culture, immunology, viral infections of man and animals, and methods and techniques. Short indicative abstracts. Each issue also has notices of proceedings and new books. Indexes for each issue include virus names, author, patentee and assignee. There are annual author and subject indexes.

9-46   **Zentralblatt fuer Bakteriologie, Parasitenkunde, Infectionskrankheiten und Hygiene.** Abteilung 1: Referate. v. 1- . Stuttgart, West Germany: 1975- . 4v. of 7 nos. per year.
Abstracting service providing international coverage of information on microbiology and hygiene. Arranged by broad subject categories. Also includes a section on book reviews.

Encyclopedias

9-47   **Encyclopedia of Biochemistry.** By Roger John Williams and Edwin M. Lansford. New York: Van Nostrand Reinhold, 1967; repr. New York: Krieger, 1977. 876p. tables.
Some 370 expanded, signed articles and 270 brief unsigned articles dealing with biochemical matters. Bibliographic references are provided. Cross-references and index.

9-48   **Encyclopedia of Microscopic Stains.** By Edward Gurr. Baltimore, MD: Williams and Wilkins, 1960. 498p. illus.
Useful guide to the properties and applications of materials used for staining microscopic tissue preparations. Arranged in alphabetical order by names of dyes. Entries give the code name and number, molecular weight, formula, solubility, structural formula, and use.

9-49   **Encyclopedia of Microscopy and Microtechnique.** By Peter Gray. New York: Van Nostrand Reinhold, 1973. 638p. illus. bibliog. index.
Emphasis is on microscopy in the biological sciences. The nearly 175 articles, prepared by over 180 authorities, are illustrated with tables and formulas; most have bibliographies appended. Comprehensive coverage of microscopes and methods. Analytic index.

Dictionaries

9-50   **Dictionary of Genetics.** By Robert C. King. 2nd ed. rev. New York: Oxford University Press, 1974. 375p. illus.

Concentrates on words routinely used by geneticists but normally excluded from collegiate and general biological dictionaries. Covers 5,000 terms; many are newly coined words in molecular biology as well as species and genera useful for genetic research. Appended is a chronological list of persons and events, and a list of available teaching aids such as films and film loops.

9-51   **Dictionary of Microbiology.** By Paul Singleton and Diana Sainsbury. New York: Wiley, 1978. 481p. illus. (A Wiley-Interscience Publication).
This is one of the most extensive dictionaries of biological terms relating to microorganisms and their environment available today. It covers terms, concepts, techniques, tests, and microbial taxa.

9-52   **Enzyme Nomenclature.** By the Standing Committee on Enzymes of the International Union of Biochemistry. New York: Elsevier, 1965. 219p.
"Recommendations (1964) of the International Union of Biochemistry on the Nomenclature and Classification of Enzymes, together with their Units and the Symbols of Enzyme Kinetics." Subject and enzyme indexes.

9-53   **Glossary of Bacteriological Terms.** By P. Samson. London: Butterworths, 1975. 155p.
A handy source of medical laboratory terms in bacteriology. Useful for students and professionals.

9-54   **Glossary of Genetics in English, French, Spanish, Italian, German, Russian.** Comp. by Francoise Biass-Ducroux. New York: American Elsevier, 1970. 436p. index.
A comprehensive multilingual, technical glossary of 3,000 genetic terms, for professional and beginning interpreters. Entries are listed in the alphabetical order of the English terms; each term has an identifying number and each is followed by the French, Spanish, Italian, and German equivalents. Alphabetical indexes follow for each of the non-English languages, referring to the equivalent English word by number. The Russian entries are presented in a separate section and are also coded to the equivalent English word.

9-55   **Glossary of Molecular Biology.** By A. Evans. New York: Wiley, 1974. 55p.
"Prepared for the use of those reading the specialized literature of molecular biology." Bibliography and index are included.

Handbooks and Laboratory Guides

9-56   **Biological Stains: A Handbook on the Nature and Uses of the Dyes Employed in the Biological Laboratory.** By Harold Joel Conn. 8th ed. Baltimore, MD: Williams and Wilkins, 1969. 498p. diagrams. tables.
This handbook, sponsored by the Commission on Standardization of Biological Stains, gives explanations of biological dyes and stains, their standardization, chemical structure, and use. Lengthy bibliography. Author index.

9-57   **Electron Microscopy: A Handbook for Biologists.** By E. H. Mercer and M. S. C. Birbeck. 3rd ed. rev. and expanded. London: Blackwell Scientific Publications; distr. Philadelphia, PA: F. A. Davis, 1972. 145p. illus. bibliog. index.

Provides a brief and practical set of instructions to facilitate biologists' preparation of material for examination by the electron microscope. The third edition has an expanded section on autoradiography and additional sections on immunomethods and scanning electron microscopy. Appendixes include data on buffers, fixatives, embedding media, processing schedules, stains, autoradiography, cytochemistry, and immuno-techniques.

9-58 **Guidebook to Biochemistry: A New Edition of a Guidebook to Biochemistry by K. Harrison.** By Michael Yudkin and Robin Offord. New ed. New York: Cambridge University Press, 1971. 195p. illus. index.

The object of the guidebook is to introduce crucial concepts in biochemistry and macromolecular chemistry. Section 1 outlines the relationship between structure and function in macromolecules. Section 2 outlines cellular metabolic reactions. Section 3 covers molecular genetics and the control of protein synthesis.

9-59 **Guidebook to Microscopical Methods.** By A. V. Grimstone and R. J. Skaer. New York: Cambridge University Press, 1972. 134p. illus. index.

Written to systematize the many methods reported in the literature. Subjects covered include bacteria, protozoa, insects, blood, plants, and mammalian tissues. More commonly used techniques, such as fixation, staining and mounting are described in sufficient detail to enable them to be carried out without further information. Additional references are provided for less frequently used techniques such as freeze drying and freeze substitution. Special explanations for autoradiography, narcotisation, glassware cleaning, and SI units.

9-60 **Handbook of Biochemistry and Molecular Biology.** Ed. by Gerald D. Fasman. 3rd ed. Cleveland, OH: Chemical Rubber Co., 1976. 9v.

A comprehensive source providing information in the fields of biochemistry and molecular biology. It is divided into four sections: proteins; nucleic acids; lipids, carbohydrates, steroids; and physical and chemical data. Directed toward advanced graduate students, post-doctoral fellows, and research workers. Extensive index. Another older compilation with excellent tabulation is Cyril Norman Hugh Long's *Biochemists' Handbook* (New York: Van Nostrand Reinhold, 1961. 1192p.), which is organized into sections on chemical data, enzyme data, metabolic pathways, chemical composition of animal and plant tissue, and related data.

9-61 **Handbook of Intermediary Metabolism of Aromatic Compounds.** By Brian L. Goodwin. London: Chapman and Hall, 1976. 1v. (various paging).

Provides quick access to the metabolic reactions of aromatic compounds, and cites the original literature for each reaction. The book is divided into two sections. The first section is a review of metabolic enzymes and reactions, and the second section, which is the main portion of the work, deals with metabolic reactions compound by compound.

9-62 **Handbook of Microbiology.** Ed. by Allen I. Laskin and Hubert A. Lechevalier. 2nd ed. Cleveland, OH: Chemical Rubber Co., 1977. 4v. index.

Provides microbiologists and allied scientists with current data on the properties of microorganisms, their composition, products, and activities. Volumes 1 and 2 provide systematic information on bacteria, fungi, algae, protozoa and viruses. Volumes 3 and 4 cover amino acids and proteins, carbohydrates, lipids, nucleic acids, and minerals as they pertain to microbiology.

9-63 **Handbook of Micromethods for the Biological Sciences.** By Georg Keleti and William H. Lederer. New York: Van Nostrand Reinhold, 1974. 166p. illus. index.

A detailed compilation of laboratory micromethods used in the U.S. and Europe. The book is divided into three sections: 1) preparation of material, 2) microanalytical methods, and 3) biological characterization. For each method, a step-by-step procedure is stated without further discussion or explanation. Reagents are listed and include preparation suggestions. Original literature references are cited.

9-64 **International Code of Nomenclature of Bacteria and Viruses.** Bacteriological Code, June 1958. Ed. by the Editorial Board of the International Committee on Bacteriological Nomenclature. Ames, IA: Iowa State University Press, 1959. 186p. index.

Presents the 1958 revised edition of the International Code of Nomenclature of Bacteria and Viruses. Appended are lists of conserved and rejected names.

9-65 **Microtomist's Formulary and Guide.** By Peter Gray. London: Constable, 1954; repr. Huntington, NY: Krieger, 1975. 794p. diagrams.

Compendium for methods, techniques, stains, and formulas for the preparation of microscopic slides in biological research. Bibliography and analytic index.

9-66 **Staining Procedures Used by the Biological Stain Commission.** Ed. by George Clark. 3rd ed. Baltimore, MD: Williams and Wilkins, 1973. 418p.

Lists the staining methods used in the laboratories of the commission members. "It is not intended to be a complete treatise in regard to staining . . . nor are the procedures given intended to be considered as standard or official." Divided into three parts: 1) animal histology, 2) botanical sciences, and 3) microbiology. Includes a bibliography and index.

Treatises

9-67 **Comprehensive Biochemistry.** Ed. by M. Florkin and E. H. Stotz. New York: Elsevier, 1962- .

Distinguished treatise in biochemistry. To be completed in 34 volumes. Main sections are: I, Physico-Chemical and Organic Aspects of Biochemistry (v. 1-4); II, Chemistry and Biological Compounds (v. 5-11); III, Biochemical Reaction Mechanisms (v. 12-16); IV, Metabolism (v. 17-21); V, Chemical Biology (v. 22-29); VI, History of Biochemistry (v. 30-33); and Index (v. 34). Extensive bibliographic references.

9-68 **Enzymes.** Ed. by P. D. Boyer. 3rd ed. New York: Academic Press, 1970-1976. 13v.

This authoritative treatise is the first source to consult for detailed information on enzymes and enzyme catalyses at the molecular level. Recommended for scholarly documentation and clear presentation. Bibliographic references and author and subject indexes.

9-69   **Methods in Enzymology: A Multivolume Work.** Ed. by Sidney P. Colowick and Nathan O. Kaplan. New York: Academic Press, 1955-1978. 51v. indexes.

Presents a comprehensive and scholarly collection of the methods used in the study of enzymes. Serves as a complement to *Enzymes* (9-68). Each volume covers a separate topic and is edited by a different person.

BOTANY AND AGRICULTURE

Abstracts, Indexes, and Bibliographies

9-70   **Abstracts of Mycology.** v. 1- . Philadelphia, PA: Biosciences Information Service, 1967- . monthly.

A compilation of all the abstracts and references pertaining to fungi and lichens which are reported in *Biological Abstracts* (9-4), *BioResearch Index* (9-7), and *Biological Abstracts/RRM* (9-5). Each issue includes author, biosystematic, CROSS, and subject indexes to more than 10,000 abstracts and references annually from about 1,200 primary literature sources.

9-71   **Agrindex.** v. 1- . Rome: AGRIS; distr. New York: Unipub, 1975- . monthly.

This index identifies agriculture related materials contributed by nine participating centers including the National Agricultural Library. Coverage includes books, journal articles, atlases, technical reports, etc. The Food and Agriculture Organization of the United Nations (FAO) serves as the coordinating body, where input from the contributing institutions is merged into a magnetic tape data base. From this data base are derived *Agrindex* and a magnetic tape service, in which all cited references in *Agrindex* are available to participating governments and NGOs in machine-readable form. Most of the materials can be obtained from Aglinet, a worldwide network of agricultural libraries, designed to provide interlibrary loan, photoduplication service, and bibliographic information.

9-72   **Asher's Guide to Botanical Periodicals.** Amsterdam: A. Asher and Co., 1973- . 15/yr.

A table-of-contents service, prepared by Asher, a well-known natural history publisher, and produced by INFONET, a computer-processing firm. Speed of announcements of selected botanical journals is admirable, a time lag of about six weeks. An important feature of each issue is the author-subject index; cumulations are also published. Conference proceedings and symposia are published once a year.

9-73   **Bibliography of Plant Viruses and Index to Research.** Comp. and ed. by Helen Purdy Beale. New York: Columbia University Press, 1976. 1495p. index.

This book provides references to over 29,000 articles on plant-attacking viruses with the entries gathered from about 6,500 perodicals.

9-74 **Bibliography of Seeds.** By Lela Viola Barton. New York: Columbia University Press, 1967. 858p.

Derived from the files at Boyce Thompson Institute for Plant Research, Yonkers, New York, the bibliography lists some 20,000 citations by author. Includes plant and subject indexes.

9-75 **Botanical Abstracts.** Baltimore, MD: Williams and Wilkins, 1918-1926. 15v. and index.

A retrospective service published in classified arrangement. Each entry gives the complete citation. Superseded by *Biological Abstracts* (9-4).

9-76 **Botanical Bibliographies: A Guide to Bibliographic Materials Applicable to Botany.** By Lloyd H. Swift. Minneapolis, MN: Burgess, 1970. 804p. index.

An excellent guide to bibliographies of botany and allied subjects for historical and current material. Part 1, General Bibliography, includes sections on library classification, periodicals, book reviews and abstracts; part 2, Background Literature, refers to mathematical, physicochemical, and general life science sources; part 3, Botanical Literature, limits itself to the pure science, its reference works and specialized sources for plant taxonomy, ecology, physiology, etc.; part 4, Literature of Applied Areas of Plant Study, lists citations on applied botany. Planned especially for beginning graduate level.

9-77 **Catalog of the Farlow Reference Library of Cryptogamic Botany, Harvard University.** Boston: G. K. Hall, 1979. 6v.

This catalog represents nearly 60,000 items in the form of books, periodicals, and reprints covering all non-flowering plants.

9-78 **Catalog of the Manuscript and Archival Collections and Index to the Correspondence of John Torrey in the New York Botanical Garden Library.** Boston: G. K. Hall, 1973. 482p.

A dictionary catalog listing 180,000 items from the late nineteenth century to the present that are available from one of the major botanical-horticultural research libraries in the United States. Among the materials listed are research notes; papers; diaries; unpublished manuscripts; and personal, administrative, and scientific correspondence, including a descriptive calendar of the Charles Darwin letters. The index to John Torrey's correspondence is in a separate index.

9-79 **Catalog of the Royal Botanic Gardens, Kew, England.** Boston: G. K. Hall, 1973. 9v. (Author Catalog, 5v.; Classified Catalog, 4v.).

This catalog is classified by the Dewey Decimal system for specific subjects; however, for systematic works on specific plant groups and species, the Bentham and Hooker Botanical Classification is used. The Kew Herbarium has its own geographical schedule for flora. The library's collection is especially rich in plant taxonomy, economic botany, early botanical works, botanical travel and exploration, as well as plant cytology, physiology, and biochemistry.

9-80 **Dictionary Catalog of the Library of the Massachusetts Horticultural Society.** Boston: G. K. Hall, 1963. 3v.

9-80a **Supplement.** 1st- . 1972- .

This library contains over 31,000 volumes on all aspects of gardening practice from the earliest times to the present. It is especially rich in American publications. Other specialties include pomology, herbals and early gardening, home landscaping, and the history of garden design.

9-81   **Excerpta Botanica.** Stuttgart: Fischer, 1959- . section A, 10 issues/yr.; section B, quarterly.

Section A, "Taxonomica et Chrologica," consists of abstracts of articles on systematic botany (i.e., studies concerning specific plants or plant groups). Annotations are in English, German, or French. The last part of this section abstracts periodical literature of herbaria, gardens, museums, congresses, biographies, and bibliographies. Author and subject indexes. Section B, "Sociologica," is an international topical bibliographic listing (without annotations) of monographic literature on plant geography and ecology. Each issue deals with selected countries, subclassified by categories of botanical interest. Author index.

9-82   **Farm and Garden Index.** v. 1- . Mankato, MN: Minnesota Scholarly Press, 1978- . quarterly.

This index covers some 119 agricultural and horticultural periodicals. Well cross-referenced and excellent coverage.

9-83   **Flowering Plant Index of Illustration and Information.** By the Garden Center of Greater Cleveland; comp. by Richard T. Isaacson. Boston: G. K. Hall, 1979. 2v.

This index lists, by both common and botanical name, thousands of the world's flowering plants, and indicates where illustrations of these plants may be found. More than 100 botanical/horticultural titles have been indexed.

9-84   **Geographical Guide to Floras of the World.** By Sidney Fay Blake. Washington: U.S. Department of Agriculture; distr. Washington: GPO, 1942-1961. 2v. (U.S. Department of Agriculture. Miscellaneous Publications No. 401, 797).

This government document is an annotated catalog of the literature of floras and floristic works. It includes periodical sources, but excludes works on local areas and popular material. The catalog is arranged by continent and subdivided by country and region. Part 1 has some 3,000 entries on Africa, Australia, North America, South America, and oceanic islands. Part 2 lists about 4,000 entries on Western Europe. Part 2 differs from part 1 by including all complete works on pteridophytes, weeds, and poisonous plants, as well as a few works on general ecology and phytogeography. Part 2 indicates library location if other than the National Agricultural Library.

9-85   **Gray Herbarium Index, Harvard University.** Boston: G. K. Hall, 1968. 10v.

9-85a   **Supplement.** 1st- . 1978- .

This index contains about 259,000 cards with single or multiple entries devoted to name and literature citations of newly described or established vascular plants of the Western Hemisphere.

9-86   **Huntia: A Yearbook of Botanical and Horticultural Bibliography.** v. 1- . Pittsburgh, PA: Hunt Botanical Library, 1964- . annual.

A bibliography reporting literature on systematic botany and horticulture, botanical voyages and explorations, early agriculture, medical botany, botanical biography, and iconography.

9-87   **Index Bergeyana.** Ed. by Robert E. Buchanan et al. Baltimore, MD: Williams and Wilkins, 1966. 1472p.

Subtitle: *An Annotated Alphabetical Listing of Names of the Taxa of the Bacteria.* It is a companion volume to Bergey's *Manual of Determinative Bacteriology* because it incorporates species that were not adequately described or placed in the classification when the *Manual* was published. The *Index* evaluates the significance of more than 20,000 names of bacteria taxa cited in the literature. It also has a lengthy bibliography arranged alphabetically by authors of the cited literature.

9-88   **Index Kewensis. Plantarum Phanerogamarum Nomina et Synonyma Omnium Generum et Specierum a Linnaeo Usque ad Annum MDCCCLXXXV Complectens.** London: Oxford University Press, 1893-1895. 2v.

9-88a   **Supplement.** 1886/1895- . every 5 years.

This repository of names for flowering plants includes identification and original source publications. A more comprehensive register compiled by computer and to be complete in 50 volumes covering 1.2 million plants, is the *International Plant Index* (New Haven, CT: The Index, 1962- .). An older standardized list of common and scientific names for American trees, shrubs, and plants is that sponsored by the American Joint Committee on Horticultural Nomenclature and prepared by Harlan P. Kelsey and William A. Dayton, *Standardized Plant Names* (2nd ed. Harrisburg, PA: J. Horace McFarland, 1942. 675p.). It is a listing of approved scientific and common names of plants and plant products in American commerce or use. A similar index for genera and species of mosses is R. van der Wijk's *Index Muscorum* (v. 1- . Utrecht: International Bureau for Plant Taxology, 1959- .).

9-89   **Index of Fungi.** Kew, England: Commonwealth Mycological Institute, 1940- . semiannual.

A register of new genera, species, and varieties of fungi as documented in the world literature. It supersedes the *Index of Fungi, 1936-1939*, compiled by F. Petrak, and Verzeichnis . . . der Pilze, 1932-1935. From 1940 to 1947 the Institute's *Index* was issued as a supplement to its *Review of Applied Mycology*, a systematically arranged abstracting journal on fungi and plant pathology, which in 1971 changed its title to *Review of Plant Pathology*. Another semiannual published by the Institute that is of interest here is its *Bibliography of Systematic Mycology* (1942- .), which lists international papers and books on the taxonomy of fungi.

9-90   **Index to American Botanical Literature, 1886-1966.** Comp. by the Torrey Botanical Club. Boston: G. K. Hall, 1969. 4v.

9-90a   **Supplement.** 1st- . (1967-1976- .). 1977- .

An author catalog of approximately 102,000 books and papers published in the Western Hemisphere since 1886. It was compiled from a card file representing the cumulated contents of the *Index*, which appears regularly as a feature of the Club's *Bulletin*. Topics covered are taxonomy, phylogeny, and floristics of the fungi, pteridophytes, bryophytes, and spermatophytes; morphology, anatomy, cytology, genetics, physiology, and pathology of the same groups; plant ecology; and general botany, including biography and bibliography. Annuals from 1959 to 1972 may be purchased as files of cards. Book-form supplements are expected every ten years.

9-91   **Index to Grass Species.** Comp. by Agnes Chase and Cornelia D. Niles at Smithsonian Institution. Boston: G. K. Hall, 1963. 3v.
Includes more than 62,000 entries of genera, species, and varieties of the world's grasses. Genera are arranged alphabetically, subdivided alphabetically by species and varieties; species and varieties are represented in the general alphabetical arrangement only to provide cross-references to the proper genera. Following the complete name of each grass is the bibliographic citation for the original publication of identification, plus such information as the location, date, and collector of the original specimen and further bibliographic references. The documentation cites grass literature from 1763 to 1962.

9-92   **Index to Plant Distribution Maps in North American Periodicals through 1972.** Comp. by W. Louis Phillips and Ronald L. Stuckey. Boston: G. K. Hall, 1976. 752p.
Indexes plant distribution maps in 267 North American periodicals. Entries are arranged by taxa with like taxa arranged chronologically. Maps that cover the entire plant kingdom are grouped together without regard to the different plant groups. Each entry includes name of taxon mapped, type of map, geographical distribution, periodical reference, and author of article. Includes a separate list of North American books containing plant distribution maps, geographically arranged according to the area covered.

9-93   **Index to the Gray Herbarium of Harvard University.** Boston: G. K. Hall, 1968. 10v.
A photoreproduction of about 259,000 cards from the Gray Herbarium catalog, a collection devoted to name and literature citations of established or newly described plants of the Western Hemisphere. Citations are international in scope and range from 1886 to the present. There is some duplication with *Index Kewensis* (9-88); however, the *Gray Herbarium Index* (9-85) provides additional access by indexing names below the species level. Includes ferns, other pteridophytes, pymosperms, and angiosperms. Extensive cross-references.

9-94   **International Bibliography of Vegetation Maps.** Comp. by A. W. Kuchler. Lawrence, KS: University of Kansas Libraries, 1965-1970. 4v. (University of Kansas Publications. Library Series, no. 36).
An international reference work for botanical, agricultural, and biogeographic studies. Each of the volumes lists about 1,500 vegetation maps by geographical area (continent, region, country) and then by date (from oldest to most recent). Data for each map include title, color, scale, author, legend transcript (generally in the original language, but translated into English if not one of the better known European tongues), and when and where published.

9-95 **Plant Science Catalog: Botany Subject Index of the U.S. National Agricultural Library.** Boston: G. K. Hall, 1958. 15v.

Photographic reproduction of the 315,000 subject cards from the *Plant Science Catalog*, compiled by the U.S. Department of Agriculture from 1903 to 1952. Arranged alphabetically by subject, the file contains references of international scope, from earliest times to 1952, to botanical books, serials, proceedings, bulletins, etc. Large categories are divided and subdivided; plant families are filed alphabetically with systematic subdivision by genera. The index also contains references to textbooks arranged chronologically, voyages and travels, biographies, and geographical botany.

9-96 **Trees and Shrubs of the United States: A Bibliography for Identification.** By Elbert L. Little, Jr., and Barbara H. Honkala. Washington: U.S. Forest Service, 1976; distr. Washington: GPO, 1977. 56p. index. (Miscellaneous Publication, no. 1336).

This is a classified bibliography of 470 books and journal articles covering popular tree guides to highly technical monographs, together with a number of federal and state publications. Separate sections on the United States as a whole, on regions, on the 50 states, Puerto Rico, the Virgin Islands, and Guam are prefaced with lists of bibliographies, check lists, atlases, and references for genera and families.

Encyclopedias

9-97 **Atlas of Plant Life.** By Herbert Edlin. New York: John Day, 1973. 128p. index.

This is a continent-by-continent presentation of plant distribution. Includes chapters on climate and the spread of plants by man. Common plant names are used in the text; Latin equivalents are provided in tabular form.

9-98 **Chilton's Encyclopedia of Gardening.** By Martin Stangl. Radnor, PA: Chilton, 1975. 206p. illus. (part col.). index.

A good first reference on gardening covering the initial planning of a garden, choosing tools, structure and functioning of plant tissues, flowers, fruits, vegetables, plant foods, and bug killers and disease curatives.

9-99 **Diseases and Pests of Ornamental Plants.** By Pascal P. Pirone. 4th ed. New York: Ronald Press, 1970. 546p. illus. index. (An Official Publication of the New York Botanical Garden).

Part 1 provides a summary of plant diseases, pests, and control. Part 2 is arranged by botanical name of host and lists the diseases and pests of ornamental plants. For each host, the diseases and pests are fully described along with their recommended control. Numerous illustrations, descriptive footnotes, and diagrams make this an extremely useful book for professional and amateur gardeners, arborists, nurseryworkers, landscape architects, floriculturists, and greenskeepers.

9-100 **Encyclopedia of Mushrooms.** Ed. by Colin Dickinson and John Lucas. 1st American ed. New York: G. P. Putnam's, 1979. 280p. illus. (part col.).

This is a combination field guide and natural history sourcebook for amateur mushroom hunters. The first half of the text covers biology; life styles of fungi; habitat; mushrooms and man; and food. The second part is a guide to identifying mushrooms.

9-101 **Illustrated Encyclopedia of Indoor Plants.** By Kenneth Beckett and Gillian Beckett. Garden City, NY: Doubleday, 1976. 192p. illus. (part col.). index.

A concise guide to some 800 genera and 2,000 species of house plants giving desirable growing conditions and methods of propagation.

9-102 **International Code of Botanical Nomenclature.** Prepared and edited by F. A. Stafleu and others at the 11th International Botanical Congress, Seattle, 1969. Utrecht, Netherlands: Oosthoek, 1972. 426p. (Regnum Vegetabile, v. 82).

Title is also in French and German; text is in English, French, and German. These rules and regulations are accepted by botanists around the world for establishing botanical names. The basis for this particular code was set forth by the Ninth International Botanical Congress (Montreal, 1959) and further refined by the Tenth Congress (Edinburgh, 1964). During the course of establishing the code and its refinements, Rogers McVaugh, Robert Ross, and Frans Anthonie Stafleu compiled a dictionary of terms to ensure international agreement on the code's interpretation.

9-103 **Living Plants of the World.** By Lorus Milne and Margery Milne. London: Newson, 1967. 336p. illus. index.

Although this account is directed toward the layman and is descriptive rather than systematic, it is an excellent botanical work containing information and uses of many species. It has superb illustrations, both color and black and white.

9-104 **McGraw-Hill Encyclopedia of Food, Agriculture and Nutrition.** Daniel N. Lapedes, ed. in chief. New York: McGraw-Hill, 1977. 732p. illus. index.

A good encyclopedia that discusses the interrelations between food, agriculture, and nutrition. It covers the world food problem and related areas of agricultural productivity as well as all other aspects of food, its production, and the nutritional vlues. The longer articles have extensive bibliographies.

9-105 **Morphologic Encyclopedia of Palynology.** By Gerhard Otto Wilhelm Kremp. Tucson, AZ: University of Arizona Press, 1968. 263p. illus. (Program in Geochronology, Contribution no. 100).

Defines 1,280 terms on spores and pollen that are quoted from original sources. The 38 plates provide illustrations to the more than 800 forms discussed. Kremp also compiled a work on this subject for palaeontologists, *Catalog of Fossil Spores and Pollen* (University Park, PA: College of Mineral Industries, Pennsylvania State University, 1957-1965. 25v. looseleaf).

9-106 **Pocket Encyclopedia of Plant Galls.** By Arnold Darlington, New York: Philosophical Library, 1969. 192p. illus. index.

Field guide to 293 galls and the parasites responsible for their growth. Section 1 provides colored illustrations of each gall. Section 2 discusses each gall and is arranged by host plant with references to the first section of illustrations. The last section summarizes the life histories of many of these organisms.

9-107 **Tanaka's Cyclopedia of Edible Plants of the World.** By Tyozaburo Tanaka. Tokyo: Keigaku Publishing Co., 1976. 924p. index. bibliog.
Lists over 10,000 edible plants by Latin name. Each entry records common name, distributions, edible parts, recipes, and literature citations. Cross-references are provided from common names to Latin names. Drugs and fodder plants are not included.

Dictionaries

9-108 **Botanical Latin: History, Grammar, Syntax, Terminology and Vocabulary.** By William T. Stearn. New York: Hafner, 1966. 566p. illus.
A manual of instruction for using Latin in describing plants. After the beginning sections on grammar and syntax, special attention is directed to plant habitats, geographical names, color terms, and symbols and abbreviations. The final part is a vocabulary and a general bibliography.

9-109 **Collegiate Dictionary of Botany.** By Delbert Swartz. New York: Ronald Press, 1971. 520p.
Lists nearly 24,000 entries gathered from more than 170 sources. Common botanical terms found in standard dictionaries have not been included. American usage is represented in terminology and spelling. Few Latin names are included.

9-110 **Dictionary of Economic Plants.** By Johannes Cornelius Theodorus Uphof. 2nd ed. New York: S-H Service Agency, 1968. 591p.
Over 9,500 plant species are identified in this comprehensive dictionary. The Latin scientific names are used for entries, with cross-references from the common names. A brief description is given including geographical distribution, products, and principal uses. Appended is a systematically arranged bibliography.

9-111 **Dictionary of the Flowering Plants and Ferns.** By John Christopher Willis. Rev. by H. K. Airy Shaw. 8th ed. New York: Cambridge University Press, 1973. 1245p. index.
First published in 1897 this comprehensive botanical dictionary provides concise entries on family and generic names. Both the seventh and eighth editions exclude common plant names, plant products, and general terms that were in previous editions; these categories are now found in the complementary work by F. N. A. Howes, *Dictionary of Useful and Everyday Plants and Their Common Names* (9-115). One new feature of the eighth edition of Willis is an alphabetical list of accepted plant family names including their equivalents in Bentham and Hooker's *Genera Plantarum* and Engler's *Syllabus*. The list of generic names in this work includes all variant spellings and intergeneric hybrids. In many cases brief characteristics of subfamilies are given. The *Genera Plantarum*, a great work of descriptive botany, was published in three volumes from 1862 to 1883 by George Bentham and Joseph Dalton Hooker. Adolf Engler's taxonomic classification, *Syllabus der Pflanzen-Familien*, was revised and enlarged in its twelfth edition in two volumes, published in Berlin by Nikolassee Gebruder Borntraiger, 1954-1964.

9-112 **Dictionary of Gardening.** Ed. by Fred J. Chittenden and Patrick M. Synge. Repr. of 2nd ed. Oxford: Clarendon Press. 4v. Suppl. 554p. illus.
A comprehensive listing of gardening and horticulture terms. The work was sponsored by the Royal Horticultural Society. Some entries are quite lengthy and there are good illustrations.

9-113 **Dictionary of Nutrition and Food Technology.** By Arnold E. Bender. New York: Chemical Publishing, 1976 (c.1975). 250p.
This small dictionary defines about 1,500 terms as found in 42 standard texts covering chemistry, biochemistry, microbiology, and mechanical engineering, as well as nutrition and food sciences.

9-114 **Dictionary of Plants Used by Man.** By George Usher. New York: Hafner Press, 1974. 619p.
Arranged alphabetically by plant genus. Each entry provides the vernacular name, country of origin, brief description of the genus, and the total number of species.

9-115 **Dictionary of Useful and Everyday Plants and Their Common Names; Based on Material Contained in J. C. Willis: A Dictionary of the Flowering Plants and Ferns.** By F. N. A. Howes. New York: Cambridge University Press, 1974. 290p.
Howes' compilation contains general botanical terminology, particularly trade and common names of plants and commercial plant products, now excluded from the seventh and eighth editions of Willis's dictionary. The material for this book was based on the sixth edition (1931) of Willis's work. Howes' authority is based on his experience as former keeper of the Museum, Royal Botanic Gardens, Kew, England. The plant entries give the Latin names, countries of origin, and practical uses. Howes expands into short essays on some general terms, such as aquarium plants, bee plants, bonsai, gourds, Bible plants, snakebite remedies, tannin, and timber, with citations to further references. Addendum: bibliography of major botanic dictionaries and manuals.

9-116 **Elsevier's Lexicon of Plant Pests and Diseases.** By Manuel Merino-Rodriquez. New York: American Elsevier, 1966. 351p. 5 language indexes.
Almost 3,000 terms in Latin, English, French, Italian, Spanish and German. The dictionary is divided into two sections: 1) zooparasites (protozoans, nematodes, molluscs, arthropods, vertebrates) and 2) phytoparasites (bacteria, actinomycetes, viruses, fungi, algae, lichens, flowering plants).

9-117 **Glossary of Botanic Terms with Their Derivation and Accent.** By Benjamin Dayden Jackson. 4th ed. rev. and enlarged. New York: Hafner, 1960. (repr. of 1928 ed.). 481p.
In spite of the publication date, this is still the standard English-language botanical glossary. Latin equivalents, derivations, and pronunciation generally accompany the 10,000 definitions; generic terms are subdivided by as many as 20 entries. Bibliography of botanical dictionaries.

9-118 **Glossary of Mycology.** By Walter H. Snell and Esther A. Dick. Rev. ed. Cambridge, MA: Harvard University Press, 1971. 181p. illus. bibliog.

Includes definitions to over 7,000 terms dealing with fungi research and its applications. A special feature provides color terms from the *Repertoire de Couleurs.*

9-119 **Hortus Third: A Concise Dictionary of Plants Cultivated in the United States and Canada.** Initially compiled by Liberty Hyde Bailey and Ethel Zoe Bailey; rev. and expanded by the staff of the Liberty Hyde Bailey Horotorium. New York: Macmillan, 1976. 1290p. illus. index.

This important work is designed to account for all the species and botanical varieties of plants in cultivation within the continental United States and Canada providing brief descriptions on uses, propagation, and cultivation. There is an index to 10,408 common plant names.

9-120 **Science Dictionary of the Plant World.** By Michael Chinery. New York: Watts, 1969. 264p. illus. (part col.).

Definitions are listed for the more common terms and major taxonomic groups in the botanical world, and vary from a single word to several pages. Illustrations are exceptionally instructive and clear for learning anatomical features of plants.

9-121 **Thesaurus of Agricultural Terms as Used in the Bibliography of Agriculture.** From data provided by the National Agricultural Library, U.S. Department of Agriculture. Scottsdale, AZ: Oryx Press, 1976. 83p.

This thesaurus contains over 25,000 terms that are the subject headings used in the *Bibliography of Agriculture.*

9-122 **Vocabularium Botanicum.** By E. F. Steinmetz. 2. druk. Amsterdam: Steinmetz, 1953. 359p.

Multilingual dictionary of 3,500 terms used in trade literature and catalogs for botany and botanical drugs. Entries are alphabetically arranged by the Latin and/or Greek form followed by equivalents in Dutch, German, English, and French.

Handbooks and Field Guides

9-123 **Atlas of United States Trees.** v. 1- . Washington: U.S. Forest Service; distr. Washington: GPO, 1971- . illus. (part col.). bibliog. index. (Miscellaneous publication, no. 1146, 1293, 1314).

This series of atlases contains large maps showing the natural distribution or range of the tree species native to the continental United States. Each volume contains an introductory text with a clear explanation of the maps, discussion of the scientific and common names, notes on ranges, and latest reports of rare and endangered species. Volume 1 covers Conifers and Termperate Hardwoods; volume 2, Alaska Trees and Common Shrubs; and volume 3, Minor Western Hardwoods. Volume 4 is to cover Minor Eastern Hardwoods and volume 5, Florida Trees.

9-124 **Audubon Society Field Guide to North American Wildflowers, Eastern Region.** By William A. Niering and Nancy C. Olmstead. New York: Alfred A. Knopf, a division of Random House, 1979. 863p. illus. (part col.). index.

9-124a **Audubon Society Field Guide to North American Wildflowers, Western Region.** By Richard Spellenberg. New York: Alfred A. Knopf, a division of Random House, 1979. 862p. illus. (part col.). index.

These two guides cover the entire North American continent. The books are divided into two divisions. The first presents photographs of the wildflowers, arranged by color, and the second is a full descriptive text arranged by families. For each species there are included vernacular and scientific name, full description with height of plant and size of flower, flowering dates, habitat, range, fascinating comments on outstanding characteristics, origin of name, historical facts, plant lore, and poisonous and edible qualities.

9-125 **Bergey's Manual of Determinative Bacteriology.** By the Society of American Bacteriologists. Ed. by Robert E. Buchanan and Norman E. Gibbons. 8th ed. Baltimore, MD: Williams and Wilkins, 1974. 1272p.

Descriptive classification of all the bacteria cited in the world literature. Descriptions of all identifying elements are provided from class to species to subspecies, with appropriate keys. The companion volume, *Index Bergeyana* (9-87), is an alphabetical listing of the bacterial taxa.

9-126 **Common Weeds of the United States.** Prepared by the Agricultural Research Service of the U.S. Department of Agriculture. Washington: GPO, 1970; repr. by Dover, 1971. 463p. index.

The original work, published by the department in March 1970, is entitled *Selected Weeds of the United States.* Describes and illustrates 224 species of weeds, with maps of their geographical distribution in the United States. The plants selected are prevalent weeds in croplands, grazing lands, non-croplands, and aquatic sites. The descriptive text is in great detail, enabling the reader to identify each plant quite readily.

9-127 **Complete Book of Mushrooms: Over 1,000 Species and Varieties of American, European, and Asiatic Mushrooms with 460 Illustrations in Black and White and in Color.** By Augusto Rinaldi and Vassili Tyndalo; trans. from the Italian by Italia and Alberto Mancinelli. New York: Crown, 1974. 332p. illus. (part col.). bibliog. index.

Comprehensive guidebook for amateur mushroom hunters, first published in 1972 in Italy as *L'Atlante dei Funghi.* Part 1 identifies about 1,000 species and varieties of mushrooms and other fungi and lists their characteristic anatomical features, odor, flavor, and whether poisonous or not. Range is indicated by time of year and special growing areas. More than half of the entries are illustrated in color. Part 2 contains a general discussion about mushrooms including characteristics, hunting, habitats, edibility, cultivation, history, and a complex identification key. A special feature is the gastronomic classification of mushrooms with information on nutrients, cooking, and preserving. Glossary; index. Because the identifications in part 1 omit geographic location, American enthusiasts will find a native guide useful, such as Alexander Hanchett Smith's *The Mushroom Hunter's Field Guide*, revised and enlarged edition (Ann Arbor, MI: University of Michigan Press, 1963. 264p.).

9-128 **Eastern North America's Wildflowers.** By Louis C. Linn. New York: E. P. Dutton, 1978. 277p. index.

Describes 372 of the common and rarer wildflowers found in the eastern United States. Each entry gives common and Latin names, plant height, description, and habitat. The plates are reproductions of watercolor paintings done on location. A color key to identification provides access by color. A glossary is included.

9-129 **Families of Flowering Plants Arranged According to a New System Based on Their Probable Phylogeny.** By John Hutchinson. 3rd ed. New York: Oxford University Press, 1973. 968p. illus. index.

The late Dr. Hutchinson, one of the outstanding botanists of the twentieth century, devoted a lifetime of research devising a classification scheme for plants based on a hierarchy of their probable evolutionary relationships. Linnaeus recognized that his system was artificial and anticipated that future scholars, benefiting by increments of knowledge, would in time provide natural alternatives. And there have been others, most notably those of Bentham and Hooker (1862-1883) and Engler and Prantl (1887-1898). Hutchinson, associated with the Royal Botanic Gardens, Kew, England, first published his system in 1934, and revised it in 1959. New features in the third edition are 67 drawings, miscellaneous notes, and additional keys.

9-130 **Flowers of Europe: A Field Guide.** By Oleg Polunin. New York: Oxford University Press, 1969. 662p. illus. 192 plates. map. bibliog.

Describes 2,600 common wildflowers in Europe. Arrangement is by family and entries provide adequate identification with information on habitat and uses. Some 1,000 plants are illustrated by color photographs in the appendix. Includes a glossary and lists of common French, German, and Italian names of well-known species.

9-131 **Guide to the Study of Lichens.** By Ursula K. Duncan. Arbroath, Angus: Buncle, 1959. 164p. plates.

Eight families and 45 genera are described systematically with keys and illustrations. Glossary; bibliography. The author published a companion work, *Lichen Illustrations* (Arbroath, Angus: Buncle, 1963. 144p.), which is composed of 142 plates of drawings and magnifications as identification aids. Another handbook on lichens is *The Lichens* edited by Vernon Ahmadjian and Mason Ellsworth Hale (New York: Academic Press, 1973. 697p.). For an introductory guide see Mason E. Hale's *The Lichens* (Dubuque, IA: W. C. Brown, 1969. 226p.).

9-132 **Living Trees of the World.** By T. H. Everett. London: Thames and Hudson, 1969. 315p. illus. index.

Excellent guide, beautifully produced and illustrated, discussing pteridophytes, gymnosperms, angiosperms, monocotyledons, and dicotyledons. To be defined as a tree and included in this text, a plant must be twenty feet high at maturity. Glossary.

9-133 **Manual of the Grasses of the United States.** By A. S. Hitchcock. Washington: GPO, 1950; repr. by Dover, 1971. 2v. 1051p. illus. index.

Includes 2,000 line drawings, identification keys, descriptions, distribution ranges, and common uses of grasses in the United States. Over 200 pages are devoted to synonyms of grasses that have appeared in botanical literature.

9-134 Mosses with Hand Lens and Microscope. By Abel Joel Grout. Ashton, MD: Lundberg, 1972 (c.1903). 416p. illus.

A standard handbook for identification in the field. Another useful handbook by Grout is *Moss Flora of North America: North of Mexico* (New York: Hafner, repr. 1872 [c.1928], 3v.).

9-135 Mushrooms of North America. By Orson K. Miller, Jr. New York: E. P. Dutton, 1977. 368p. illus. (part col.). (A Dutton Paperback).

A field guide not only to mushrooms in the strict sense but also to bracket fungi, puffballs, earthstars, tongue fungi, and many other less familiar groups. Excellent photographs, keys, and an illustrated glossary make this a great little book. Each species is described clearly with information on edibility or toxicity indicated, geographic range and habitat, and sometimes suggestions for cooking.

9-136 Mycology Guidebook. Ed. by Russell B. Stevens for the Mycological Guidebook Committee of the Mycological Society of America. Seattle: University of Washington Press, 1974. 703p. illus. bibliog. index.

Definitive and indispensable manual prepared by a team of eminent mycologists. It is intended for those who teach introductory mycology and for the beginning specialist. Contents: part 1, field observations of fungi and techniques of collecting and preserving; part 2, taxonomic groups; part 3, ecological aspects; part 4, fungi as biological tools. Appendixes: quarantine and shipment of biological materials, culture collections, stains, reagents, media, and teaching aids.

9-137 North American Trees (Exclusive of Mexico and Tropical United States): A Handbook Designed for Field Use, with Plates and Distribution Maps. By Richard J. Preston, Jr. 3rd ed. Ames, IA: Iowa State University Press, 1976. 399p. illus. maps. index.

A good field handbook for tree identification. The description of each species includes an enumeration of its habit, leaf form, floral, fruit, twig, and bark characteristics and its silvicultural value. Most descriptions are accompanied by line drawings of the flower, leaf, fruit, and occasionally twig of the tree in question, and a map of its distribution in North America.

9-138 Plant Disease Handbook. By Cynthia Westcott. 3rd ed. New York: Van Nostrand Reinhold, 1971. 843p. illus. bibliog. index.

Discusses the history of plant diseases, garden chemicals, and classification of plant pathogens. Cross-references are provided among host plant, disease, and control. A list of agricultural experiment stations, a glossary, and a bibliography complete the useful handbook.

9-139 Plants of the World. By H. C. D. de Wit; trans. by A. J. Pomerans. New York: Dutton, 1966-1969. 3v. illus. index.

Translated from the Dutch, this handbook consists of three parts. Volume 1 treats the higher plants, monocotyledons and some dicotyledons. Volume 2 lists the remainder of the dicotyledons. Volume 3 treats the lower plants, including ferns, fungi, mosses, and lichens. Arrangement is by representative families of each order. Each entry lists botanical description, distribution and habitat, cultivated types, uses, and history of the common name. Illustrations are in color and black and white. Analytical indexes for each volume.

9-140 **Research Methods in Plant Science.** By Richard M. Klein and Deana T. Klein. New York: Natural History Press, 1970. 756p. index.
This book includes information on acquisition and maintenance of plant collections; structure; environmental control; and procedures to study plant cultivation, separation and analysis of plant components, growth factors, metabolism, reproduction, and plant diseases.

9-141 **Seed Identification Manual.** By Alexander Campbell Martin and William D. Barkley. Berkeley; Los Angeles, CA: University of California Press, 1973 (c.1961). 221p. illus. index.
Contains 824 seed photographs that are arranged by general location of farmlands, wetlands and woodlands. A systematic arrangement of identification clues follows. Short bibliography.

9-142 **Trees and Shrubs Hardy in the British Isles.** By William Jackson Bean. George Taylor, general ed. 8th ed. fully rev. London: John Murray, 1970; distr. New York: Scribner's, 1976- . illus. index.
Bean's authority stems from 60 years' experience in curator positions at the Royal Botanic Gardens, Kew, England; from his travels; and from his study of library and herbarium records. The revising editorial board of similar expertise has added information on older entries, described new plants, and modernized the botanical nomenclature. The work consists of an annotated alphabetical list of the species and varieties, giving full description (without keys), common names, date of original identification, and economic uses. Includes a glossary, bibliography, and index. A similar standard work for British flowering plants is by Arthur R. Clapham, T. G. Tutin, and E. F. Warburg, *Flora of the British Isles*, 2nd ed. (Cambridge: University Press, 1962. 1269p. illus.). In addition to general identification, origin, and common names, it contains data on flowering and fruiting times, pollination, seed dispersal, and chromosome number.

9-143 **Trees of North America and Europe.** By Roger Phillips. New York: Random House, 1978. 224p.
Indexes 500 of the most common trees found in North America and Europe. Part 1 is a leaf index and provides the Latin botanical name, common name(s), leaf photograph, and page number for the main text. Part 2 is the main text and includes photographs of each tree's flowers and fruits, and a drawing of the adult tree shape. A desciption of the entry follows. There is a short glossary and an index of common names.

9-144 **Westcott's Plant Disease Handbook.** Rev. by R. Kenneth Horst. 4th ed. New York: Van Nostrand Reinhold, 1979. 803p. illus. (part col.). bibliog. index.
This book is divided into two main sections: plant diseases and their pathogens, and host plants and their diseases. For each pathogen listed, the common name of the disease, plant symptoms produced by the organism, and chemicals that are effective in alleviating the disease are mentioned. In the second section the plants are listed alphabetically by common name followed by the various diseases of these plants. These diseases are also listed alphabetically by their common names, along with their scientific names.

9-145 **Wild Flowers of the United States.** By Harold William Rickett and Wesley Niles. New York: McGraw-Hill, 1966-1975. 6v. and index. illus. (mostly col.). maps.

This set is one of the most comprehensive and authoritative guides to American wildflowers ever published. The plant descriptions are written with the layperson and amateur botanist always in mind. Each entry is described in detail with accompanying color photographs and line drawings. The volumes cover: v. 1 (in 2 parts): The Northeastern States, 1966; v. 2 (in 2 parts): The Southern States, 1967; v. 3 (in 2 parts): Texas, 1969; v. 4 (in 3 parts): The Southwestern States, 1970; v. 5 (in 2 parts): The Northwestern States, 1971; v. 6 (in 3 parts): The Central Mountains and Plains, 1973; and Complete Index for the Six Volumes of *Wild Flowers of the United States*, 1975.

Treatises

9-146 **Fungi: An Advanced Treatise.** Ed. by G. C. Ainsworth and A. S. Sussman. New York and London: Academic Press, 1965-1973. 4v. in 5. illus. diagrams. tables. indexes.

Contents include the fungal cell, organism, and population. Includes a taxonomic review with keys. There are author and subject indexes as well as fungi, lichen, and actinomycetes.

9-147 **Illustrated Reference on Cacti & Other Succulents.** v. 1- . By Edgar Lamb and Brian Lamb. Poole, England: Blandford Press; distr. New York: Sterling Publishing, 1955- . illus. (part col.). index.

This set is to be the most comprehensive survey of the world's succulents. Since the main concerns are horticultural, growing instructions are offered for each entry.

9-148 **Plant Pathology: An Advanced Treatise.** By James Gordon Horsfall. New York: Academic Press, 1959-1960. 3v.

The subject treatment is directed to disease concepts of plants rather than to specific diseases. Contents: v. 1, scope, importance and history of plant pathology; pathological processes; defense devices; predispositions of the diseased plant; and therapy; v. 2, pathogenicity and its inhibition; v. 3, diseased population of plants; epidemics and their control.

9-149 **Yeasts.** Ed. by A. H. Rose and J. S. Harrison. London and New York: Academic Press, 1969-1971. 3v. bibliog. indexes.

The contents of this detailed treatise are: v. 1, Biology of Yeasts; v. 2, Physiology and Biochemistry of Yeasts; v. 3, Yeast Technology. Indexed by author and analytical subject.

Directories

9-150 **Biographical Dictionary of Botanists.** In the Hunt Institute Portrait Collection of the Hunt Botanical Library at the Carnegie-Mellon University. Boston: G. K. Hall, 1972. 451p.

Over 11,000 entries for botanists, horticulturists, and others working with plants with biographical identification for each.

9-151 **Biographical Notes upon Botanists: Maintained in the New York Botanical Library.** Comp. by John Hindley Barnhart. Boston: G. K. Hall, 1965. 3v.

A reproduction of an annotated file of biographical information on more than 44,000 past and present botanists. Each name entry has dates, education, honors, professional positions, memberships, outstanding contributions and publications, and biographical works.

9-152 **Dictionary of British and Irish Botanists and Horticulturists Including Plant Collectors and Botanical Artists.** By Ray Desmond. London: Taylor and Francis Ltd., 1977. 747p. index. bibliog.

Retrospective biographical directory of British and Irish botanists and horticulturists. Entries provide name; dates and places of birth and death; education, awards, honors, selected publications; biographical references in books and periodicals; locations of plant collections, herbaria, manuscripts, drawings, portraits; and any plant commemorating the individual. The subject index classifies entries under profession, plants, or the country where the flora has been studied and collected.

9-153 **Great Botanical Gardens of the World.** By Edward Hyams. Photography by William MacQuilty. New York: Macmillan, 1969. 288p. illus.

A pictorial guide to the most renowned botanical gardens in the world. Information given for each of the 41 gardens, grouped by continent, consists of the history and range of the collection, specialties, and research activities. There is a general index and an index/map of the world's botanical gardens. Superb photographs.

9-154 **Index Herbariorum: A Guide to the Location and Contents of the World's Public Herbaria.** v. 1- . By Joseph Lanjouw and F. A. Stafleu. 5th ed. Utrecht, The Netherlands: International Bureau for Plant Taxonomy and Nomenclature of the International Association for Plant Taxonomy, 1964- .

An herbarium is a collection of dried plants arranged systematically by botanical classification. The index is published in two parts: part 1, Herbaria, provides directory information for about 7,000 herbaria, arranged alphabetically under cities and towns, plus publications and loan and exchange regulations. It also contains a country listing and a personal name index. Part 2, Collectors, lists some 20,000 collectors, alphabetically arranged, with dates, locations, and the range of their collections.

ZOOLOGY

Abstracts, Indexes, and Bibliographies – General

9-155 **Bibliography of Reproduction: A Classified Monthly Title List Compiled from the World's Research Literature. Vertebrates, Including Man.** v. 1- . Cambridge, England: Reproduction Research Information Service, 1963- . monthly.

Covers over 600 books and papers from the literature of biology, medicine, agriculture, and veterinary science. Each issue is divided into 35 sections and has an author and animal index. Authors' addresses are provided. Indexes are cumulated semiannually.

9-156   This number has not been used.

9-157   **Catalogue of the Library of the Museum of Comparative Zoology, Harvard University.** Boston: G. K. Hall, 1967. 8v.

9-157a **Supplement.** 1st- . 1976- .
This catalog contains author and added entry access to some 250,000 volumes and more than 2,000 periodicals in the fields of zoology, paleozoology, geology, and palaeontology at the Museum of Comparative Zoology. The older periodical holdings are especially valuable; all important monographic serials are analyzed. Subject entries are excluded except for biographical works and publications on institutions and expeditions.

9-158   **Nomenclator Zoologicus: A List of the Names of Genera and Subgenera in Zoology from the Tenth Edition of Linnaeus 1758 to the End of 1955.** By Sheffield Airey Neave. London: Zoological Society of London, 1939-1966. 6v.
A comprehensive list of accepted scientific zoological names from 1758 to 1955. Arranged alphabetically, each of the 250,000 entries cites the namer and the original reference. Together with *Zoological Record* (9-161) or *Biological Abstracts* (9-4) it forms a complete record from 1758.

9-159   **Research Catalog of the Library of the American Museum of Natural History: Authors.** Boston: G. K. Hall, 1977. 13v.

9-159a **Research Catalog of the Library of the American Museum of Natural History: Classed Catalog.** Boston: G. K. Hall, 1978. 12v.
Contains entries for a collection of 325,000 volumes that covers mammalogy, geology, general natural history, general zoology, animal behavior, ornithology, anthropology, entomology, paleontology, herpetology, ichthyology, mineralogy, invertebrates, and peripheral biological sciences. Extensive holdings of 17,000 serial titles add to its strength and comprehensiveness.

9-160   **Zoobooks: A Bibliography of New and Forthcoming Books: Veterinary Medicine, Zoology . . . Books, Series, Proceedings, Journals.** Basel, Switzerland: Karger Libri, 1969- . annual.
A classified list with an author index. Since publisher and price information is given, it is a good purchasing guide. Text is in English, German, French, Italian, and Spanish.

9-161   **Zoological Record . . . Being the Record of Zoological Literature Relating to the Year. . . . v. 1- . 1864- .** London: Zoological Society of London, 1865- . annual.
A comprehensive retrospective bibliography of the world's zoological literature. Because all information on new names is sent to the Zoological Society and is incorporated into the bibliography, *Zoological Record* is the authoritative source of taxonomic references. Volumes 43 to 52 were issued concurrently as Section N

of the *International Catalogue of Scientific Literature* (1906-1915) (4-15) with bindings to match either set. *Zoological Record* is divided into 20 taxonomic sections, each with its own alphabetical author listing of terms, subject index, and systematic index. Section 1 is designated for general zoological works that have appeared in the year covered; section 20 is a list of new genera and subgenera reported in the year covered. There is no overall index for the 20 sections. Publication lag time is about three years.

Abstracts, Indexes, and Bibliographies – Invertebrates

9-162 **Abstracts of Entomology.** v. 1- . Philadelphia, PA: Biosciences Information Service, 1970- . monthly.
Covers pure and applied studies of insects and spiders and other arachnids.

9-163 **Catalogue of the Library of the Royal Entomological Society of London.** Boston: G. K. Hall, 1979. 5v.
The library's catalogue records some 9,000 monographs and reference works, 50,000 pamphlets, and 600 journals, plus correspondence, manuscripts, and drawings on entomology.

9-164 **Entomological Nomenclature and Literature.** By Willard Joseph Chamberlin. 3rd ed. rev. and enlarged. Westport, CT: Greenwood Press, 1970 (c.1952). 141p.
Chronological bibliography of insect nomenclature. Includes separate annotated bibliographies of serials, and general works.

Abstracts, Indexes, and Bibliographies – Vertebrates

9-165 **Aquatic Sciences and Fisheries Abstracts.** v. 1- . London: Information Retrieval, Ltd., 1971- . monthly.
Reviews over 3,000 journals in physical and chemical oceanography, limnology, aquatic biology, ecology, and pollution effects on fishes. Arranged by broad subject categories. Includes author, taxonomic, and geographic indexes in each issue. Cumulative author indexes are issued semiannually. Originally published as *Aquatic Biology Abstracts* (1969-1971), the change of title occurred when the *Abstracts* merged with *Current Bibliography for Aquatic Sciences and Fisheries* (1958-1971), compiled by the Food and Agriculture Organization of the UN and published by Taylor and Frances, London.

9-166 **Bibliography of Birds.** By Reuben Myron Strong. Chicago: Natural History Museum, 1939-1959. 4v.
"Special reference to anatomy, behavior, biochemistry, embryology, pathology, physiology, genetics, ecology, aviculture, economic ornithology, poultry, culture, evolution and related subjects" is the subtitle and adequately explains the coverage of this book. Author catalog with subject index. Coverage is comprehensive to 1926 and selective to 1938; for ornithological literature after 1926, consult *Biological Abstracts* (9-4). Although entries were largely taken from the *Zoological Record* (9-161), all references were checked against the original works to verify accuracy.

9-167 **Bibliography of Fishes.** By Bashford Dean. Enlarged and ed. by C. R. Eastman. New York: American Museum of Natural History, 1916-1923. Repr. New York: Stechart Hafner, 1972. 3v.

Contains 50,000 references on the habitats, structure, development, physiology, pathology, and distribution of fishes. Volumes 1 and 2 contains the bibliography, which is arranged by author, and volume 3 is the subject index (excludes species, genus, and family names). Special features: pre-Linnaean publications; references to general bibliographies with ichthyology coverage; voyages and expeditions; and list of periodicals.

9-168 **Catalogue of the Ellis Collection of Ornithological Books in the University of Kansas Libraries.** v. 1- . Comp. by Robert M. Mengel. Lawrence, KS: University of Kansas Libraries, 1972- . bibliog. (University of Kansas Publications, Library Series, 33).

A comprehensive catalog of ornithology. Each entry has a complete citation and annotation plus information on illustrations and artists, contents, chapter titles, and other major bibliographies that cite the entry.

9-169 **Dean Bibliography of Fishes, 1968 and 1969.** Comp. by James W. Atz. New York: American Museum of Natural History, 1971-1973. 2v. index.

Basic tool for ichthyological research. Since 1971 the work has appeared as section 15, Pisces, of the *Zoological Record* (9-161). Volume 2 lists some 3,500 items; volume 2 lists 5,600 entries. Each volume includes the following indexes: systematic, one arranged alphabetically and another by taxonomic groups; classified subject; geographic, arranged regionally; author; serial sources; and an alphabetic index to the classified subject and geographic indexes.

9-170 **Dictionary Catalog of the Blacker-Wood Library of Zoology and Ornithology, McGill University, Montreal.** Boston: G. K. Hall, 1966. 9v.

A card catalog reproduction of some 60,000 volumes in zoology and ornithology. The manuscript holding of 20,000 pieces of notes and correspondence of famous naturalists, as well as the older serial items, contributes to the library's distinction as a research center. Includes bound reprints and serials; analytic entries are provided for the older serials.

9-171 **Dolphins and Porpoises: A Comprehensive Annotated Bibliography of the Smaller Cetacea.** Comp. by Deborah Truitt. Detroit, MI: Gale Research Co., 1974. 582p. index.

The bibliography contains some 3,500 entries, ranging chronologically from 560 B.C. through 1972, citing books, report literature, and journal articles on dolphins and porpoises. Most of the titles are scientific works, but fiction, mythology, and even children's stories are included. The arrangement is by subject. Titles in the Roman alphabet are cited in the original language; titles in non-Roman alphabets are translated into English. Each entry includes a concise but complete bibliographic description; most include a brief descriptive annotation as well. Works not seen by the compiler are indicated. Indexes include author, taxonomic, and subject access. A comprehensive and well-organized tool.

9-172 **Guide to the Taxonomic Literature of Vertebrates.** By Richard E. Blackwelder. Ames, IA: Iowa State University Press, 1972. 259p.

Emphasizes twentieth century literature. Entries are abbreviated, giving place of publication but not publisher; conflicts among variant forms of authors' names are not resolved.

9-173 **Introduction to the Literature of Vertebrate Zoology.** By Casey Albert Wood. London: Oxford University Press, 1931; repr. New York: Arno Press, 1974. 643p.

"Based chiefly on the titles in the Blacker Library of Zoology, the Emma Shearer Wood Library of Ornithology, the Bibliotheca Osleriana, and other libraries of McGill University." Section 1 provides an introduction to the literature of vertebrate zoology; section 2 provides a geographical and chronological author-title index to section 2; and the last section lists a partially annotated catalog on vertebrate zoology of McGill University.

9-174 **Laboratory Animal Science: A Review of the Literature.** Argonne, IL: Argonne National Laboratory, Biological and Medical Research Division, 1966- . quarterly.

Brief abstracts of articles on laboratory animal technology and medicine from some 155 international journals. Classified arrangement with species index.

9-175 **Laboratory Animals: An Annotated Bibliography of Informational Resources Covering Medicine — Science (Including Husbandry) — Technology.** Ed. by Jules Cass. New York: Hafner, 1971. 1v. (various paging).

Extensive guide to the literature pertaining to the care and use of laboratory animals. Major sections include normal anatomy, physiology and psychology; disease, abnormalities, and injuries; nutrition and diet; breeding programs and colony design and maintenance; procurement and use of animals; and periodical and other publication sources. Citations are followed by descriptive and critical annotations. Emphasizes the vertebrates.

Encyclopedias

9-176 **Atlas of Animal Migration.** By Cathy Jarman. New York: John Day, 1972. 124p. illus. (col.). index.

Discusses the reasons and mechanisms of migration and animal navigation. Principal migratory groups are considered individually. Both migration and emigration are included, though the emphasis is on the former. Two final maps locate migratory bird refuges in the United States and bird migration watchpoints in Europe.

9-177 **Atlas of Wildlife.** By Jacqueline Nayman. New York: John Day, 1972. 124p. illus. (col.). index.

This book begins with a short chapter on zoogeography, giving special attention to the concept of continental drift. Following chapters cover major regions with the final chapter discussing island faunas. Appended is a world map locating major wildlife parks and refuges.

9-178 **Encyclopedia of Animal Care.** Ed. by Geoffrey P. West. 12th ed. Baltimore, MD: Williams and Wilkins, 1977. 867p. illus.

This is the English edition of a British title, *Black's Veterinary Dictionary*

(London: A & C Black, 1976. 867p.). It emphasizes first-aid and preventive medicine, veterinary techniques of interest to the farmer, and public health matters.

9-179 **Grzimek's Animal Life Encyclopedia.** By Bernhard Grzimek. New York: Van Nostrand Reinhold, 1972-1975. 13v. illus. (part col.). index.
Originally published in Germany in 1967, the encyclopedia is now published in an English edition. The set is arranged by animal groups: four volumes on mammals, three on birds, two on fishes and amphibians, one on reptiles, one on insects, one on mollusks and echinoderms, and one on lower animals. Material in each volume is arranged by animal orders and familiers. The text discusses the evolution of each group of animals, indicates physical description, range and habitat, feeding and mating habits, and other behavioral notes. The writing is descriptive, presenting a substantial amount of information in concise form. At the end of each volume is a systematic classification index. There is a language animal name glossary in English, German, French, and Russian.

9-180 **Hyman Series in Invertebrate Biology: Zoological Sciences.** By Libbie Henrietta Hyman. New York: McGraw-Hill, 1940-1967. 6v. illus. (McGraw-Hill Publications in the Zoological Sciences).
This is an advanced treatise on invertebrate zoology covering the morphology, physiology, and embryology of the invertebrates. References are to groups rather than to individual species. Bibliographic references.

9-181 **Illustrated Encyclopedia of Birds: All the Birds of Britain and Europe in Color.** New York: Marshall Cavendish, 1979. 5v. illus. (col.). index.
This is a translation of a 1971 Italian edition. It covers 642 different species of birds giving their common names in English, French, Italian, Spanish, and German, plus the Latin name. Descriptions include habitat, identification, call, reproduction, food, distribution and movements, and subspecies. Excellent photos, drawings, and distribution maps make this a very good reference work.

9-182 **The International Wildlife Encyclopedia.** Ed. by Maurice Burton and Robert Burton. New York: Marshall Cavendish Corporation; distr. New York: Purnell Library Service, 1969-1970. 20v. illus. index.
Published in England and the United States, this 20-volume encyclopedia covers animal life throughout the world. Lavishly illustrated with color photos and drawings, each article has information on geographic distribution and is often accompanied by a small map. Articles are subdivided by rubrics such as breeding, conservation, feeding habits, habitat, enemies and defense, life cycles, and distribution, and conclude with classification according to class, order, family, genus, and species. The text is informative and enjoyable.

9-183 **Larousse Encyclopedia of the Animal World.** New York: Larousse, 1975. 640p. illus. (part col.). index.
A richly colored general reference book with the arrangement classified, beginning with simple unicellular life and culminating with complex animals. Each chapter generally covers a phylum with information on such topics as structure, locomotion, communication, reproduction, feeding, adaptations, habitats and classification discussed.

9-184 **The Living Sea: An Illustrated Encyclopedia of Marine Life.** By Robert Burton, Carole Devaney and Tony Long. New York: Putnam's, 1976. 240p. illus. (col.). index.

Divided into five main sections—invertebrates, fish, reptiles, seabirds, and mammals—this reference work is useful for its excellent illustrations rather than its rather brief and general text.

9-185 **Traite de zoologie: Anatomie, systematique, biologie.** By Pierre Paul Grasse. paris: Masson, 1948- . illus. (17v. each issued regularly in parts).

According to Malcles' *Les Sources du travail bibliographique*, this work is the only major treatise on zoology worthy of the name, authoritative and of uniformly high standard and remarkable unity even though there is some variation in extent of bibliographies. The comparable German work is *Handbuch der Zoologie: eine Naturgeschichte der Stamme des Tierreiches*, v. 1- . Gegrundet von Willy Kukenthal, hrsg. von J. G. Helmcke, D. Starck und H. Wermuth. 2. Aufl. (Berlin: W. de Gruyter, 1968- . illus. in progress). The standard British work is an excellent treatise though much older: *The Cambridge Natural History*, edited by S. F. Harmer and A. E. Shipley (London: Macmillan, 1895-1909; repr. Weinheim, Engelman, 1958. 10v. illus. maps.).

Dictionaries

9-186 **Dictionary of Entomology.** By A. W. Leftwich. London: Constable; New York: Crane, Russak, 1976. 360p.

Over 3,000 species are defined in this dictionary for the amateur entomologists, naturalists with an interest in insects and students of zoology. Definitions have been omitted that are of general biological terms except where these have special significance in relation to insects. Overall there are in excess of 4,000 definitions.

9-187 **Dictionary of Herpetology.** By James Arthur Peters. New York: Hafner, 1964. 391p. illus.

Scientific terminology of snakes is explained with brief definitions; conflicting meanings are also considered. Periodical and monographic sources are cited. Several cross-references.

9-188 **Dictionary of Zoology.** By A. W. Leftwich. 2nd ed. London: Constable and Co., Ltd.; distr. New York: Crane, Russak, 1973. 478p.

A student's dictionary covering about 5,500 zoological terms. Appended materials include classification and nomenclature, translation of Greek words, and a short bibliography.

9-189 **Entomologisches Worterbuch mit besonderer Berucksichtigung der morphologischen Terminologie.** By S. von Keler. 3. durchgesehene und erw. Aufl. Berlin: Akademie-Verlag, 1963. 788p. illus. plates. tables.

A scholarly dictionary of about 10,000 terms. Includes discussions of conflicting meanings, cross-references, and 27 pages of drawings accompanied by explanations. A bibliography of glossaries and handbooks is provided.

9-190 **Glossary of Entomology: Smith's "An Explanation of Terms Used in Entomology."** By Jose Rollin De La Torre-Bueno. Completely rev. and rewritten. Brooklyn, NY: Brooklyn Entomological Society, 1937. 336p. illus.

Brief definitions of about 8,000 words found in the literature of insects. Two appendixes offer lists of abbreviations and signs and symbols. A bibliography is provided.

9-191 **Glossary of Some Foreign-Language Terms in Entomology.** By R. O. Ericsson. Washington: GPO, 1961. 59p. illus.

Issued by the Entomology Division of the U.S. Agricultural Research Service. The languages are Czech, Danish, Dutch, French, German, Polish, Russian, and Swedish. The terms are interfiled alphabetically with language designations and the English equivalent.

9-192 **New Dictionary of Birds.** By A. L. Thomson. London: Nelson, 1964. 928p. illus. plates.

Sponsored by the British Ornithologists' Union, this text includes articles on bird groups, definitions of ornithological terms, and explanations of American, British, and other native bird names. There is an index of generic names.

9-193 **Systematic Dictionary of Mammals of the World.** By Maurice Burton. 2nd ed. London: Museum Press, 1965; New York: Apollo Editions, 1968. 307p. illus.

Covers between 4,000 and 5,000 different mammals. Each entry describes characteristics, habits, habitat, distribution, breeding, longevity, and food. Arranged systematically by families and subgroups.

Handbooks — General

9-194 **Animal Life of Europe: The Naturalist's Reference Book.** By Jakob Graf. English version prepared by Pamela and Maurice Michael. New York: Warne, 1969. 595p. illus. (part col.). index.

Identification manual of insects, birds, fish, mammals, and reptiles of Europe, with over 2,000 illustrations. Well written; designed for popular use.

9-195 **Biological Research Method: A Practical Guide.** By H. H. Holman. 2nd ed. New York: Hafner, 1969. 280p. illus.

Directed to researchers who work with animals or their products. Chapters on data collection, records and their interpretation, mathematical and statistical methods, and the organization and preparation of scientific papers. Bibliographical references.

9-196 **Code Internationale de Nomenclature Zoologique.** By the International Commission on Zoological Nomenclature. Adopté par le XVe Congres Internationale de Zoologie. **International Code of Zoological Nomenclature.** Adopted by the XVth International Congress of Zoology. London: published for the International Commission on Zoological Nomenclature by the International Trust for Zoological Nomenclature, 1961. 176p.

The draft of the code was discussed and modified in London in 1958 by the Colloquium on Zoological Nomenclature. In its final form, the code sets forth the rules and regulations for assigning scientific names to animals and groups of animals with proper form of citation. Glossaries: English and French.

9-197 **Handbook of Sensory Physiology.** v. 1- . Ed. by Werner R. Loewenstein. New York: Springer-Verlag, 1971- . illus. index.

A comprehensive tool on the physiology of nervous tissue in higher and lower life forms. Topics include: general receptors, enteroreceptors, muscle type receptors, electroreceptors and other unusual lower animal senses, auditory system, vistibular system, photoreceptors, visual photochemistry and photophysics, and central processing of visual information.

9-198 **New York Aquarium Book of the Water World: A Guide to Representative Fishes, Aquatic Invertebrates, Reptiles, Birds, and Mammals.** By William Bridges. New York: published for the New York Zoological Society by American Heritage Press, 1970. 287p. illus. (part col.).

A selection of 462 species of water creatures is examined in detail, accompanied by 120 full-color illustrations. The emphasis is on animals of the seas, such as jellyfish, clams, shrimp, corals, octopuses, penguins, sea turtles, sea snakes, crocodilians, etc.

9-199 **Poisonous and Venomous Marine Animals of the World.** By Bruce W. Halstead. Sections on Chemistry by Donovan A. Courville. Washington: GPO, 1965-1970. 3v. illus.

A useful sourcebook of technical data on toxic marine animals from antiquity to the present. Contents: v. 1, Invertebrates; v. 2-3, Vertebrates. Copious illustrations, including color plates. References cited are well documented. General and personal name indexes. A similar publication produced by members of the U.S. Department of the Navy, Office of Naval Intelligence, is *Poisonous Snakes of the World. A Manual for Use by the U.S. Amphibious Force* (Washington: GPO, 1968. 212p. illus.). This manual is limited, however, to snakes found in geographic areas adjacent to the oceans. Includes a chapter on treatment of snake bites.

Handbooks—Invertebrates

9-200 **Butterflies of the World.** By H. L. Lewis. Chicago: Follett, 1973. 312p. illus. (part col.). index.

A detailed guide identifying the main species of butterflies throughout the world. The plates are arranged by geographical areas. Butterflies of a family are grouped together, the genera alphabetically within the family and the species alphabetically within the genera. The following text provides generic and specific names of each butterfly, author of the name, and locality where it is most frequently seen. Brief notes on points of variance are given. Color photographs of more than 6,500 specimens from the collections of the British Museum are included. The index refers to both the plate number and insect's number on the plate.

9-201 **Destructive and Useful Insects: Their Habits and Control.** By Clell L. Metcalf and W. P. Flint. 4th ed. rev. by R. L. Metcalf. New York: McGraw-Hill, 1962. 1087p. illus. tables.

A compendium of information on more than 400 insect pests found in North America. Line drawings and halftone illustrations complement the textual descriptions as aids to identification, but the work's most important contribution is the control methodology presented for the various pests. Includes bibliographic references and an analytic subject index.

9-202 **Insects of the World.** By Walter Linsenmaier. Trans. from the German by Leigh E. Chadwick. New York: McGraw-Hill, 1972. 392p. illus. (col.). index.

A useful guide arranged sytematically with the two exceptions of the social and aquatic insects, which are discussed in separate chapters. The descriptions stress the insect as a living organism. The illustrations are prominent.

9-203 **Oxford Book of Insects.** By John Burton and others. New York: Oxford University Press, 1969. 208p. illus. bibliog.

A useful guide designed for the layperson describing the insects of Great Britain.

9-204 **Oxford Book of Invertebrates: Protozoa, Sponges, Coelenterates, Worms, Mollusks, Echinoderms and Arthropods (Other Than Insects).** Text by David Nichols and John A. L. Cooke (Arthropods); illus. by David Whiteley. London: Oxford University Press, 1971. 218p. illus.

A useful handbook consisting of alternating pages of color illustrations (drawn to a single scale where practical) with facing pages of textual descriptions. A glossary and bibliography of sources provide further information. Designed for popular use.

9-205 **Seashells of North America: A Guide to Field Identification.** By Tucker R. Abbott; illus. by George F. Sandstrom. New York: Golden Press, 1968. 280p. illus.

Introductory field guide to the seashells of North America. Includes about 160 color plates. A short bibliography and index of common and scientific names are provided.

9-206 **Shell Book.** By Julia Ellen Rogers. Rev. ed. Boston: Bransford, 1951. 503p. illus. index.

A long-standing favorite handbook to identifying shells, first published in 1908. Includes illustrative plates and a list of modern names. Designed for popular use.

Handbooks—Birds

9-207 **Check-List of Birds of the World.** By James Lee Peters. Cambridge, MA: Harvard University Press, 1931-1968. 15v.

A scholarly and comprehensive list of ornithological general, species, and subspecies with bibliographic sources.

9-208 **Checklist of the World's Birds: A Complete List of the Species, with Names, Authorities and Areas of Distribution.** By Edward S. Gruson with the assistance of Richard A. Forster. New York: Quadrangle/New York Times Book Co., 1976. 212p. bibliog. index.

A well-organized book with the names obtained from the standard, recognized reference works on birds of different areas of the world. Number codes in the book indicate the source books from which the names were obtained. Letter codes indicate which of the world's great faunal regions each species occurs.

9-209 **Field Guide to the Nests, Eggs, and Nestlings of North American Birds.** By Colin Harrison. Cleveland, OH: William Collins Publishers, 1978. 416p. illus. index.

An excellent guide to the eggs, nests, and nestlings. Color plates are used throughout to help in identifying the item. No other guide has attempted to be this comprehensive.

9-210 **Handbook of North American Birds.** Ed. by Ralph S. Palmer. New Haven, CT: Yale University Press, 1962-1976. 3v. illus. (part col.). index.

These encyclopedic volumes are the result of the concerted efforts of 32 authors, who have compiled and synthesized the literature and data, both published and unpublished. The descriptions of both sexes of each species cover plumage at all ages and in all seasons, measurements, weight, hybrids, and geographical variations. Field identification, voice, habitat, distribution, migration, banding status, reproduction, survival, habits, and food are also included.

9-211 **North American Birds.** By Lorus Milne and Margery Milne. Englewood Cliffs, NJ: Prentice-Hall, 1969. 340p. illus. (part col.). index.

Covers 300 species grouped by habitat. Full-color paintings by Marie Bohlen and prose style make this publication particularly appealing to amateur ornithologists. For a more scholarly handbook, consult that edited by Ralph S. Palmer, *Handbook of North American Birds* (9-210). The British equivalent is *The Handbook of British Birds*, ed. by H. F. Witherby (London: Witherby, 1938-1941. 5v. illus. maps.).

9-212 **Waterfowl of North America.** By Paul A. Johnsgard. Bloomington, IN: University Press, 1975. 575p. illus. (part col.). bibliog. index.

This book is intended to complement F. H. Kortright's *The Ducks, Geese and Swans of North America* (Washington: Wildlife Management Institute, 1943. 476p.). It emphasizes biology, behavior, and management of waterfowl where Kortright emphasized description and systematics. For each species information is given on common and scientific names, range, subspecies, measurements and weights, identifying features in hand and in the field, aging and sexing data, distribution and habitat, biology, ecology, behavior, a black and white drawing, and a range map.

Handbooks – Fish

9-213 **Fishes of North America.** By Earl Stannard Herald. Garden City, NY: Doubleday, 1972. 254p.

A useful handbook for fish identification in North America. Includes beautiful underwater photographs.

9-214 **Fishes of the World: A Key to Families and a Checklist.** By Georgii U. Lindberg. New York: Wiley, 1974. 545p. illus. bibliog. index.

Identification manual and guide designed for the researcher. Originally published by Nauka (Leningrad) in 1971.

9-215 **Freshwater Fishes of the World.** By Gunther Sterba. Trans. and rev. by Denys W. Tucker. London: Vista Books, 1962. 878p. illus. plates. tables.

A major reference work, originally published in Germany in 1959, written by a scholar who has served as director of the Zoological Institute of the University of Leipzig. The entries, systematically arranged, concentrate on basic data for identification of 1,300 species of fishes and on conditions necessary for their maintenance in aquaria. The excellent illustrations consist of photographic plates, some in color, and hundreds of line drawings. Annotated bibliography; analytic index, with references to illustrations.

9-216 **Handbook of Freshwater Fishery Biology.** By Kenneth D. Carlander. Ames, IA: Iowa State University Press, 1969-1977. 2v. index.
Volume 1 covers Life History Data on Freshwater Fishes of the United States and Canada, Exclusive of the Perciformers, and volume 2 covers Life History Data on Centrarchid Fishes of the United States and Canada. This handbook is mainly a collection of highly specialized published life history and statistical data intended primarily for the conservationist and biologist working in the field and in the laboratory. The information is in tabular form and is arranged by species.

9-217 **Living Fishes of the World.** By Earl Stannard Herald. Garden City, NY: Doubleday, 1961. 303p. illus.
Arranged systematically by classification of fishes. Notable for its excellent underwater photographs. The author's experience as curator of the Steinhart Aquarium of the California Academy of Sciences brings authority to the text, which is informative for student and amateur naturalists.

9-218 **Tropical Fish Identifier.** By Braz Walker. New York: Sterling, 1971 (c.1968). 256p. illus. (col.). index.
A concise introductory manual covering 120 of the more popular tropical aquarium fishes for the amateur fancier. For each fish the manual gives order, family, scientific and common name, native range and habitat, physical description, size, care, feeding, breeding data, and a color photograph.

Handbooks — Amphibians and Reptiles

9-219 **Living Amphibians of the World.** By Doris M. Cochran. Garden City, NY: Doubleday, 1961. 199p. illus.
An introductory survey with useful illustrations, including 77 color plates. Includes general discussion of amphibia as well as descriptions of representative members and groups. An appendix gives directions for the care of amphibia as pets.

9-220 **Reptiles of the World: The Crocodiles, Lizards, Snakes, Turtles and Tortoises of the Eastern and Western Hemisphere.** By Raymond Lee Ditmars. London: Lane, 1933. 321p. illus.
A popular survey of long-standing interest. The descriptive information is supplemented by almost 90 plates of illustrations. The author has contributed a similar work for the American continent, *The Reptiles of North America: A Review of the Crocodilians, Lizards, Snakes, Turtles and Tortoises Inhabiting the United States and Northern Mexico* (New York: Doubleday, 1936. 476p. illus.), illustrated with 135 plates, some in color. For field guide information, the Putnam series has a fairly recent addition directed to the beginning student and amateur: Doris M. Dochran and J. Goin Coleman, *The New Field Book of Reptiles and Amphibians* (New York: Putnam, 1970. 359p.). It is comprehensive, concise, and accurate for the species and subspecies, native and naturalized in the United States.

9-221 **Turtles of the United States.** By Carl H. Ernst and Roger W. Barbour. Lexington, KY: University Press of Kentucky, 1973. 347p. illus. (part col.). maps. bibliog. index.

A comprehensive work on U.S. turtles. A short introduction on general characteristics precedes the detailed accounts of the species, in which behavior, ecology, and conservation are emphasized. Good photographs and distribution maps accompany the text. Appendixes: a discussion of turtle evolution; care of turtles in captivity; a list of the parasites, symbionts, and commensals reported for the species included; a glossary; and an extensive bibliography emphasizing the period 1950 through 1970. Index of subjects and common and scientific names.

Handbooks — Mammals

9-222 **Field Guide to the Larger Mammals of Africa.** By Jean Dorst. Boston: Houghton Mifflin, 1970. 287p. illus. (part col.). maps. bibliog. index.

For each species there is a concise description emphasizing identifying characteristics of physical appearance, habitat, and behavior; a range map; and a colored painting with indicative field marks. It covers the game animals, carnivores, and primates found in mainland Africa South of the Tropic of Cancer. For hard-to-separate species, distinguishing features are carefully noted.

9-223 **Handbook of Living Primates: Morphology, Ecology and Behavior of Nonhuman Primates.** By John Russell Napier and Prue H. Napier. London: Academic Press, 1967. 456p.

A scholarly compilation in three parts: part 1, functional morphology of primates; part 2, profiles of primate genera; part 3, supplementary and comparative data including taxonomic features, habitats, appendages, and their uses. Bibliography of references arranged alphabetically by authors. Index of animals, including illustration notations.

9-224 **Hoofed Mammals of the World.** By Ugo Mochi and T. Donald Carter. New ed. New York: Scribner's 1971. 268p. illus. bibliog. index.

A useful handbook of the ungulates covering all living species, many subspecies, and a few recently extinct forms. Excellent illustrations.

9-225 **Mammals of North America.** By Eugene Raymond Hall and Keith R. Kelson. New York: Ronald, 1959. 2v. 1083p. illus. maps.

A major sourcebook in mammalogy. It is arranged by order-family-genus-subgenus. The exhaustive description for each entry is complete with distribution maps, literature citations, and illustrations. Extensive bibliography.

9-226 **Mammals of the World.** By Ernest Pillsbury Walker. 3rd ed. (vol. 3 published in 1968 by the same press, is still in 2nd ed.). Baltimore, MD: Johns Hopkins Press, 1975. 3v. illus.

The first two volumes are a comprehensive account of recent genera of mammals arranged by taxonomic classification. The illustrations are excellent. Volume 2 has subject index to the first two volumes. The third volume contains an exhaustive, classified bibliography on mammalogy, with a separate section on reference works and checklists and another on periodicals.

Directories

9-227 Animals Next Door: A Guide to Zoos and Aquariums of the Americas.
By Harry Gersh. New York: Fleet Academic, 1971. 170p.
Comprehensive directory to zoos and aquariums, public and private, in the
Western Hemisphere. Arrangement is geographic. Gives institutional information, such as name, address, hours, fees, collection statistics, publications, educational programs, directors, etc. Preliminary chapters explain the origin, history,
and management of these animal collections.

9-228 International Zoo Yearbook. v. 1- . London: Zoological Society, 1960- .
illus. annual.
International sourcebook for animal populations in zoological gardens and
aquaria. There is a section that discusses special animal groups in captivity,
another that reports new developments in the zoo world, and a reference section
that lists federations, societies, new institutions, censuses of captive animals, and
breeding activities. Taxonomic authorities consulted for preparation of the yearbook are also listed. Subject index.

9-229 World Directory of National Parks and Other Protected Areas. By the
International Union for Conservation of Nature and Natural Resources.
Distr. New York: Unipub, 1979. 2v. (looseleaf).
Provides comprehensive information on conservation activities in national parks
and other protected areas. Listings are by country and give the following information: name, type, biotic province, legal protection, date established,
geographical location, altitude, total area, land tenure, physical features, vegetation, fauna, zoning, disturbances or deficiencies, and tourism.

9-230 World of Zoos: A Survey and Gazetteer. Ed. by R. Kirchshofer. Trans.
by Hilda Morris. 1st English language ed. London: Batsford, 1968.
327p. illus.
The survey section consists of introductory articles followed by an extensive collection of illustrative plates, many in color, of captive animals and their zoo
habitat and management. The concluding gazetteer of zoological gardens around
the world by country gives directory information for each zoo with brief indication of size, staff, hours, tours, demonstrations, photographic facilities, and
research projects.

**9-231 World Wildlife Guide: A Complete Handbook Covering All the World's
Outstanding National Parks, Reserves, and Sanctuaries.** Ed. by Malcolm
Ross-Macdonald. London: Threshold Books, 1971; New York: Viking
Press, 1972. 416p. illus. maps.
A directory and illustrated catalog of 179 national parks, 264 reserves and sanctuaries, 206 state parks, and other accessible wildlife areas in 66 countries. Information for each entry includes fauna, flora, landscapes, facilities for accom-
modation and transportation, and location directions. Special features: 25 maps
and 10 species' lists.

**9-232 Zoos and Aquariums in the Americas; Including Roster of Membership,
Association History, Purposes and Objects.** Wheeling, WV: American
Association of Zoological Parks and Aquariums, 1930- . biennial.
The directory section is divided into three parts: United States (state-city arrange-
ment), Canada, and Latin America. Besides the address, basic information is
given on hours, fees, property holdings, animal population, and budget.

10

Geoscience

Geoscience or geology investigates the physical features of the earth. It studies rocks, minerals, landforms, oceans, rivers, streams, fossils, and atmospheric conditions. The broad scope of geology ties it to other scientific disciplines. For example, the study of the earth's form, size, and physical condition is a branch of astronomy; the study of ancient life forms is related to botany and zoology; and the study of physical and chemical changes from the earth's past and present involves physics, chemistry, archaeology, and anthropology.

The rise of geology (geoscience is a modern term) began with Abraham Gottlob Werner (1749-1817) who used empirical and experimental methods to study the earth. Werner was a Neptunist, believing that mountains and other landforms on the earth were formed from depositions of primeval oceans that covered the earth. James Hutton (1726-1797) and John Playfair (1748-1819) refuted Werner's theories. They were Vulcanists, believing that volcanic eruptions and internal stresses created the land forms. Between 1830 and 1833, Sir Charles Lyell published his *Principles of Geology* where he stated that the present is the key to the past; that current processes acting on the earth must be studied, analyzed, and interpreted to reconstruct the geologic past. Although much has been learned since Lyell wrote his *Principles*, his theory that geology is a systematic science remains unchanged.

The subdisciplines of geoscience are many. Cosmology studies the early history of the earth, sun, planets, stars, and other heavenly bodies in the universe. Petrology is the study of the physical, chemical, and geologic conditions that produce rocks and rock units. Structural geology is closely related to physics. It deals with the structural relationships between rocks and geologic formations. Physiography, or geomorphology, studies the surface of the earth and the formation of mountains, valleys, and plains. Paleontology, the science of fossils, reveals historical events recorded in the earth's crust. The study of paleontology includes paleobotany, paleozoology, invertebrate paleontology, and micropaleontology. It also contributes to investigations in archaeology and anthropology.

Investigations concerning the birth of the earth and all living forms supported by earth is termed historical geology. A branch of historical geology is stratigraphy, which studies the layers of strata in regional formations and establishes age and time sequence with other formations in the world. The study of stratigraphy is dependent on much of the knowledge of paleontology. Economic geology covers the occurrence, origin, and distribution of minerals that are beneficial to man, and includes studies in mineralogy, crystallography, mining geology, petroleum geology, and engineering geology. Meteorology is the study of the atmosphere and its effects on the geologic features of earth. Oceanography deals with the composition, physical forces, life, and sediments of ocean water, including the development of coastlines and ocean floors. The study of other bodies of water on or within the earth, is called hydrology.

The science librarian must be aware of the many areas of geoscience, a science in which basic research can be as detailed, as complicated, and as abstruse as any other scientific discipline. Perhaps this can be best exemplified in geophysics, which combines the most advanced, mathematical methods with pertinent geological, physical, and chemical principles.

GUIDES TO THE LITERATURE

10-1   **Geologic Reference Sources: A Subject and Regional Bibliography to Publications and Maps in the Geological Sciences.** By Dederick C. Ward and Marjorie W. Wheeler. 2nd ed. Metuchen, NJ: Scarecrow Press, 1972. 453p.
Probably the best available guide to recent geological materials with most listed items having been published since 1950. Its purpose is to introduce the reader to the literature of the geologic discipline – including maps which deal with the general geology of countries and regions. Selected publications range from introductory to highly technical texts, treatises, and serials. Occasional annotations.

10-2   **Guide to Information Sources in Mining, Minerals, and Geosciences.** By Stuart R. Kaplan. New York: Wiley, 1965. 599p. illus. (Guides to Information Sources in Science and Technology, v. 2).
A conveniently organized volume providing guidance to reference sources in general geology, geophysics, geography, and pure and applied earth sciences. The second part annotates more than 600 reference publications in these fields; the first lists and describes approximately 1,000 national and international organizations. Although dated, it is still useful.

10-3   **Guide to U.S. Government Maps: Geologic and Hydrologic Maps.** By Laurie Andriot and Donna Andriot. Preliminary ed. (updated through August 1976). McLean, VA: Documents Index, 1977. 703p. illus. maps. index.
This is a cumulation of the map entires previously published in the catalogs entitled *Publications of the (U.S.) Geological Survey, 1879-1961* and *1962-1980*, and the annual supplements and monthly issues through August 1976. It includes an area-subject index, a subject-area index and a coordinate index of latitude and longitude.

10-4   **Map Librarianship.** By Harold Nichols. London: Clive Bingley; distr. Hamden, CT: Linnet Books, 1976. 298p. bibliog. index.
A good book on map librarianship with bibliographical notes on selected mapping services and samples of map cataloging. One has to overlook the author's style of writing.

10-5   **Map Librarianship.** By Mary Larsgaard. Littleton, CO: Libraries Unlimited, 1978. 330p. bibliog. index.
Provides guidance in selection, classification, and computer applications. Appendixes include sample policies, directories, map sources, and a glossary.

10-6   **Sources of Information in Water Resources: An Annotated Guide to Printed Materials.** By Gerald J. Giefer from the Water Resources Center

Archives at the University of California, Berkeley. Port Washington, NY: Water Information Center, 1976. 290p. index.
This guide includes a collection of current secondary sources pertinent to many areas of water resources such as flood, runoff, ground water, pollution and dams. It is intended to guide the user to places to begin a thorough literature search.

10-7   **Use of Earth Sciences Literature.** Ed. by D. N. Wood. Hamden, CT: Archon Books, 1973. 459p. index. (Information Sources for Research and Development).
Systematically covers stratigraphy (historical geology), structural geology, mineralogy and petrology, hydrology, glaciology, meteorology, oceanography and geomorphology, soil science, and several aspects of applied geology (e.g., geology of metalliferous ore deposits, exploration for metalliferous ore deposits, coal, oil and natural gas, industrial metals, etc.). Introductory chapters discuss libraries and their use, primary literature, secondary literature (reference works), translations, geological maps, and literature searches.

ABSTRACTS, INDEXES, AND BIBLIOGRAPHIES

10-8   **Abstracts of North American Geology.** Washington: U.S. Geological Survey; distr. Washington: GPO, 1966-1971. 6v.
Contains abstracts of technical papers and books and citations to maps on geology of North America including Greenland, West Indies, Guam, and other island possessions. No limited circulation material is included. It is arranged by author. Continued by *Bibliography and Index of Geology* (10-14).

10-9   **Annotated Bibliographies of Mineral Deposits in Africa, Asia (Exclusive of the USSR), and Australia.** By John Drew Ridge. New York, Pergamon, 1976. 545p.
A comprehensive bibliography of mineral deposits for the continents named. For each bibliography the following information is given: 1) the location of the deposit, 2) probable age of formation, 3) the metals or minerals for which it is being mined, and 4) the Lindgren classification category chosen for it by the author. Arrangement is by continent and then country. In addition to the author index, indexes are by deposits, deposits according to age of mineralization, deposits according to metals or minerals produced, and deposits according to categories of the modified Lindgren classification.

10-10   **Annotated Bibliographies of Mineral Deposits in the Western Hemisphere.** By John D. Ridge. Boulder, CO: Geological Society of America, 1972. 681p. index. (G.S.A. Memoir no. 131).
A revised and expanded version of the author's *Selected Bibliographies of Hydrothermal and Magmatic Mineral Deposits*, published by G.S.A. as *Memoir 75* in 1958. At the head of each bibliography is given: 1) the location of the deposit, 2) probable age of formation, 3) the metals or minerals for which it is being mined, and 4) the Lindgren classification category chosen for it by the author. Basic arrangement of the bibliography is by continent and country, and, for Canada and the United States, by state or province. In addition to the author index, indexes are by deposits, deposits according to age of mineralization, deposits according to metals or minerals produced, and deposits according to categories of the modified Lindgren classification.

10-11   **Annotated Bibliography of Economic Geology.** Lancaster, PA: Economic Geology Publishing Co., 1929-1966. 39v.
This bibliography covers all aspects of the geology of mining and petroleum. It is arranged by subject, with author index in annual volumes. The entries in each volume are complete for that year except for a few earlier titles inadvertently missed.

10-12   **Antarctic Bibliography.** v. 1- . Washington: U.S. Library of Congress. Science and Technology Division; distr. Washington: GPO, 1962- .
The bibliography, international in scope, lists significant materials in biological sciences, expeditions, geological sciences, logistics, equipment and supplies, atmospheric physics, and political geography. Indexed by author, title, and geographic area.

10-13   **Arctic Bibliography.** Prepared by the Arctic Institute of North America; ed. by Maret Martna and Maria Tremaine. Washington: GPO (v. 1-12); Montreal: McGill-Queen's University Press (v. 13-21), 1947-1975. 26v. illus. map (col.). index.
This comprehensive bibliography covers thousands of books and papers published in 40 languages and in all pertinent fields of science, technology, and the arts that pertain to the Arctic. Resources of libraries in Canada and the United States as well as collections abroad have been utilized.

10-14   **Bibliography and Index of Geology.** v. 32- . Boulder, CO: Geological Society of America in Cooperation with the American Geological Institute, 1968- . monthly with annual indexes.
This index is photocomposed from citations in the GeoRef data base. Prior to this publication, there were 31 volumes published as *Bibliography and Index of Geology Exclusive of North America.* An international index to books, monographs, papers, and maps on geology, each issue of which is divided into 29 broad categories, with author and subject indexes.

10-15   **Bibliography of Fossil Vertebrates.** New York: Geological Society of America, 1902- .
These five-year bibliographies are published as issues of the Society's *Memoirs* numbers 37, 57, 84, 92, 117, 134, and 141. Two earlier ones were published in the *Special Papers* numbers 27 and 42. They are arranged alphabetically by author with a detailed subject and systematic index. Very comprehensive.

10-16   **Bibliography of North American Geology.** Washington: U.S. Geological Survey; distr. Washington: GPO, 1923-1971. 49v.
This bibliography is a subseries within the U.S. Geological Survey *Bulletin.* Each year covered the geologic literature for that year. It is arranged by author, with a subject index. Each entry gives author, title, periodical, and pagination. Continued by *Bibliography and Index of Geology* (10-14).

10-17   **Bibliography of Theses in Geology, 1965-1966.** By Dederick C. Ward and T. C. O'Callaghan. Washington: American Geological Institute, 1969. 255p.

10-17a **Bibliography of Theses in Geology, 1967-1970.** Ed. by Dederick C. Ward. Boulder, CO: Geological Society of America, 1973. 160p.

Continuation of a series that was inaugurated in 1958 with the *Bibliography of Theses in Geology through 1957*, by John and Halka Chronic, and continued by the same authors in 1965, with the *Bibliography of Theses in Geology, 1958-1963*. Both are available from the American Geological Institute. For theses in 1964 consult "Bibliography of Theses in Geology, 1964," by Dederick C. Ward, in *Geoscience Abstracts*, v. 7, no. 12, supplement (1965). For geologists the advantages of the above geology bibliography over *Dissertation Abstracts International* (3-147) is that the former is indexed in great depth, using an index compatible with those in the two main English language geology bibliographies, *Abstracts of North American Geology* (10-8) and *Bibliography and Index of Geology* (10-14).

10-18 **British Geological Literature (New Series).** v. 1- . Worthington, England: Bibliographic Press Ltd., 1972- . quarterly.

This is a bibliography and index of geology and related topics of the British Isles and adjacent sea areas. It is arranged by broad subject areas with author index. There are short annotations.

10-19 **Catalog of the Arctic Institute of North America.** Boston: G. K. Hall, 1968. 4v.

10-19a **Supplement.** 1st- . 1971- .

This is the catalog of a library of over 9,000 volumes and 20,000 pamphlets dealing with all aspects of the polar regions of the Arctic.

10-20 **Catalog of the U.S. Geological Survey Library.** Boston: G. K. Hall, 1964. 25v.

10-20a **Supplement.** 1- . 1972- .

This important catalog indexes by author, title, and subject more than one-half million pieces, mostly bound volumes, pamphlets and, to some extent, maps. The library's principal subject interests are geology, paleontology, petrology, mineralogy, ground and surface water, cartography and mineral resources, plus mathematics, engineering, certain fields of physics, chemistry, soil science, botany, zoology, oceanography and natural resources.

10-21 **Catalogs of the Glaciology Collection in the Department of Exploration and Field Research of the American Geographical Society of New York.** Boston: G. K. Hall, 1971. 3v.

This important library catalog contains 18,900 author, 17,000 subject, and 13,400 regional entries. Subjects covered are geography, geophysics, geology, and earth science. Maintained in combination with the World Data Center A: Glaciology, the collection itself comprises journals, books, reprint photographs, maps, and unpublished reports from the mid-1800s to the present. The subject classification is based on the *Universal Decimal Classification for Use in Polar Libraries*, published by the Scott Polar Research Institute, Cambridge, England, with the cooperation of the British Standards Institution, London, 1963. The alphabetical list of the main subjects, with their subject numbers, is included at the beginning of the catalog. Guide cards, giving the topic covered by each subject number, are provided within the catalog for the convenience of the researcher.

10-22 **Catalogue of Published Bibliographies in Geology, 1896-1920.** By Edward Bennett Mathews. Washington: National Research Council, 1923. 228p. index. (NAS-NRC Bulletin 36).

A retrospective but still useful index to bibliographies on geology. International in coverage and scope, it is arranged in three parts: general bibliographies, special bibliographies, author bibliographies and necrologies.

10-23 **Climatological Data.** Washington: U.S. National Oceanic and Atmospheric Administration; distr. Washington: GPO, 1959- .

10-23a **Monthly Weather Review.** v. 1- . Washington: U.S. National Oceanic and Atmospheric Administration; distr. Washington: GPO, 1872- .

The *Climatological Data* is a summary of the reports of U.S. weather stations arranged by state. The *Monthly Weather Review* is a periodical containing articles on meteorology and summaries of weather phenomena such as tornadoes, hurricanes, and other forms of turbulence that have occurred in immediate past months.

10-24 **Dictionary Catalog of the Map Division, New York Public Library.** Boston: G. K. Hall, 1971. 10v.

A catalog of a library containing depository maps of the U.S. Army Map Service, foreign maps, manuscript and early printed maps, atlases, and some 11,000 volumes on all aspects of cartography.

10-25 **Dictionary Catalog of the U.S. Department of the Interior Library.** Boston: G. K. Hall, 1967. 37v.

10-25a **Supplement.** 1- . 1968- .

This is a dictionary catalog of a highly specialized collection on mines, minerals, petroleum, and coal, as well as a good working collection on geology. Extensive holdings on fish, fisheries, and wildlife; conservation; land management; land reclamation; public land policy; irrigation; water; and power are also included.

10-26 **Geography and Earth Sciences Publications, 1968-1972; An Author, Title and Subject Guide to Books Reviewed, and an Index to the Reviews.** Comp. by John Van Balen. Ann Arbor, MI: Pierian Press, 1978. 313p. index.

10-26a **Geography and Earth Sciences Publications, 1973-75; An Author, Title and Subject Guide to Books Reviewed, and an Index to the Reviews.** Ann Arbor, MI: Pierian Press, 1978. 232p. index.

These two volumes provide bibliographic coverage of books published internationally in geography, geology, and related earth sciences. The first part is an author section with citations listed alphabetically by author's last name, and with each author numbered. The author section is followed by subject, geographical, and title indexes, which give both page and number of citations on each page for each site.

10-27 **Geological Abstracts.** New York: Geological Society of America, 1953-1958. 6v.

Consisted of abstracts from U.S. and translated Russian periodicals. Author and subject indexes. Superseded by *Geoscience Abstracts* (10-29).

10-28   **Geophysical Abstracts: Abstracts of Current Literature Pertaining to the Physics of the Solid Earth and to Geological Exploration.** Washington: U.S. Geological Survey; distr. Washington: GPO, 1929-1971. 299nos. index.

Arrangement is alphabetic by broad subject headings ranging from age determination to volcanology. Coverage is international, but all abstracts are in English. There are annual subject and author indexes.

10-29   **Geoscience Abstracts.** Washington: American Geologic Institute, 1959-1966. 8v.

This abstracting service replaced the *Geological Abstracts* (10-27). It is arranged by subject with an author index. No foreign material is included except translated Russian material. Entries include author, title, periodical, and pagination with abstract.

10-30   **Geotitles Weekly.** v. 1- . London: GeoServices, 1969- . weekly.

The only international current awareness service for geoscience. Consists of three sections: 1) news, 2) subject classification, and 3) source and author.

10-31   **Index of Generic Names of Fossil Plants, 1820-1965.** By Henry N. Andrews, Jr. Washington: U.S. Geological Survey; distr. Washington: GPO, 1970. 354p.

Revises and updates USGS *Bulletin* no. 1013, which surveyed the field for the years 1850-1950. It is based on the Geological Survey's *Compendium Index of Paleobotany*, which omits diatoms, spores, and pollens. For each genus a type species or one that is representative is cited. Also a brief notation is given concerning the age, geographic origin, and taxonomic status of most of the fossils.

10-32   **Index of State Geological Survey Publications Issued in Series.** By John Boyd Corbin. Metuchen, NJ: Scarecrow Press, 1965. 667p. index.

Includes all numbered monographic publications in series issued by the state geological surveys or their designated equivalents through 1962. All unnumbered monographs and annual or biennial administrative reports have been omitted. It is arranged by state, then alphabetically by title of series, then numerically; entries give title, authors, and date. There is an author and a subject index.

10-33   **Meteorological and Geoastrophysical Abstracts.** v. 1- . Washington: American Meteorological Society, 1950- . monthly.

The first ten volumes were called *Meteorological Abstracts and Bibliography.* Devoted to current abstracts on important meteorological and geoastrophysical articles, papers and other primary documents in every language. The emphasis is on atmospheric science and boundary layer problems. Each citation includes the author, decimal classification, title, unabbreviated periodical title, volume, number, month, year, pages, presence of maps, abstract, and subject headings. All titles are in the language of the article, with a translated title following. Annual cumulative author and subject indexes.

10-34   **Mineralogical Abstracts.** v. 1- . London: Mineralogical Society of Great Britain, 1920- . quarterly.

An international abstracting service covering books, pamphlets, reports, and periodical literature in mineralogy. It is arranged by subject with annual topographical index and a detailed author and subject index.

10-35 **Publications of the Geological Survey, 1879-1961.** Washington: U.S. Geological Survey; distr. Washington: GPO, 1964. 457p.

10-35a **Publications of the Geological Survey, 1962-1970.** Washington: U.S. Geological Survey; distr. Washington: GPO, 1971. 586p.

10-35b **New Publications of the Geological Survey, 1971- .** Washington: U.S. Geological Survey; distr. Washington: GPO, 1971- . monthly with annual cumulations.

These three items serve as a checklist to the many publications published by the U.S. Geological Survey, one of the most important publishers of geological literature in the United States. All series, old and new, are listed with all numbers, parts, and chapters. The publications also list map series.

10-36 **Selected Bibliography on Urban Climate.** By T. J. Chandler. Geneva, Switzerland: World Meteorological Organization; distr. New York: Unipub, 1970. 383p.

This bibliography contains more than 2,000 citations listed under 25 broad subject headings, ranging from air pollution to atmospheric electricity.

10-37 **Zentralblatt fuer Geologie und Palaeontologie. Teil 1: Allgemeine, Angewandte, Regionale und Historische Geologie.** Stuttgart: E. Schweizerbart'sche Verlagsbuchhandlung, 1950- . 13 issues per year.

10-37a **Zentralblatt fuer Geologie und Palaeontologie. Teil 2: Palaeontologie.** Stuttgart: E. Schweizerbart'sche Verlagsbuchlandlung, 1950- . 7 issues per year.

10-37b **Zentralblatt fuer Mineralogie. Teil 1: Kristallographie, Mineralogie.** Stuttgart: E. Schweizerbart'sche Verlagsbuchhandlung, 1950- . 7 issues per year.

10-37c **Zentralblatt fuer Mineralogie. Teil 2: Petrographie, Technische Mineralogie. Geochemie und Lagerstaettenkunde.** Stuttgart: E. Schweizerbart'sche Verlagsbuchhandlung, 1950- . 13 issues per year.

This is a comprehensive index to the world's literature on geology, paleontology, and mineralogy arranged by broad subject categories with several appropriate indexes. The four titles were formed by the merger of teils 1-3 of the *Zentralblatt fuer Mineralogie, Geologie und Palaeontologie.*

ENCYCLOPEDIAS

10-38 **Climates of the States.** New material by James A. Ruffner. Detroit: Gale Research Co., 1978. 2v. illus. maps.

This is a reprint of two U.S. Government documents: *Climates of the States*, a 52-part series originally issued from 1959-1960 and *Local Climatological Data*, originally issued in 1975. A brief introduction is given to each state followed by the various tables that give data for major weather stations in that state.

10-39 **Color Encyclopedia of Gemstones.** By Joel E. Arem. New York: Van Nostrand Reinhold, 1977. 147p. illus. (col.). bibliog. index.

Arranged alphabetically according to mineral species, this encyclopedia lists basic mineralogical and gemological data on every known species and variety of gemstone giving chemical formula, crystal structure, colors, luster, hardness, density, cleavage, optics, spectral data, luminescence, and size.

10-40   **Deserts of the World: An Appraisal of Research into Their Physical and Biological Environments.** Ed. by William G. McGinnies and others. Tucson, AZ: University of Arizona Press, 1969. 788p. maps. bibliog. index.
Prepared with the assistance of the Office of Arid Lands Studies, University of Arizona, this guide evaluates existing arid land studies, numbering approximately 5,000 items. Its scope includes climate, flora and fauna, surface and ground-water hydrology for 13 major desert regions.

10-41   **Encyclopedia of Marine Resources.** Ed. by Frank E. Firth. New York: Van Nostrand Reinhold, 1969. 740p. illus. index.
Analyzes more than 125 topics concerned with the most significant aspects of the ocean's resources. Some brief summaries are included on topics in oceanography and marine engineering. The latest information is presented on such topics as marine botany, nutrition, pollution, minerals, and ecology. Each article is signed and concludes with a bibliography.

10-42   **Encyclopedia of Minerals and Gemstones.** Ed. by Michael O'Donoghue. New York: Putnam's, 1976. 304p. illus. (part col.). bibliog. index.
This encyclopedia covers crystal morphology, basic and economic geology, mineral identification, gem cutting, and mineral display. Over 1,000 minerals are systematically described for the non-specialist.

10-43   **Encyclopedia of Oceanography.** New York: Van Nostrand Reinhold, 1966. 1021p. illus. (Encyclopedia of Earth Sciences Series, v. 1).

10-43a **Encyclopedia of Atmospheric Sciences.** New York: Van Nostrand Reinhold, 1967. 1200p. illus. (Encyclopedia of Earth Science Series, v. 2).

10-43b **Encyclopedia of Geomorphology.** New York: Van Nostrand Reinhold, 1968. 1295p. (Encyclopedia of Earth Sciences Series, v. 3).

10-43c **Encyclopedia of Geochemistry and Environmental Sciences.** New York: Van Nostrand Reinhold, 1972. 1321p. illus. (Encyclopedia of Earth Sciences Series, v. 4a).

10-43d **Encyclopedia of Sedimentology.** Stroudsburg, PA: Dowden, Hutchinson & Ross; distr. New York: Academic Press, 1978. 901p. illus. maps. index. (Encyclopedia of Earth Sciences, v. 7).
This series of encyclopedias edited by Rhodes Whitmore Fairbridge is published for a diverse group of potential users, ranging from high school students to graduate students. On the average, the volumes are keyed to the college level. The subject of each volume is covered in great detail.

10-44   **Geography and Cartography: A Reference Handbook.** By C. B. Muriel Lock. 3rd ed. rev. and enlarged. London: Clive Bingley; Hamden, CT: Linnet Books, 1976. 762p. index.

This handbook has about 30 pages devoted to map librarianship; the rest to various topographical and thematic maps and atlases published around the world. It gathers together much information, generally on geographical publications and organizations and on prominent deceased geographers.

10-45　**Illustrated Encyclopedia of the Mineral Kingdom.** Alan Woolley, consultant ed. New York: Larousse, 1978. 240p. illus. (part col.). bibliog. index.

This beautifully illustrated book contains several sections of which the main one is the Mineral Kingdom. It contains descriptions of 300 of the most common minerals. Other sections include Minerals, Rocks, and Their Geological Environment; Crystals; Properties and Study of Minerals; Gemstones; Economic Minerals; Building a Collection; and A Guide to the Literature on Minerals.

10-46　**McGraw-Hill Encyclopedia of Ocean and Atmospheric Sciences.** Sybil P. Parker, ed. in chief. New York: McGraw-Hill, 1980. 580p. illus.

Covers all aspects of ocean and atmospheric sciences including pollution, satellites, climate modification and sea diving. The articles are signed and arranged alphabetically with some coming from the *McGraw-Hill Encyclopedia of Science and Technology* (3-48).

10-47　**McGraw-Hill Encyclopedia of the Geological Sciences.** Daniel N. Lapedes, ed. in chief. New York: McGraw-Hill, 1978. 915p. illus. index.

This encyclopedia provides a comprehensive treatment of the geological sciences, including geology, geochemistry, geophysics, and related aspects of oceanography and meteorology. Includes a table of accepted mineral species giving chemical formula, crystallography class, hardness, and specific gravity of each.

10-48　**Maps for Books and Theses.** By A. G. Hodgkiss. New York: Pica Press; distr. New York: Universe Books, 1970. 267p. illus.

Thoroughly describes the preparation of thematic maps for books, theses, and journals. It will serve as a useful reference source for authors who need to prepare their own maps or for beginners in map-making offices. It covers selection of a base map, changing scale, tools and equipment, lettering and drawing, presentation of statistical data, map design and layout, specialized maps, and reproducing maps. A classified list of readings is provided, plus several appendixes and a list of suppliers.

10-49　**Oilfields of the World.** By E. N. Titratsoo. Beaconsfield, England: Scientific Press; distr. New York: Geology Press, 1973. 376p. illus. index.

A well-researched volume describing the geology and geography of the world's major oil fields. More than 50 maps and diagrams are presented depicting the occurrence of hydrocarbons in approximately 70 countries. Crude oil production and reserve statistics are summarized in about 100 tables. The final chapter of the book examines 1) the known world oil reserves, 2) the exploration potential, and 3) possible future replacement. As a reference work this book should find wide readership extending beyond those engaged in the exploration and production of oil.

10-50 **Planet We Live On: Illustrated Encyclopedia of the Earth Sciences.** Ed. by Cornelius S. Hurlbut, Jr. New York: Harry N. Abrams, 1976. 527p. illus. (part col.).

An excellent general encyclopedia on the earth sciences with signed articles that discuss earth history and the generalities of the subjects making up earth sciences.

10-51 **Standard Encyclopedia of the World's Mountains.** By Anthony Julian Huxley. New York: Putnam, 1969. 383p. illus. (part col.). index.

A standard reference for factual information on mountain ranges, including their geology, special fauna and flora, historical importance, first discovery, and first explorers or climbers. A gazetteer is included with some feature articles.

10-52 **Standard Encyclopedia of the World's Oceans.** By Anthony Huxley. New York: Putnam, 1969. 383p. illus. index.

A standard reference for information on the world's oceans, including their special fauna and flora, historical importance, first discovery, and first explorers. A gazetteer is included along with some feature articles.

10-53 **Standard Encyclopedia of the World's Rivers and Lakes.** By R. Kay Gresswell and Anthony Huxley. New York: Putnam, 1966. 384p. illus. (part col.).

Comprehensive coverage of all major rivers and lakes of the world. Each entry gives the river's location, source, outlet, and length, or the lake's length, breadth, and area. All entries are keyed to a map. A short narrative discusses the river's or the lake's geographic importance.

10-54 **Treatise on Invertebrate Paleontology.** Prepared under the guidance of the Joint Committee on Invertebrate Paleontology. New York: Geological Society of America and University of Kansas Press, 1953- . illus. index.

The most comprehensive treatise on invertebrate paleontology available. Each entry provides a full description and contains illustrations, drawings, and bibliographical citations.

10-55 **Weather Almanac: A Reference Guide to Weather, Climate, and Air Quality in the United States and Its Key Cities, Comprising Statistics, Principles, and Terminology.** . . . Ed. by James A. Ruffner and Frank E. Bair. 2nd ed. Detroit: Gale Research Co., 1977. 728p. illus. index.

This is a comprehensive compilation of weather and climatic data throughout the United States comprised of numerous tables, charts, maps, and diagrams that provide such information as severe weather conditions, health and safety rules, yearly weather records for selected U.S. cities, air quality data, marine weather, retirement and health weather, and round-the-world weather.

DICTIONARIES

10-56 **Dictionary of Earth Sciences.** Ed. by Stella E. Stiegler. New York: Pica Press; distr. New York: Universe Books, 1977. 301p. illus.

This dictionary defines roughly 3,000 words or phrases of common or historical usage in geology, paleontology, geophysics, meteorology, and several other related sciences.

10-57 Dictionary of Geography: Definitions and Explanations of Terms Used in Physical Geography. By W. G. Moore. Rev. ed. New York: Barnes and Noble, a Division of Harper and Row, 1978. 260p. illus.

In addition to physical geography terms this dictionary gives definitions of terms in the related fields of climatology, geology, astronomy, and other earth sciences.

10-58 Dictionary of Geological Terms. By American Geological Institute. Rev. ed. Garden City, NY: Anchor Press Doubleday, 1976. 472p.

This is an abridgement of the *Glossary of Geology* (10-62) containing 8,500 of the more common definitions found in the glossary, plus 1,000 new definitions that update the basic volume.

10-59 Dictionary of Geology. By John Challinor. 5th ed. Cardiff: University of Wales Press; New York: Oxford University Press, 1978. 365p. index.

A small, concise dictionary of 1,500 terms that are not only defined but also placed in relation to other terms and concepts. The book probes geology "by examining the meaning and usage of names and terms that stand for the more significant things, facts and concepts of the science."

10-60 Dinosaur Dictionary. By Donald F. Glut. Secaucus, NJ: Citadel Press, 1972. 218p. illus.

An alphabetical listing of nearly all genera of dinosaurs currently known to paleontologists, with cross-references to similar genera. Each entry describes the characteristics of the genus, tells where it is known to have existed, and lists the skeletal remains by which it has been identified. In some cases the date and place of discovery are noted. Drawings are provided to illustrate bone fragments, skulls, and reconstructions.

10-61 Elsevier's Dictionary of Hydrogeology: In Three Languages; English, French, German. Comp. and arranged by Hans-Olaf Pfannkuch. New York: American Elsevier, 1969. 168p. bibliog. index.

Contains 3,040 entries in English, with definitions and French and German equivalents. This dictionary covers the terms used in ground water geology and hydrology to describe the occurrence of ground water, its infiltration into the geological framework, its subsurface movement, and its final discharge into the atmosphere or the sea. Properties and qualities of the water are dealt with only insofar as they relate to the above topics. The geological terms included are those that describe the specific interaction between the geological environment and water. Finally, technological terms used in exploration, drilling, and production of ground water, and in the working of water wells, are incorporated.

10-62 Glossary of Geology and Related Sciences. Ed. by Margaret Gary, Robert McAfee, Jr., and Carol L. Wolf. 3rd ed. Washington: American Geological Institute, 1972. 805p.

A glossary of preferred current meanings, encompassing all of the subdivisions of geology described in the introduction to this chapter.

10-63 Glossary of Meteorology. Boston: American Meteorological Society, 1959. 638p.

Includes those important meteorological terms likely to be found in the literature. Three or more experts had to agree on a definition before it was admitted to the glossary.

10-64   Illustrated Dictionary of Place Names, United States and Canada. Ed. by Kelsie B. Harder. New York: Van Nostrand Reinhold, 1976. 631p. bibliog. illus. (A Hudson Group Book).

This quasi-gazetteer lists over 15,000 populated places, natural features, and historic sites located north of the Rio Grande. Brief entries include information on general location plus derivation of name, along with variants and earlier names. In the case of rivers, direction of flow, principal tributaries, and termini are indicated. Short biographical sketches appear throughout for persons after whom sites have been named.

10-65   International Dictionary of Metallurgy, Mineralogy, Geology, and the Mining and Oil Industries: English, French, German, Italian. By Angel Cagnacci-Schwicker. Milan: Technoprint International, 1968; repr. New York: McGraw-Hill, 1970. 1550p.

This dictionary contains 27,000 terms arranged alphabetically in English, followed by the French, German, and Italian equivalents covering metallurgy, mining and oil, mineralogy, and geology. A symbol beside each English term indicates the industry in which the word is used.

10-66   Nomenclature of Petrology: With References to Selected Literature. By Arthur Holmes. London: Allen and Unwin, 1920; repr. New York: Hafner, 1971. 284p.

A facsimile edition of an old but very useful geological dictionary. The words used by American and British geologists are often not mutually understood, so "translation" has sometimes been necessary.

10-67   Oil Terms: A Dictionary of Terms Used in Oil Exploration and Development. By Leo Crook. New York: International Publications Service, 1976 (c. 1975). 160p. illus. bibliog.

This dictionary presents clear and concise definitions of the terminology and equipment used in the fields of oil exploration and production.

10-68   Penguin Dictionary of Geology. By D. G. A. Whitten. Baltimore, MD: Penguin Books, 1976. 493p. illus.

A small alphabetically arranged dictionary that includes not only geologic terms but also stratigraphic terms and brief biographies of famous geologists.

10-69   Petroleum Dictionary. By David F. Tver and Richard W. Berry. New York: Van Nostrand Reinhold, 1980. 374p. illus.

This is "the first comprehensive combination dictionary-handbook that covers virtually all aspects of the petroleum industry, with comprehensive coverage of geology, geophysics, seismology, drilling, gas processing, production, a detailed analysis of various refining operations and processes, offshore technology, and description of various materials and supporting techniques used in drilling and production of oil and gas."

10-70   Standard Dictionary of Meteorological Sciences: English-French/French-English. By Gerard J. Proulx. Montreal and New York: McGill-Queen's University Press, 1971. 307p. bibiog.

An outstanding dictionary on meteorology; the definition for each entry is in the same language as the entry itself.

HANDBOOKS, MANUALS, AND FIELD GUIDES

10-71   **Audubon Society Field Guide to North American Rocks and Minerals.**
New York: Knopf, 1978. 850p. illus. (col.). index.

An excellent field guide to the rocks and minerals of North America. For each listed mineral the following is given: mineralogical name, chemical name, formula, color and luster, hardness, cleavage, other data, crystals, best field marks, similar species, environment, occurrence and color plate. For each rock the following is given: name, texture, structure, color, luster, hardness, other data, field features, metamorphism, parent rock, mineralogy, and environment. Appendixes include a glossary, bibliography, rock-forming minerals, chemical elements, and localities.

10-72   **Dana's Manual of Mineralogy.** By Cornelius S. Hurlbut, Jr. 18th ed.
New York: Wiley, 1971. 579p. illus.

Although designed as a text for beginning courses in mineralogy, this work is excellent as a reference book. Divided into several sections, it encompasses crystallography; physical mineralogy; chemical mineralogy; descriptive and determinative mineralogy. This book is based on *System of Mineralogy* (10-81).

10-73   **Field Guide to Landforms in the United States.** By John A. Shimer. New
York: Macmillan, 1972. 272p. illus. index.

A well-organized manual of basic surface geology. Part 1 is devoted to the major geologic provinces of the nation. Descriptions are brief and informative without being pedantic. Part 2 will be of greatest reference use because of its lavishly illustrated descriptions of landscape features.

10-74   **Field Guide to Rocks and Minerals.** By Frederick H. Pough. 4th ed.
Boston: Houghton Mifflin, 1976. 317p. illus. (part col.). bibliog. index.
(The Peterson Field Guide Series, no. 7).

The largest part of this excellent guide covers 270 minerals grouped according to chemical composition. Excellent color photographs add to the usefulness of this book, which also includes chapters describing the physical and chemical attributes to minerals, the crystal classification, and home laboratory techniques.

10-75   **Handbook of Ocean and Underwater Engineering.** Ed. by John J. Myers
and others. Prepared under the auspices of North American Rockwell
Corporation. New York: McGraw-Hill, 1969. 1070p. illus. index.

Written in technical language for engineering personnel, this handbook contains 12 sections dealing with such topics as oceanography, hydrodynamics, underwater instrumentation, tools and machinery, cables, power, materials and testing, etc. Articles are accompanied by tables, graphs, and bibliographical citations. Well illustrated and indexed.

10-76   **Handbook of World Salt Resources.** By Stanley J. Lefond. New York:
Plenum, 1969. 384p. maps.

A detailed listing of all salt deposits, solar salt operations, and salt springs. Presentation is by continent and then alphabetically by country. Detailed maps, charts, diagrams, and tabulations help to fix the various deposits geographically. In most cases, a full historical account of the deposit is provided, along with a detailed bibliography. Statistics on mining and reserves are liberally cited.

10-77　**Larousse Guide to Minerals, Rocks, and Fossils.** By W. R. Hamilton, A. R. Woolley, and A. C. Bishop. New York: Larousse, 1977. 320p. illus. (part col.). index.

A beautifully illustrated guide to 220 minerals, 90 rocks, and approximately 300 fossils. A typical entry for a mineral includes chemical formula, crystal system, specific gravity, hardness, cleavage, fracture, color and transparency, streak, luster, distinguishing features, and occurrence. Rocks are identified in terms of color, grain size, texture, structure, mineralogy, and field relations. Fossil descriptions include geologic period, occurrence, and physical characteristics.

10-78　**Lexicon of Geologic Names of the United States for 1961-1967.** By Grace C. Keroher. Washington: U.S. Geological Survey; distr. Washington: GPO, 1970. 848p. (USGS Bulletin 1350).

An alphabetical listing of 2,860 names of geologic formations and stratigraphic classifications and nomenclatures, providing for each location variant names, periodical citation where first described, and where it outcrops. The geographic area includes the United States, its possessions, the Panama Canal Zone, and the trust Territory of the Pacific Islands. For coverage of previous years, see: Fred B. Weeks, *North American Geologic Formation Names*, USGS Bulletin 191 (1902); M. Grace Wilmarth, *Names and Definitions of the Geologic Units of California*, USGS Gulletin 826 (1931); M. Grace Wilmarth, *Lexicon of Geologic Names of the U.S. (Including Alaska)*, USGS Bulletin 896 (1938); Druid Wilson, *Geologic Names of North America Introduced in 1936-1966*, USGS Bulletin 1056-A (1957); Druid Wilson, *Index to the Geologic Names of North America*, USGS Bulletin 1956-B (1959); Grace Keroher, *Lexicon of Geologic Names of the United States for 1936-1960*, USGS Bulletin 1200 (1966).

10-79　**Simon and Schuster's Guide to Rocks and Minerals.** Ed. by Martin Prinz, George Harlow, and Joseph Peters. New York: Simon and Schuster, 1978. 607p. illus. (col.). index. (A Fireside Book).

This is a field guide, handbook, and reference work on mineralogy, containing 276 entries divided into minerals and rocks. It covers methods of identification, principles of classification, etc. The entries for minerals include chemical composition, crystal form, physical properties, rock environment, geographical distribution, major uses, and rarity. For rock entries, information includes chemistry, origin, classification, type, grain size, environment in which formed, and pressure and temperature of formation.

10-80　**Smithsonian Meteorological Tables.** 6th ed. rev. Washington: Smithsonian Institution; distr. Washington: GPO, 1951. 527p. (Smithsonian Miscellaneous Collections, 114).

A desk handbook for the meteorologist, this handbook contains all the various tables that would be needed in interpreting and compiling meteorological reports. Supplementally, there are conversion tables of statute miles to nautical miles; hours, minutes, and seconds to decimals of a day; wind tables, barometric tables, and atmospheric and altimetry tables.

10-81　**System of Mineralogy.** 7th ed. New York: Wiley, 1944-1962. 3v. illus.

This is the most comprehensive classification of minerals available. All minerals are listed with complete descriptions of physical and chemical characteristics.

ATLASES

10-82　**Atlas of Economic Mineral Deposits.** By Colin J. Dixon. Ithaca, NY: Cornell University Press, 1979. 143p. illus. (part col.). maps. index.

This atlas covers 48 major economic mineral deposits. For each deposit, the diagrams include a location map, a map showing the regional geological setting of the deposit, and maps of sections of the deposit itself. The text includes paragraphs on location, geographical setting, history, and mining, geological setting, geology of the deposit, size and grade, and geological interpretations.

10-83 **Atlas of World Physical Features.** By Rodman E. Snead. New York: Wiley, 1972. 158p. illus. maps. bibliog. index.

A descriptive atlas of a wide variety of physical features showing where in the world the features are particularly numerous. The 103 small-scale maps show the overall distribution of the landforms but do not give their exact numbers, sizes, or shapes.

10-84 **Climatic Atlas of Europe. Maps of Mean Temperature and Precipitation.** Paris: World Meteorological Organization, 1970; distr. New York: Unipub, 1971. 27 detachable map sheets. Title page and text in English, French, Spanish, and Russian.

Sponsored by the World Meteorological Organization and based on WMO specifications, this atlas is the first volume in the projected series sponsored jointly by WMO and Unesco. The maps of the distribution of monthly and annual mean values of temperature, of the annual range of temperature, and of monthly amounts of precipitation have been published on the scale 1:10,000,000. There are two separate maps, on the scale of 1:5,000,000, showing the distribution of the annual amounts of precipitation for the western half and the eastern half of Europe. In all monthly maps the areas between isotherms and isohyets, with the same values of temperature and amount of precipitation, respectively, are shown in the same color, so that in the series of monthly maps the annual variation of the changes in the distribution of temperature or precipitation is shown distinctly. Based on temperature data collected by more than 2,500 stations and precipitation data collected by more than 7,000 stations, this is one of the most authoritative sources on the subject.

10-85 **International Cloud Atlas.** Abridged Atlas. Geneva: World Meteorological Organization; distr. New York: Unipub, 1969. 1v. (various paging).

Composed of two parts, the first of which consists of descriptive and explanatory text, the second of a collection of plates in black and white and in color, designed to illustrate the text.

10-86 **Landforms and Topographic Maps Illustrating Landforms of the Continental United States.** By Williams Gayly Upton, Jr. New York: Wiley, 1970. 134p. maps. (col.)

A collection of significant topographic maps with some accompanying textual description of each. It is an outgrowth of the USGS *Portfolio of One Hundred Maps.* It contains only portions of the full quadrangles that were selected by 60 well-known geologists as representative of the particular land forms.

10-87 **National Atlas of the United States.** Washington: U.S. Geological Survey; distr. Washington: GPO, 1970. 417p. illus. index.

This magnificent reference work contains 756 maps under the following headings: General Reference, Physical, Historical, Economic, Sociocultural, Administrative, Mapping and Charting, and the World. Science librarians will be the most interested in the physical category, which includes expertly drawn maps in every domain of geoscience insofar as these pertain to the national territory of the United States.

10-88 **Oceanographic Atlas of the Pacific Ocean.** Comp. by Richard A. Barkley. Honolulu, HI: Press of the University of Hawaii, 1969. 176p. charts. tables. maps.

Embodies the results of observations made throughout the Pacific during a period of over 50 years, as supplied by more than 50,000 oceanographic and fisheries stations in Australia, Canada, France, French Oceania, Great Britain, Japan, and the United States, many of which have never before been used in an ocean-wide cartographic study. The component maps derive from about three million separate measurements of temperature, salinity, dissolved oxygen, and sampled depths.

10-89 **Rand McNally Atlas of the Oceans.** New York: Rand McNally, 1977. 208p. illus. (part col.). index.

This is actually an encyclopedia of the world's oceans. In addition to bathymetric maps of each of the oceans and the principal seas there are maps of mineral and living resources, and wind and water circulation. Numerous color photographs help to make this a very useful reference book.

10-90 **Space Age Photographic Atlas.** Comp. by Ken Fitzgerald. New York: Crown, 1970. 288p. illus. maps.

Contains more than 200 photographs of the earth as seen by astronauts and from aircraft, covering every type of climate and topography. Explanatory maps and text accompany each photo.

10-91 **Volcanic Landforms and Surface Features: A Photographic Atlas and Glossary.** Ed. by Jack Green and Nicholas M. Short. New York: Springer-Verlag, 1971. 418p. illus. bibiog.

A compilation of approximately 400 photographs of nearly all volcanic surface structures and features described during the development of volcanology in this century. Each photo is accompanied by a caption identifying the feature, giving its location, and elaborating on its particular aspect. Terms or concepts introduced in a caption are defined in the glossary or in the introductory chapter, which treats volcanoes as landforms.

10-92 **Water Atlas of the United States.** By James J. Geraghty and others. 3rd ed. Port Washington, NY: Water Information Center, 1973. 1v. (unpaged). illus. maps.

A good general visual guide to the water situation in the United States, including Alaska and Hawaii. The information is presented in 122 maps arranged on a common base, so that the data can be compared from map to map. The data cover all aspects of precipitation, surface water, and ground water, plus new material on water pollution, water quality, water conservation/recreation, and water law. The text that accompanies each map defines the subject (e.g., water hardness) and gives facts and statistics necessary to cover the subject.

DIRECTORIES

10-93 **Geologic Field Trip Guidebooks of North America: A Union List Incorporating Monographic Titles.** By the Guide Book and Ephemeral Materials Committee of the Geoscience Information Society. 2nd ed. Houston, TX: Phil Wilson, 1971. 152p.

A long-needed reference work, since geologic field trip guidebooks are an important but elusive part of the literature of geology. Indispensable to catalogers who must trace the history of field conferences. This edition contains guidebook holdings of 59 participating libraries, including important libraries in the

Southwest (particularly the University of Texas collection). Arranged alphabetically by the organization sponsoring the field trip, with a location index.

10-94   **Map Collections in the United States and Canada.** Comp. by David K. Carrington. 3rd ed. New York: Geography and Map Division of Special Libraries Association, 1978. 230p. index.

A directory of 745 Anglo-American cartography collections housed in academic, public, or special (but not private) libraries. Arrangement is alphabetical by state or province and within each, alphabetical by city. Each entry lists staff, number of maps, books and other materials, annual accessions, area, subject and chronological specializations, special cartographic collections, cataloging and classification, depositories, public served, reproduction facilities, when established, interlibrary loan, name of map librarian, and telephone number when available.

10-95   **World Directory of Map Collections.** Comp. by the Geography and Map Libraries Sub-Section; ed. by Walter W. Ristor. Munich: Verlag Dokumentation; distr. New York: Unipub, 1976. 326p. bibiog. (IFLA Publications 8).

This directory lists about 285 collections in more than 45 countries. Arrangement is alphabetical by country, then by city with information given including name and address of the collection, date of establishment, telephone number, director, number of professional staff members, size of collection, annual additions, geography, subject and chronological range of the collection, classification scheme, reference service provided, exchange policy, reproduction facilities, storage and preservation facilities, and publications.

11
Energy and Environment

This chapter is new to this edition. In the second edition it was part of geosciences for lack of a better place to locate it. The literature in the areas of energy and environment has greatly expanded. The numbers of sources are now as varied as in any of the other science disciplines.

Environment can be defined as the assemblage of material factors and conditions surrounding the living organisms and its component parts. It includes both external forces on the organism and internal forces. Physical environment is made up of the inanimate objects and the forces associated with them while the organic environment is the living things and their derivatives with which the animal may be associated. The internal environment is determined by organization within the body to which all of its parts respond directly, whether or not they also have external contacts.

Energy is generally defined as the ability or capacity to do work. In this chapter we are interested in all the factors related to power or energy technology. The chapter covers coal and gas technology; nuclear energy; and petroleum, geothermal, solar, and wind energy.

Because this discipline is young, it does not have a developed history. Instead, the reader is directed to the histories of other disciplines where references are made to pollution, solar energy, contamination, ecological balance, environmental quality, fuel resources, and exploration of resources. The effects of environmental and energy research have far reaching ramifications in all the sciences. Physicists and chemists are concerned with developing new sources of energy. Geoscientists are renewing their search for oil and minerals. Biologists are watching and studying the effects of pollution and contamination on living organisms. Engineers are experimenting with new modes of transportation and power. A librarian who has to work with sources of information on energy and environment has a monumental task because sources from all disciplines need to be used.

GUIDES TO THE LITERATURE

11-1   **Environmental Pollution: A Guide to Current Research.** New York: CCM, 1971. 851p. index.
This guide was produced from data gathered by Science Information Exchange, Smithsonian Institution. According to the preface, this guide represents a first attempt to describe current research projects in environmental pollution. Its contents comprise several broad subject categories: ecological systems, effects of population, air and water pollution, and pollution causes. The following information is usually provided for each project: name, organization and address, brief description of objectives, progress, and supporting organization.

11-2   **Information Sources in Power Engineering: A Guide to Energy Resources and Technology.** By Karen S. Metz. Westport, CT: Greenwood Press, 1975. 114p. bibliog. index.

A useful subject guide to information sources on energy resource development, conversion to electric power, transmission from power plants to areas of consumption, and electricity distribution to users. It covers periodicals, textbooks, conference publications, abstracting and indexing sources, and other reference tools. It also includes directory information related to organizations in power engineering, information centers, libraries, and commercial services.

11-3   **Noise Pollution: A Guide to Information Sources.** By Clifford R. Bragdon. Detroit, MI: Gale Research Co., 1979. 524p. bibliog. index.

This bibliography comprises over 3,000 references drawn from periodicals, documents, books, and nonprint sources. Material is arranged into the following topics: physiological effects, behavioral effects, abatement, community noise, environmental impact, acoustics, and education.

11-4   **Sourcebook on the Environment: A Guide to the Literature.** Ed. by Kenneth A. Hammond, George Macinko, and Wilma B. Fairchild. Chicago: University of Chicago Press, 1978. 613p. index.

This comprehensive guide to environmental literature is arranged in three major sections: environmental perspectives, case studies, and major elements of the environment. A bibliography follows each essay and is current through 1976. A fourth section contains further study aids: a briefly annotated list of 100 pertinent periodicals; a list of environmental organizations; a review of federal environmental legislation since 1785; plus notes on helpful bibliographies, abstracts, and indexes.

11-5   **Toxic Substances Sourcebook: The Professional's Guide to the Information Sources, Key Literature and Laws of a Critical New Field.** Ed. by Steve Ross and Monica Pronen. New York: Environment Information Center, 1978. 550p.

Summarizes the literature dealing with the regulations and dangers posed by chemical applications on the environment. It contains 2,000 abstracts of reports, research studies, journal articles, and technical papers from 1974 to 1977. Supplementary information includes a directory of toxic information centers, data banks, and magazines; 50 pages of statistics; an analysis of the new Toxic Substances Control Act and projections on EPA enforcement; and book and film listings. Indexes provide keyword, author, and geographic access.

11-6   **Wastewater Management: A Guide to Information Sources.** By George Tchobanoglous, Robert Smith, and Ronald Crites. Detroit, MI: Research Gale Co., 1976. 202p. bibliog. index. (Man and the Environment Information Guide Series, v. 2).

Provides quick introductory information on the field of wastewater management, including the engineering of wastewater collection, treatment, disposal and reuse systems. Addresses of newsletters, periodicals, abstracts, and digests are given, as are those of government agencies and nearly 500 publishers.

ABSTRACTS, INDEXES, AND BIBLIOGRAPHIES

11-7   **Abstracts on Health Effects of Environmental Pollutants.** v. 1- . Philadelphia, PA: Biosciences Information Service, 1972- . monthly.
This publication results from the scanning of two data bases: BIOSIS of Biosciences Information Service, and MEDLARS of the National Library of Medicine. It covers about 1,000 articles each month on occupational health and industrial medicine with emphasis on their effects on human health. Indexes provide subject, author, genus-species, biosystematic, and general concept access to the abstracts.

11-8   **Air Pollution Abstracts.** v. 1- . Research Triangle Park, NC: Air Pollution Technical Information Center; distr. Washington: GPO, 1970- . monthly.
This publication abstracts information on the methods used for measuring and controlling emissions, air quality, and atmospheric interactions; effects of airborne pollutants on plants, materials, animals, and men; air pollution legislation and standards; and the effects of air pollution on industry and the economy. Each issue includes author and subject indexes.

11-9   **Air Pollution Publications: A Selected Bibliography with Abstracts.** 1955- . Washington: U.S. Public Health Service; distr. Washington: GPO, 1964- . annual.
This bibliography consists of references with abstracts to journal articles, books, conference papers, and some report literature originating with NAPCA staff and grantees. Arranged in 12 broad subject categories such as emission sources, measurements, and legal aspects. Author and subject index are included. The bibliography is compiled by the Science and Technology Division of the Library of Congress for the National Air Pollution Control Administration.

11-10   **Bibliography of the Urban Modification of the Atmospheric and Hydrologic Environment.** By John F. Griffiths and M. Joan Griffiths. Washington: U.S. Environmental Data Service, 1974. 100p.
Selectively provides a cross section of the literature, limiting its coverage to city climates that contain comparisons with adjacent rural climates. Separate sections have the following headings: City, Cloud, Cooling Power, Dust and Nuclei, Electricity and Ions, Humidity, Light, Models, Pollution, Precipitation, Radiation, Temperature, Visibility, Wind, Floods, Ground Water, Runoff, Sedimentation, Stream Temperature, and Water Quality.

11-11   **Biological Indicators of Environmental Quality: A Bibliography of Abstracts.** By William A. Thomas, William H. Wilcox, and Gerlad Goldstein. Ann Arbor, MI: Ann Arbor Science Publishers, 1973. 254p. index.
This book is the only source book presently available that lists research using biological indicators for testing of living organisms in their native habitats. The purpose of this kind of research is to provide an accurate assessment of environmental quality in situations where organisms are subjected to the combined effects of all naturally occurring stresses — as opposed to the mechanical or chemical measurements of artificially selected pollutants on organisms tested under laboratory conditions. An excellent index.

11-12   **Book Catalog of the Environmental Conservation Library, Minneapolis Public Library.** Chicago: American Library Association, 1974. 201p.
This catalog is a collection of more than 2,200 titles from the Environmental Conservation Library of the Minneapolis Public Library. It is divided into three separate sections: author/main entry, title, and subject. In addition to the standard environmental titles, this collection has everything from children's books to specialized technical reports. Two appendixes include pamphlet file subject headings and periodicals.

11-13   **Catalog of the Conservation Library, Denver Public Library.** Boston: G. K. Hall, 1974. 6v.

11-13a   **Supplement.** 1- . 1978- .
This specialized catalog is particularly strong in fish and wildlife biology and management, the history of the environmental movement and its leaders, pollution control, and the economics of ecology. The collection is international in scope and includes manuscripts, serials, photographs, and state and foreign documents.

11-14   **Catalogues of the Library of the Marine Biological Association of the United Kingdom, Plymouth, England.** Boston: G. K. Hall, 1977. 16v.
This catalog covers 13,000 books, 40,000 bound periodical volumes, 1,450 current periodicals and serials, and over 50,000 reprints and pamphlets on marine biology, oceanography, fisheries, and related subjects. The main catalog is an alphabetical author/name listing. The subject index consists of some 14,500 entries arranged under the following headings: Environmental Conservation; General Water Pollution; Detection; Analysis; Removal; Specific Chemicals; Oil; Dispersants and Detergents; Metals; Pesticides; Radioactivity; PCBs; Domestic Sewage; Pulp and Paper; Food Processing Effluent; Heat; and Solids.

11-15   **Coastal/Estuarine Pollution: An Annotated Bibliography.** By Evelyn Sinha. LaJolla, CA: Ocean Engineering Information Service, 1970. 87p. (Ocean Engineering Information Series, v. 3).
Contains 631 informative abstracts on the detection, identification, measurement, and analysis of pollution parameters; sources of pollution; coastal and estuarine processes; effects of pollution; water quality management; and waste heat utilization.

11-16   **Dictionary Catalog of the Water Resources Center Archives, University of California, Berkeley.** Boston: G. K. Hall, 1970. 5v.

11-16a   **Supplement.** 1- . 1971- .
This is a library catalog of 80,000 documents relating to water as a natural resource and its utilization; municipal and industrial water uses and problems; flood control; reclamation; waste disposal; coastal engineering; sediment transport; water quality; water pollution; water law; and water resources development and management. Emphasis is on report literature.

11-17   **Energy Bibliography and Index.** v. 1- . Comp. by the Texas A & M University Libraries. Houston, TX: Gulf Publishing, 1978- .

This index consists of the bibliography or main entry section and four supporting indexes. The coverage includes "production, utilization, and conservation of all types of fuels . . . ; energy storage and conversion; alternative energy sources . . . ; power plants and transmission systems; and . . . various nontechnical aspects. . . ." Each main entry gives the standard descriptive data plus an abstract, the source of which is indicated. It covers vertical file items; technical reports; government documents; and books, maps, serials, and periodicals.

11-18   **Energy Index.** v. 1- . New York: Environment Information Center, 1971- . annual.

Intended to be used either separately or as the annual cumulative index to *Energy Information Abstracts* (11-19). The *Index* provides access to the past year's energy literature and serves as an annual review to the major events, legislation, and statistics in the energy field. It also lists the major energy meetings, conferences and symposia held each year, as well as new energy books and films.

11-19   **Energy Information Abstracts.** v. 1- . New York: Environment Information Center, 1976- . monthly.
This monthly abstracting service specializes in policy and planning issues, new energy sources, energy alternatives, energy economics, conservation, as well as the traditional energy categories. The abstracted literature is gathered from special reports, task force reports, conference papers, government documents, research reports, and 1,000 energy-related international periodicals. It also lists upcoming energy conferences and announces newly published books. Each issue includes author and subject indexes, which are cumulated annually and published as the *Energy Index* (11-18).

11-20   **Environment Abstracts.** v. 1- . New York: Environment Information Center, 1971- . monthly.
This monthly abstracting service summarizes over 1,000 articles and reports gathered from technical periodicals, government documents, conference proceedings, and popular magazines covering food and drugs, population planning, weather modification, and wildlife, as well as the more common environmental topics. It also lists upcoming environmental conferences and announces new books in print. Each issue includes author and subject indexes, which are cumulated annually and published as the *Environment Index* (11-21).

11-21   **Environment Index: A Guide to the Key Environmental Literature of the Year.** New York: Environment Information Center, 1971- . annual.
This index is the cumulative annual index to the documents listed in *Environment Abstracts* (11-20). It provides subject, author, and geographic access to scientific and technical journals, government documents, conference papers and proceedings, general magazines, books, and films. It also lists pollution control device patents. Supplementary features include pollution control officials, and a synopsis of the major environmental bills in the U.S. Congress.

11-22   **Environmental Effects on Materials and Equipment.** v. 1- . Washington: National Academy of Sciences-National Research Council, Prevention of Deterioration Center, 1961- . monthly.

In two sections; section A continues *Prevention of Deterioration Abstracts* (1946-1962); section B continues *Environmental Effects on Materials and Equipment* (1961). Abstracts cover the deteriorating effects of environmental forces on materials and equipment. Also includes environmental testing and aerospace technology. Each section has its own monthly and annual subject and author indexes.

11-23   **Environmental Quality Abstracts.** v. 1- . Louisville, KY: Data Courier, 1975- . quarterly.
This abstract searches some 200 periodicals for significant, new information relevant to the environmental sciences. The abstracts are arranged under the following divisions: public policy, population and health, conservation and endangered species, resources and recycling, and energy resources.

11-24   **Man and the Environment: A Bibliography of Selected Publications of the United Nations System, 1946-1971.** Comp. and ed. by Harry N. M. Winton. New York: Unipub; New York: R. R. Bowker, 1972. 305p. index.
Includes some 1,200 numbered entries arranged under broad subject categories, such as, natural resources and the earth sciences, geography, geology, geophysics, seismology, mineral resources, and oceanography, which are subdivided into more specific topics. In addition to reference works, monographic works, proceedings, surveys, and periodical literature are covered. Most entries have brief descriptive annotations, and the volume is well indexed with separate title and subject indexes.

11-25   **Pollution: A Selected Bibliography of U.S. Government Publications on Air, Water, and Land Pollution, 1965-1970.** Comp. by Louis Kiraldi and Janet L. Burk. Kalamazoo, MI: Institute of Public Affairs, Western Michigan University, 1971. 78p.
The 800 publications listed in this selective bibliography are arranged into three major sections: air, water, and land pollution. Within each section entries are arranged by Superintendent of Documents call numbers. Each entry indicates the kind of reference work (abstracts, bibliographies, directories, guides, and laws), which is followed by names of departments, agencies, and commissions. Entries give title, date, pages, and issuing agency. Annotations are provided for reference titles but not for the other entries.

11-26   **Pollution Abstracts.** v. 1- . Louisville, KY: Data Courier, Inc., 1970- . bimonthly.
Screens over 2,500 international primary sources each year that deal with air pollution, water pollution, solid wastes, noise, pesticides, radiation, and general environmental quality. Each issue provides indexing access by subject keyword, permuted keyword, and author. The permuted keyword and author indexes are cumulated annually; five-year cumulations are also available.

11-27   **Selected Water Resources Abstracts.** v. 1- . Washington: U.S. Office of Water Research and Technology, Water Resources Scientific Information Office, U.S. Department of the Interior; distr. Washington: GPO, 1968- . semimonthly.

This publication contains abstracts of current and earlier monographs, journal articles, reports and other publications covering the water-related aspects of the life, physical, and social sciences as well as related engineering and legal aspects of the characteristics, conservation, control, use or management of water. Full bibliographical citations are given.

11-28 **Water Pollution Abstracts.** v. 1- . London: published for the Water Pollution Research Laboratory of the Department of the Environment by Her Majesty's Stationery Office, 1972- . monthly.

Abstracts all primary forms of scientific documents on a worldwide basis. Specific topics include water quality and analysis, water conservation, sewage and wastewater, and polluting effects on water resources.

11-29 **Water Publications of State Agencies: A Bibliography of Publications on Water Resources and Their Management Published by the States of the United States.** Ed. by Gerald J. Giefer and David K. Todd. Port Washington, NY: Water Information Center, 1976. 2v.

11-29a **Supplement.** 1- . 1976- .

A retrospective bibliography of water resource publications of 335 state agencies in the 50 United States as supplied by lists from the agencies to the editors. Arrangement is by state, then by issuing agency. There are no subject or author indexes, but in most cases the simple arrangement of the book obviates them. Federal government reports (except "cooperative" ones) are omitted, as are reports of agricultural and interstate agencies. Publications of the water resources research centers established by Public Law 88-376 are included. Since state libraries are depositories for these reports, a list of the libraries and their addresses is appended. Price and out-of-print information are included if supplied by the agency.

11-30 **Water Resources Abstracts.** v. 1- . Urbana, IL: American Water Resources Association, 1968- . looseleaf. monthly.

Seeks to achieve comprehensive coverge of all phases of water resources, from planning and allocation to hydrology. Indicative abstracts are arranged according to 46 categories. A related work, less comprehensive but still valuable, is *Selected Water Resources Abstracts* (11-27).

11-31 **Water Resources Research Catalog.** v. 1- . Washington: U.S. Office of Water Resources Research; distr. Washington: GPO, 1965- .

Presents summary descriptions of current research on water resources problems. Each item is entered under a broad subject, and the entry gives name of project, investigator, place, and abstract. There are indexes by subject, investigator, contractor, and supporting agency.

ENCYCLOPEDIAS

11-32 **Encyclopedia of Energy.** Daniel N. Lapedes, ed. in chief. New York: McGraw-Hill, 1976. 785p. illus. index.

Divided into two sections, Energy Perspectives and Energy Technology, this book is intended to provide a better understanding of the issues surrounding energy and its use. Includes an excellent overview of the energy crisis.

11-33 **Encyclopedia of Environmental Science and Engineering.** Ed. by James R. Pfafflin and Edward N. Ziegler. New York: Gordon and Breach, 1976. 2v. illus. index.

This impressive work attempts to combine an overview of environmental areas with an in-depth treatment of specific subjects within these fields. It provides an insight into disciplines that are closely related to the environmentalist and the practicing engineer.

11-34 **Grzimek's Encyclopedia of Ecology.** Bernhard Grzimek, ed. in chief. New York: Van Nostrand Reinhold, 1977 (c.1976). 705p. illus. (part col.). bibliog. index.

A comprehensive work covering in part 1, the environment of animals, and in part 2, the environment of man. All articles are signed; supplementary readings are suggested at the end of the text. This encyclopedia is a translation of the Swiss work published in 1973.

11-35 **McGraw-Hill Encyclopedia of Environmental Science.** New York: McGraw-Hill, 1974. 754p. illus. charts. graphs. index.

This encyclopedia is authoritative, complete, and easy to use. Each article is signed, sometimes by several authors, and treats a different aspect of environmental science. Some of the articles are repeated from the *McGraw-Hill Encyclopedia of Science and Technology* (3-48); however, most of the material is new.

11-36 **New York Times Encyclopedic Dictionary of the Environment.** By Paul Sarnoff. New York: Quadrangle Books, 1971. 352p. illus.

This is a useful dictionary of environmental terms written in easy-to-read style and suited for high school and college undergraduate students.

11-37 **Water Encyclopedia: A Compendium of Useful Information on Water Resources.** Ed. by David Keith Todd. Port Washington, NY: Water Information Center, 1970. 550p.

Based on the valid premise that data in this field are scattered in numerous inaccessible sources. Nine major chapters treat climate and precipitation; hydrologic elements; surface water; ground water; water use; water quality and pollution control; water resources management; agencies and organizations involved in the area of water resources; and constants and conversion factors relating to water configurations.

11-38 **World Energy Book: An A-Z Atlas and Statistical Source Book.** David Crabbe and Richard McBride, consultant eds. and principal contributors. New York: Nichols Publishing, 1978. 259p.

An encyclopedic dictionary to the terminology of energy and related fields, it lists over 1,500 alphabetically arranged entries. Includes extensive cross-referencing. The appendix includes 40 pages of tables, diagrams, graphs, classification scales, and conversion charts relating to energy sources.

DICTIONARIES

11-39 **Dictionary of the Environment.** By Michael Allaby. New York: Van Nostrand Reinhold, 1977. 532p.

This dictionary contains about 6,000 words and phrases currently in use gathered from the many fields that treat environmental concerns. Includes acronyms, proper names, organizations, and laws with an appendix that lists organizations concerned with the environment in the United Kingdom.

11-40 **Dictionary of the Environmental Sciences.** Comp. by Rouert W. Durrenberger. Palo Alto, CA: National Press Books, 1973. 282p. illus.
This dictionary covers a wide range of topics, providing concise definitions with illustrations and diagrams.

11-41 **Energy Dictionary.** By V. Daniel Hunt. New York: Van Nostrand Reinhold, 1979. 518p. bibliog.
Contains definitions of some 4,000 terms related to the production, conservation, and environmental aspects of energy. Definitions are generally brief. There is also a brief overview of the energy situation, a glossary of acronyms, conversion factors, and an extensive bibliography.

11-42 **Standard Terms of the Energy Economy: A Glossary for Engineers, Research Workers, Industrialists and Economists Containing Over 600 Standard Energy Terms in English, French, German and Spanish.** Elmsford, NY: Pergamon, 1978. 134p. index.
This book contains close to 600 terms relating to energy, with definitions in English, French, German, and Spanish. The terms are presented in one of eight sections — general, electricity industry, water power, mining and processing of solid fuels, extraction and refining of liquid fuels, gas industry, nuclear power technology, and impact of energy industries on the environment.

HANDBOOKS

11-43 **Citizen's Guide to Air Pollution.** By David V. Bates; sponsored by the Canadian Society of Zoologists. Montreal and New York: McGill-Queen's University Press, 1972. 140p. illus. index. (Environmental Damage and Control in Canada, no. 2).
This book provides a brief explanation of the many-sided problem of air pollution. After a careful analysis of the various pollutants, their chemical composition and sources, it describes the general effects of each on man. Examples and statistical information are drawn from worldwide sources.

11-44 **Energy Handbook.** By Robert L. Loftness. New York: Van Nostrand Reinhold, 1978. 741p. illus. maps. index.
This comprehensive handbook presents in graphic and narrative form a well-selected portion of the published information on energy and covers energy and man; fossil and mineral energy resources; renewable energy resources; energy consumption trends; energy consumption projections; recovery of fossil fuels; nuclear power; geothermal energy; solar energy; energy conversion and storage; energy efficiency and conservation; energy transport; environmental aspects of energy use; environmental control of energy; energy costs; and energy futures.

11-45 **Energy Sourcebook.** Ed. by Alexander McRae and Janice L. Dudas; Howard Rowland, consulting ed. Germantown, MD: Aspen Systems, 1977. 724p. illus. index. (A Publication of the Center for Compliance Information).

This comprehensive sourcebook covers all forms of energy, both renewable and non-renewable, and quantitatively describes in detail the origin of the energy crunch, forecasts of the supplies of the various alternative resources, and projections of consumption. Fifteen individual sources are discussed, each describing the processes and economics of utilization.

11-46 **Energy Technology Handbook.** Douglas M. Considine, ed. in chief. New York: McGraw-Hill, 1977. 1v. (various paging). illus. index.
This handbook concentrates on fundamental technologies that relate to energy sources, energy reserves, energy conversion, and energy transportation and transmission. It specifically covers trends in the coal, gas, petroleum, chemical fuels, nuclear energy, solar energy, geothermal energy, hydrogen, and power technologies.

11-47 **Environment Regulation Handbook, Nineteen Eighty.** Ed. by Steve Ross. New York: Environment Information Center, 1979. 2500p. monthly updates.
Provides the basic set of federal environmental laws and regulations, and guides the user to the proper law with flow charts, a master index, and major chapter headings. Covers all environmental areas including: air pollution, land use, NEPA, pesticides, radioactive materials, solid waster, toxic substances, and water pollution.

11-48 **Environmental Impact Data Book.** By Jack Golden and others. Ann Arbor, MI: Ann Arbor Science Publishers, 1979. 864p. illus. index.
This book has brought together information that can be difficult and time consuming to collect and will be an invaluable aid to those involved in the preparation of environmental impact statements. The chapters cover techniques for aiding in the assessment process; data bases; models; legal frameworks; air quality; water resources; noise; physical resources; ecosystems; toxic chemicals; cultural; energy; and transportation.

11-49 **Handbook of Environmental Control.** Ed. by Richard G. Bond and Conrad P. Straub. Cleveland, OH: CRC Press, 1973-1978. 6v.
The six volumes cover: v. 1–Air Pollution (1973); v. 2–Solid Waste (1973); v. 3–Water Supply and Treatment (1973); v. 4–Waste Water Treatment and Disposal (1974); v. 5–Hospital and Health-Care Facilities (1975); and v. 6–Series Index (1978). Designed primarily for the specialist, these volumes can be used profitably by all those with more than an elementary knowledge of chemistry. Each volume contains analyses of the several categories of pollutants. The emphasis is on data rather than discussion. The Series Index provides access by subject and chemical substance.

11-50 **Handbook of Environmental Data on Organic Chemicals.** By Karel Verschueren. New York: Van Nostrand Reinhold, 1978. 659p. bibliog.
This reference volume contains information on more than 1,000 organic chemicals, gathered from approximately 350 references published as recently as 1976. Four types of information are included: physical and chemical data, data on air pollution, data on water pollution, and information on the biological effects of organic chemicals on microorganisms, plants, animals, and man. The chemicals are arranged alphabetically. A long introductory chapter provides definitions of terms as well as methods of sampling and measurement.

11-51   **Handbook of Industrial Noise Control.** Ed. by L. L. Faulkner. New
         York: Industrial Press, 1976. 584p. illus. bibliog. index.
This book presents practical engineering solutions to noise reduction. It includes
fundamentals of sound and vibration appropriate for industrial noise control,
fundamentals of sound measurement necessary for effective noise control, and
standardization activities. It is illustrated with fundamental concepts, case
studies, and examples from established engineering practice. Includes extensive
calculations.

11-52   **Handbook of Toxicity of Pesticides to Wildlife.** By Richard K. Tucker
         and D. Glen Crabtree. Washington: U.S. Sport Fisheries and Wildlife
         Bureau; distr. Washington: GPO, 1970. 131p.
A compendium of pesticide toxicity data for wildlife species. Pesticides are
arranged alphabetically by the pesticide's most common name. Listed under each
chemical are variant common names, chemical name, primary use, and purity of
the samples tested. A summary of oral toxicity values follow. Also included are a
glossary of terms and a bibliography.

11-53   **Handbook of Water Resources and Pollution Control.** Ed. by Harry W.
         Gehm and Jacob I Bregman. New York: Van Nostrand Reinhold, 1976.
         840p. illus. index.
A comprehensive guide to the engineering and management problems of
municipal and industrial water supply. All major innovations in water pollution
technology are covered, along with specific data of nearly 200 public water
systems used in the United States. A section-by-section analysis of current
amendments to the federal Water Pollution Control Act is also included.

11-54   **Industrial Pollution Control Handbook.** Ed. by Herbert F. Lund. New
         York: McGraw-Hill, 1971. 1v. (various paging). illus. index.
Intended to be "a communication bridge between the best practical experts on
industrial pollution problems and all of the responsible industrial managers." The
first three chapters present an excellent summary of what the pollution problem is
and how federal, state, and local legislation has come to grips with it. The next six
chapters cover current programs of control and research. Ten chapters then go
into detail and describe pollution control in the steel industry, foundry and
plating operations, metal fabricating plants, the chemical industry, textile mills,
food industries, pharmaceutical industries, the pulp and paper industry, and the
aerospace and electronics industry. There are five detailed chapters on control
equipment and one on operating costs.

11-55   **Water and Water Pollution Handbook.** Ed. by Leonard L. Ciaccio. New
         York: Marcel Dekker, 1941. 4v. illus. bibliog. index.
This four-volume treatise covers the range of water pollution problems. Part 1
treats environmental systems and includes discussions on the chemical, physical,
and biological characteristics of water resources; estuaries, irrigation and soil
waters; and wastes and waste effulents. This gives the reader background for part
2, which deals with chemical, physical, bacterial, viral, instrumental, and
bioassay techniques needed in analyzing water. The handbook is intended for
those interested in water environment, including personnel in waste treatment
and water purification plants; the transportation industry; service industries;
government enforcement agencies; government, industrial, and university
research laboratories; and sanitary, civil, and consulting engineers.

DIRECTORIES

11-56 **Conservation Directory: A Listing of Organizations, Agencies, and Officials Concerned with Natural Resource Use and Management.** Washington: The National Wildlife Federation, 1956- . annual.

This annual directory covers organizations and agencies in the field of conservation. Its four major sections cover the United States government; international, national, and interstate organizations and commissions; state and territorial agencies and citizens' groups; and government agencies and citizens' groups in Canada. In addition to listing name, address, and telephone number of each agency, it also provides the names of the top officials and statement of purpose.

11-57 **Energy: A Guide to Organizations and Information Resources in the United States.** 2nd ed. Claremont, CA: Public Affairs Clearinghouse, 1978. 221p. index.

This guide describes more than 1,500 organizations concerned with energy policy, research, production, and use. In addition to a short description, each entry lists full name, address, and telephone number. Some of the entries include names of key officers, organization structure, activities, and publications.

11-58 **Energy Atlas: A Who's Who Resource to Information.** Washington: Fraser/Ruder and Finn, 1976. 1v. (various paging).

A directory to some 50 federal units and agencies, 100 congressional committees, state executive and legislative bodies, and 150 non-governmental groups in the field of energy. Each entry provides name and address, statement of jurisdiction, and key personnel.

11-59 **Energy Directory Update.** New York: Environment Information Center, 1975- . bimonthly.

A comprehensive guide to energy organizations, decision-makers and information sources. Covers state and federal government; trade, professional, and research organizations; oil, gas, coal, electric, and nuclear companies; information centers, libraries, and publications. Entries provide names, titles, addresses, phone numbers, statements of purpose, programs, projects, publications, advisory board members, and maps and organization charts.

11-60 **Energy Information Locator.** New York: Environment Information Center, 1975- . annual.

This is chapter 05 of the *Energy Directory Update* (11-59). It focuses on organizations that provide a formal energy-related information service or system, and on publications, such as journals, directories, newsletters, binder services, and abstracting and indexing services, that deal in some way with energy. Address and telephone number are provided for each entry, along with a description of objectives, activities, type of collection, services, publications, and the name of an individual designated the energy information contact.

11-61 **Federal Energy Information Sources and Data Bases.** By Carolyn C. Bloch. Park Ridge, NJ: Noyes Data Corporation, 1979. 115p. index.

The directory is divided into four sections: "Cabinet Departments," "Administrative Agencies," "Quasi Government Agencies," and "Congressional Offices." For each department or agency included in these sections, there is information given on libraries, data bases, information centers, projects, special services, and publications that deal with energy. Each entry includes a description and an address to contact the service.

11-62 **Guide to Ecology Information and Organizations.** By John Gordon Burke and Jill Swanson Reddig. New York: H. W. Wilson, 1976. 292p. index.

An annotated guide to information on ecology, including citizen action guides, indexes, reference books, histories, monographs, government publications, nonprint media, periodicals, organizations, and government officials.

11-63 **Leaders of American Conservation.** Ed. by Henry Clepper. New York: Ronald Press, 1971. 353p.

Presents 300 biographical sketches of contributors to the preservation of our natural resources. The entries average a page in length, are carefully edited, and, in general, are prepared by writers who knew the leaders and are themselves active in the field.

11-64 **NFEC Directory of Environmental Information Sources.** Ed. by Charles E. Thibeau. 2nd ed. Boston: National Foundation for Environmental Control; distr. Boston: Cahners, 1972. 457p.

Lists government agencies, trade organizations, citizens' organizations, professional and occupational organizations dealing with environmental concerns. A separate chapter on educational institutions describes interdisciplinary environmental programs offered by universities and colleges, but listings in this particular chapter are far from complete. The second part of the work lists information sources—that is, abstracting service, directories, and indexes; conference and symposium proceedings; documents and reports; serials; some books and films.

11-65 **Nuclear Research Index: A Guide to World Nuclear Research.** 5th ed. Guernsey, British Isles: F. Hodgson, 1976. 682p.

Lists organizations, institutes, laboratory facilities, etc., in all countries with nuclear involvement. Provides details on such particulars as staff, program, funding, primary research interests. Fourth edition was called *World Nuclear Directory: An International Reference Book* (New York: Van Nostrand Reinhold, 1971. 764p.).

11-66 **Solar Directory.** Ed. by Carolyn Pesko. Ann Arbor, MI: Ann Arbor Science, 1975. 1v. (various paging). illus. index.

A computer-produced directory listing corporations, individuals, academic institutions, and government agencies dealing with solar radiation. Each entry provides address, telephone, contact personnel, and statement of purpose.

11-67 **Solar Energy Update.** New York: Environment Information Center, 1975- . annual.

This is chapter 07 of the *Energy Directory Update* (11-59). It covers more than 200 organizations and publications in the solar energy field including congressional committees, federal and state government agencies, and trade and professional associations. Factual data include names, titles, addresses, phone numbers, statements of purpose, programs, and projects. Indexes provide title, subject, and geographic access.

12

Engineering

Engineering is the application of data that has been gathered from other branches of theoretical science. The sophisticated elaborations of today's engineering sciences are based and developed from the pure sciences. Mathematics and physics underlie all engineering disciplines. Astronomy provides the theoretical base for much that is done in aeronautical engineering. Chemistry, combined with physics, is at the heart of chemical industrial engineering. The life sciences contribute to the engineering of medical technology. Finally, mathematics, logic, and electronics are the primary disciplines on which computer engineering is based. The history of engineering is therefore a composite history of all its component science disciplines.

Since engineering is broad in its coverage, this chapter is divided into the following sections: General Engineering, Mechanical and Electrical Engineering, Production and Processing Engineering, Construction Engineering and Transportation Engineering. The literature could be divided several other ways, but this arrangement works the best for this edition. It will soon become apparent to the reader that the engineering literature is highly specialized. Each field has its own core of literature; handbooks are prolific. This chapter selectively lists the most important sources.

In today's sophisticated technological fields there is an engineer for every application. In the construction fields there are construction, bridge, railroad, municipal, hydraulic, sanitary, highway, water-supply, harbor, structural, architectural, tunnel, and agricultural engineers. In production and processing one could find geological, petroleum, quarry, chemical, ceramic, nuclear, and metallurgical engineers; in mechanical operations there are power, ordnance, refrigeration, illuminating, radio, electronics, telephone, machine, and control engineers. Engineers associated with transportation include automotive, marine, aeronautical, electric, railway, naval, aerospace, highway, railroad, tunnel, and nuclear. Engineering is, clearly, a complex discipline.

The literature of engineering is equally as complicated and diverse. The forms of literature essential to engineering are similar to those of the pure sciences. Like the medical practitioner, the engineer needs a wide variety of specialized handbooks. Standards, specifications, and report literature are particularly important to the engineer. Because the field of engineering is broad, the materials selected for inclusion in this chapter are selective.

GENERAL ENGINEERING

Guides to the Literature

12-1   **Guide to Basic Information Sources in Engineering.** By Ellis Mount. New York: Wiley, 1976. 196p. index. (A Halsted Press Book; Information Resources Series).

A small book for the engineering student and researcher covering in four broad categories: Technical Literature—What It Is, Where to Find It; Books; Periodicals and Technical Reports; and Other Sources of Information. Under each category there are entries for bibliographies, dictionaries, encyclopedias, handbooks, guides to literature, histories, etc.

12-2   **How to Find Out about Engineering.** By S. A. J. Parsons. New York: Pergamon, 1972. 271p. index.

A volume in the Pergamon series detailing general sources of information and use of libraries and organizations. Includes several sections documenting the sources of mechanical, electrical, nuclear, mining, civil, and aeronautical engineers.

12-3   **Use of Engineering Literature.** Ed. by K. W. Mildren. Woburn, MA: Butterworths, 1976. 621p. illus. index. (Information Sources for Research and Development).

This handbook represents a survey of the fields of engineering, covering electronics, communications, control engineering, aeronautics and astronautics, chemical engineering, production engineering, and soil engineering. Each of the chapters includes information on classification and indexing, journals, conferences and theses, translations, reports, patents, standards, product information, and selected coverage of publications of government and international organizations.

Abstracts, Indexes, and Bibliographies

12-4   **Applied Science and Technology Index.** New York: H. W. Wilson, 1913-. Monthly (except August) with quarterly and annual cumulations. From 1913 to 1957 known as *Industrial Arts Index.*

A subject index to about 225 English language periodicals covering both the theoretical sciences and their engineering applications. It includes pure physics, chemistry and geology, mathematics, metallurgy and computer science insofar as these subjects are represented by some of the major scientific periodicals in their fields. It also provides substantial coverage of science applications, that is, the various subfields of engineering.

12-5   **Classed Subject Catalog of the Engineering Societies Library, New York.** Boston: G. K. Hall, 1963. 12v.

12-5a   **Index to the Classed Subject Catalog.** 1963. 356p.

12-5b   **Supplements.** 1964-1974. 10v.

12-5c   **Bibliographic Guide to Technology.** 1975- . annual.

The original 12 volumes include the library's 185,000 volume collection of books, journals, films, technical reports, and unpublished manuscripts in all fields of engineering. The supplements, now called *Bibliographic Guide to Technology*, appear on an annual basis, making this a prime source of current bibliography inasmuch as the Engineering Societies Library collects comprehensively in its fields of interests. These interests extend beyond traditional engineering into the geoscience subfields of geology and geophysics.

12-6   **Engineering Index.** v. 1- . New York: Engineering Index, Inc., 1884- .
        monthly with annual cumulation.

This is an English language abstracting service in engineering, including all of the subdivisions. It indexes more than 2,050 journals and all pertinent reports, symposium papers, patents, books, and miscellaneous serials published in 20 or more languages. Indicative abstracts are arranged in classified subject order. Each entry provides a complete citation followed by an abstract. Engineering Index, Inc., also maintains a machine-readable data base with partial funding by the National Science Foundation. Its services, in addition to the printed *Engineering Index* include:

COMPENDEX, a computerized magnetic tape version of *Engineering Index*, issued well in advance of the printed versions.

Monthly, a monthly set of magnetic tapes that can be machine searched immediately for current awareness, or later in a retrospective mode.

Engineering Index Microfilm Edition.

Card-A-Lert, a weekly current awareness service issued on 3x5-inch cards.

12-7   **New Technical Books: A Selected List on Industrial Arts and Engineering Added to the New York Public Library.** v. 1- . New York: New York
        Public Library, 1915- . 10 issues per year.

Covers those books in the pure and applied physical sciences, mathematics, engineering, industrial technology, and related fields published in the United States, plus some noteworthy foreign books. Arranged by DDC, entries are numbered in consecutive order through the ten issues for the year. A serviceable classified list.

Encyclopedias

12-8   **Encyclopaedic Dictionary of Mathematics for Engineers and Applied Scientists.** By I. N. Sneddon. New York: Pergamon, 1976. 800p. illus.

"The aim has been to select those mathematical concepts and techniques which are most widely and frequently used in engineering and by an extensive cross-reference system to bind together a vast amount of information giving easy access to the fundamental definitions and main results of each of the major branches of mathematics." Entries are from two or three lines to several pages and most have a brief bibliography.

12-9   **Encyclopedia/Handbook of Materials, Parts and Finishes.** Ed. by
        Henry R. Clauser. Westport, CT: Technomic, 1976. 564p. illus. index.

This handy encyclopedia brings together in one volume concise information on the many thousands of materials, material forms and parts, and finishes used in modern industry. Information on production, size and shape characteristics, and design capabilities is given for materials parts and forms.

12-10   **Encyclopedia of Engineering Materials and Processes.** Ed. by H. R.
        Clauser. New York: Reinhold, 1963. 787p. diagrams. tables. graphs.

Materials treated range from ABS plastics to zirconium; processes discussed include inter alia, heat treating, welding, electroplating, and tanning. Excellent cross-references connect the 300 articles, written by more than 200 contributors.

12-11   **Engineering Encyclopedia: A Condensed Encyclopedia and Mechanical Dictionary for Engineers, Mechanics, Technical Schools, Industrial Plants and Public Libraries, Giving the Most Essential Facts about 4,500 Important Engineering Subjects.** By Franklin D. Jones and Paul B. Schubert. 3rd ed. New York: Industrial Press, 1963. 1431p. illus.

An excellent work for students and for practitioners needing information in a subfield not their own.

Dictionaries

12-12   **Engineering Index Thesaurus.** New York: CCM Information Corporation, 1972. 402p.

Includes more than 11,800 terms or descriptors covering plastics, electrical/electronics, aeronautics, astronautics, chemical, civil and mechanical engineering, fluid and solid mechanics, mathematics, nuclear, plasma and solid-state physics, optics, statistics, and systems engineering. Terms are listed in sets, each consisting of a main term followed by cross-reference to broader, narrower, or related terms. This is the thesaurus used by Engineering Index, Inc. in the maintenance of its data base.

12-13   **SHE: Subject Headings for Engineering, 1972.** New York: Engineering Index, 1972. 149p.

12-13a   **Supplement.** 1977. 30p.

"SHE (Subject Headings for Engineering) is an alphabetical list of terms currently in use by EI technical editorial specialists as a controlled vocabulary for indexing transdisciplinary literature of engineering and related sciences." This is the basic tool for indexers in the area of engineering. Includes many cross-references and scope notes.

12-14   **Trilingual Dictionary for Materials and Structures: Dictionnaire trilingue des materiaux et des constructions; Dreisprachiges Worterbuch der Werkstoffe und Konstruktionen.** Oxford: Pergamon, 1971. 947p.

An exception to the general practice of multilingual dictionaries, this work provides definitions. It appeared under the sponsorship of the International Union of Testing and Research Laboratories (RILEM).

Handbooks

12-15   **Book of ASTM Standards.** Philadelphia, PA: American Society for Testing Materials, 1939- . annual. illus. index.

The 48 parts "contain all current formally approved ASTM standard and tentative specifications, methods of test, recommended practices, definitions, and other related material such as proposed methods." All parts are revised annually, and each new edition supersedes the previous edition. An ASTM standard represents a common viewpoint of those parties concerned with its provisions — namely, producers, consumers, and general interest groups. It is intended to aid industry, government agencies, and the general public. The use of an ASTM standard is purely voluntary. Those standards approved by the American National Standards Institute and those adopted by or under consideration for adoption by the Boiler and Pressure Vessel Committee of the American Society of Mechanical Engineers are so marked.

12-16   **CRC Handbook of Tables for Applied Engineering Science.** Ed. by
Ray E. Bolz and George L. Tuve. 2nd ed. Cleveland, OH: CRC Press,
1973. 1166p.
Like all other CRC handbooks, this one follows the rule of comprehensive
coverage. Each table is well documented and has supporting text where necessary.
This edition has many new tables for such subjects as lasers, radiation,
cryogenics, ultrasonics, semiconductors, high-vacuum techniques, eutectic
alloys, and organic and inorganic surface coatings. The subjects treated are electrical science and radiation, chemistry and applications, nuclides and nuclear
engineering, energy engineering and transport, mechanics structures and
machines, environmental and bioengineering, environmental protection and
human safety, communication and computation, measurement and instrumentation, and processes and control. An appendix lists engineering organizations and
publishers.

12-17   **Directory of Engineering Document Sources.** Comp. by D. P. Simonton.
Newport Beach, CA: Global Engineering Documentation Service, 1972.
1v. (various paging).
A consolidated cross-index of document initialisms assigned by governmental
and industrial organizations to technical reports, specifications, standards, and
related publications in engineering.

12-18   **Engineering Formulas.** By Kurt Gieck. 2nd ed. New York: McGraw-Hill,
1976. 1v. (various paging). illus. index.
A pocket-sized guide providing engineers and scientists with important technical
and mathematical formulas in a brief and handy format. It covers tables for
units, areas, solid bodies, arithmetic, functions of a circle, analytical geometry,
hyperbolic functions, differential calculus, integral calculus, statics, kinematics,
dynamics, hydraulics, heat, strength, machine parts, machine tools, electrical
engineering, optics, and chemistry.

12-19   **Engineering Manual: A Practical Reference of Design Methods and Data
in Building Systems, Chemical, Civil, Electrical, Mechanical, and
Enrivonmental Engineering and Energy Conversion.** Ed. by Robert H.
Perry. 3rd ed. New York: McGraw-Hill, 1976. 1v. (various paging). illus.
index.
An excellent survey of all fields of engineering specialization. Sets forth most of
the key concepts, facts, formulae, and tables needed in the practice of engineering
today. It is not a replacement of the individual handbooks of each discipline.

12-20   **Engineering Materials Handbook.** Ed. by Charles L. Mantell. New
York: McGraw-hill, 1958. 1892p. diagrams. tables. (McGraw-Hill
Handbooks).
"Intended to supply the practicing engineer, the designer, the student, the archi-
tect, the purchasing agent with an authoritative work."

12-21   **Engineering Mathematics Handbook: Definitions, Theorems, Formulas,
Tables.** By Jan J. Tuma. New York: McGraw-Hill, 1971. 334p. illus.
index.
Contains a wealth of useful mathematical formulas and data that the practicing
engineer is likely to need. It is arranged into 20 sections or chapters, each of

which treats a specific mathematical topic in a wide spectrum of topics – for example, algebra, geometry, differential and integral calculus. The specific engineering application of each is clearly described.

12-22   **Engineering Tables and Data.** By A. M. Howatson, P. G. Lund, and J. D. Todd. London: Chapman and Hall; distr. New York: Halsted Press/Wiley, 1972. 167p. illus. bibliog.

Directed primarily to engineering students, but also of use to practicing engineers, the work includes material on mathematics, properties of matter, thermodynamics and fluid mechanics, elasticity and structures, mechanics, and electricity.

12-23   **Engineer's Companion: A Concise Handbook of Engineering Fundamentals.** By Matt Souders. New York: Wiley, 1966. 426p. tables.

Directed both to beginning students and to those practitioners in need of refresher instruction in one or more subfields of engineering.

12-24   **Handbook of Dimensional Measurement.** By Francis T. Farago. New York: Industrial Press, 1969. 416p. illus. tables. index.

Presents the methods and equipment of metrology for modern industrial production, with more than 480 tables with diagrammatic illustrations and brief synopses for rapid scanning and review. Contents include: line-graduated instruments, fixed gauges, gauge blocks, comparative length measurements with mechanical indicators, pneumatic gauging, electronic gauges, engineering microscopes, optical projectors, angle measurements, the systems and applications of measuring machines, profile measurements, the measurement of roundness and circular contours, and surface texture measurements.

12-25   **Handbook of Engineering Fundamentals.** Ed. by Ovid Wallace Eshbach and Mott Souders. 4rd ed. New York: Wiley, 1975. 1562p. illus. index. (Wiley Engineering Handbook Series).

A ready reference to the basic ideas of engineering sciences and mathematics for beginning students as well as those requiring refresher training.

12-26   **Handbook of Tables for Applied Engineering Sciences.** 2nd ed. Cleveland, OH: Chemical Rubber Co., 1973. 1166p. illus.

A comprehensive handbook of useful tables for the engineer.

12-27   **Industrial Safety Handbook.** Ed. by William Handley. 2nd ed. New York: McGraw-Hill, 1977. 480p. illus. (part col.). index.

This practical guide for managers and supervisors emphasizes teaching safety practices and includes topics on conducting safety assessments of potentially hazardous materials, controlling exposure to all kinds of hazards, and instructions on how to create and maintain safe working conditions.

12-28   **Maintenance Engineering Handbook.** Ed. by Lindley R. Higgins and L. C. Morrow. 3rd ed. New York: McGraw-Hill, 1977. 1v. (various paging). illus. index.

This basic work covers, in addition to the general material on maintenance, pollution control equipment, plant facility security, fuel conservation management, and the selection, installation, and upkeep of equipment and services.

12-29 **Material Handbook: An Encyclopedia for Managers, Technical Professionals, Purchasing and Production Managers, Technicians, Supervisors, and Foremen.** By George S. Brady and Henry R. Clauser. 11th ed. New York: McGraw-Hill, 1977. 1011p. illus. index.

This well-known handbook describes 14,000 commercially available substances in terms of their composition, methods of production, major properties, characteristics, and uses. Among the types of materials covered are industrial materials, chemicals, food stuffs, plastics and rubbers, and naturally occurring substances.

12-30 **Practicing Scientist's Handbook: A Guide for Physical and Terrestrial Scientists and Engineers.** By Alfred J. Moses. New York: Van Nostrand Reinhold, 1978. 1292p. index.

This handbook is intended to provide a comprehensive source of materials' property data to practicing physicists, chemists, engineers and designers by pulling together numerous tables, charts, and diagrams covering properties of the elements, organic compounds, inorganic compounds, alloys, glasses and ceramics, composites, polymers and adhesives, semiconductors, superconductors, environmental, and miscellaneous materials.

12-31 **Standard Handbook of Engineering Calculations.** Ed. by Tyler Hicks. New York: McGraw-Hill, 1972. 1v. (various paging). illus. index.

Contains more than 2,000 step-by-step calculation procedures for solving almost all routine, and many non-routine, problems met in everyday engineering practice. Each of the 12 sections covers those calculations necessary for solving problems in each of 12 fields of engineering, from aeronautical to sanitary. Problem solutions are exhibited with all diagrams and formulas needed for their full comprehension.

Directories

12-32 **American Engineers of the Nineteenth Century: A Biographical Index.** By Christine Roysdon and Linda A. Khatri. New York: Garland, 1978. 247p. (Garland Reference Library of Social Science, v. 53).

Listed alphabetically, each entry gives birth and death dates, plus a statement of field of interest. References to obituaries or biographies follow.

12-33 **Directory of Engineering Societies and Related Organizations.** By the Engineers Joint Council. 9th ed. New York: Engineers Joint Council, 1979. 236p.

A directory of basic information on over 300 national, regional, state, and local organizations primarily engaged in engineering. Each entry gives purposes of the society, address, telephone number, administrators, current officers, date of founding, number of staff, library services, geographic regions, local sections, student chapters, annual budget, dues and membership, publications, membership requirements and qualifications.

12-34 **Directory of Testing Laboratories: Commercial, Institutional.** 6th ed. Philadelphia, PA: American Society for Testing and Materials, 1975. 59p.

Lists U.S. laboratories with the capability of testing materials and commercial products. Arranged alphabetically, geographically, and by specific commodity tested.

12-35   **Who's Who in Engineering.** 3rd ed. New York: Engineering Joint Council, 1977. 605p.

This is a who's who type of directory providing reasonably complete biographic data on some 4,500 engineers. Appended are sections on engineering societies and related organizations, a society index, and an awards index. It was formerly called *Engineers of Distinction.*

12-36   **Who's Who of British Engineers, 1974-1975.** Ed. by R. A. Baynton. Athens, OH: Ohio University Press, 1974. 526p.

This edition contains new names plus all active entrants from previous editions. Each entry lists full name, title, publications, professional interests, address, birth date, and degrees.

12-37   **World Space Directory, Including Oceanology.** v. 1- . Washington: Ziff-Davis, 1962- . illus.

"A complete reference handbook of government agencies, major and component space/oceanology manufactures, consultants and special services, industrial representatives, colleges/universities, nonprofit research organizations and publications in the space/oceanology industry. A special international section lists the same information for foreign countries."

MECHANICAL AND ELECTRICAL ENGINEERING

Guides to the Literature

12-38   **Guide to Literature on Mechanical Engineering.** By J. K. K. Ho. Washington: American Society for Engineering Education, 1970. 18p.

An effective little guide in ten sections, dealing with the secondary and primary literature forms discussed in chapter 2.

12-39   **How to Find Out in Electrical Engineering: A Guide to Sources of Information Arranged According to the Universal Decimal Classification.** By Jack Burkett and Philip Plumb. Oxford: Pergamon, 1967. 234p. illus. index.

This work gives the reader good control over the literature. An introductory chapter on careers is followed by sections on the organization, acquisition, and selection of materials.

12-40   **Mechanical Engineering: The Sources of Information.** By Bernard Houghton. Hamden, CT: Archon, 1970. 311p.

A handy guide to sources of information on mechanical engineering and related disciplines, specifically research in the United Kingdom, trade associations in the United Kingdom, research in the United States, periodical literature, subject directories, documents, basic reference tools, and bibliographic control.

Abstracts, Indexes, and Bibliographies

12-41 **Applied Mechanics Review, Assessment of World Literature in Engineering Science.** v. 1- . Easton, PA: American Society of Mechanical Engineers, 1948- . monthly.

A critical review of the world's primary and secondary literature in applied mechanics and related sciences. Arranged by subject. Most abstracts are short, but those for books may be two or more pages in length. If the original article is not in English, the original language is indicated. Many of the abstracts for Russian articles have been translated from *Referativnyi Zhurnal* (3-17). In addition to the abstracts, there is a feature article in each issue and a section on book reviews. All articles abstracted may be obtained from the Linda Hall Library in Kansas City, Missouri.

12-42 **Current Papers in Electrical and Electronics Engineering.** v. 1- . London: Institution of Electrical Engineers, 1964- . monthly.

Contains references to every item abstracted in *Electrical and Electronics Abstracts* (12-43), listing items in the same classification scheme. As a listing, rather than an abstracting service, it appears much sooner and hence serves the purpose of current awareness.

12-43 **Electrical and Electronics Abstracts.** v. 1- . London: Institution of Electrical Engineers. 1898- . monthly. (Science Abstracts/Series B).

Groups electrical and electronics engineering into general topics, circuits and electronics, electron devices and materials, electromagnetics and communications, instrumentation and special applications, and power and industry. For each entry complete bibliographic data are given. More than 40,000 items in the primary and secondary literature categories per year are abstracted. Includes author and subject indexes.

12-44 **Electronics and Communications Abstracts Journal.** v. 1- . Riverdale, MD: Cambridge Scientific Abstracts, 1967- . 10 issues per year.

An international abstracting service covering electronic physics, electronic systems, electronic circuits, electronic devices, and communications. Indexed sources are periodicals, government reports, conference proceedings, books, dissertations and patents. Approximately 5,000 items per annum are abstracted.

12-45 **Fluid Power Abstracts: Worldwide Coverage of Hydraulics and Pneumatics Literature.** v. 1- . Cranfield, Bedford, England: British Hydromechanics Research Association, 1960- . monthly.

Devoted exclusively to abstracts of hydraulics and pneumatics taken from more than 400 technical journals.

12-46 **Fluidics Feedbacks; Abstracts, Reviews, Industrial News, Products, Patents: A Monthly Review of the World of Fluidics.** v. 1- . Cranfield, Bedford, England: British Hydromechanics Research Association, 1967- . monthly.

An excellent specialized bibliographic and reporting service in hydraulic engineering, surveying the world's primary literature and producing about 800 abstracts each year. Annual subject, personal, and corporate author indexes. Its several sections cover circuits and circuit components, general topics, devices of moving parts, non-moving part devices, applications, and new products.

12-47 **ISMEC Bulletin.** v. 1- . London: Institution of Mechanical Engineers, 1973- . monthly.

The acronym ISMEC designates Information Service in Mechanical Engineering. As the full title emphasizes, the thrust of this service is speed of dessemination. The *Bulletin* concentrates on presenting a classified list of current papers, documents, articles, reports and patents published throughout the world in all phases of mechanical engineering. Semiannual author and subject indexes are available.

12-48 **Laser Literature: An Annotated Guide.** By Keyo Tomiyasu. New York: Plenum Press, 1968. 172p. index.

From 1965 to 1967, a series of bibliographies were published in the *IEEE Journal of Quantum Electronics*. The present is an enriched derivation from them. Organized in 27 sections, the work lists about 4,000 papers, reports, articles, and monographs.

12-49 **Semiconductor Abstracts: Abstracts of Literature on Semiconducting and Luminescent Materials and Their Applications.** Comp. by the Battelle Memorial Institute, Solid State Devices Div. and sponsored by the Electro-Chemical Society, Inc. New York: Wiley, 1951-1959. 7v.

Though ceased, these seven volumes remain valuable for retrospective searching.

Encyclopedias

12-50 **Encyclopaedia of Hydraulics, Soil and Foundation Engineering.** By Ernst Vollmer. New York: American Elsevier, 1967. 398p. illus. tables.

Subjects are hydraulic engineering, hydromechanics, river training weirs and dams; water resources; soil engineering, soil mechanics, soil excavation, soil deposition, and foundation excavation in dry soil; foundation materials behavior in soil symbols, abbreviations and conversion factors.

12-51 **Encyclopedia of Instrumentation and Control.** Ed. by Douglas M. Considine. New York: McGraw-Hill, 1972. 788p. illus. index.

A fairly comprehensive and well-organized compilation of material on instrumentation and control. Conceived as a bridge between current knowledge and projected developments.

12-52 **Water Supply Engineering.** By Harold E. Babbitt. 6th ed. New York: McGraw-Hill, 1962. 672p. tables. index.

A text for undergraduate students, a reference work for graduate students, and a handbook for the practicing engineer. Contains 27 chapters ranging from methods of water analysis to the management of waterworks.

Dictionaries

12-53 **Dictionary of Electrical Engineering, Telecommunications and Electronics.** By W. Goedecke. London: Pitman, 1966. 3v.

This is a multilingual dictionary covering all the terms associated with the fields of electrical engineering, telecommunications, and electronics. The three volumes cover: v. 1 — German-English-French; v. 2 — French-English-German; and v. 3 — English-German-French.

12-54   Dictionary of Electronics. By Harley Carter. 2nd ed. Blue Ridge Summit, PA: TAB Books, 1972. 410p. illus.
A broad view of definitions in the fields of radio, TV, communications, radar electronic instrumentation, broadcasting, and industrial electronics. The terms are alphabetically arranged and the definitions aim at providing more acceptable explanations of many familiar electronic phenomena, and at helping readers understand some less familiar electronics terminology.

12-55   Dictionary of Mechanical Engineering. By J. L. Nayler. 2nd ed. London: Butterworths, 1978. repr. of 1975 ed. 410p.
The most commonly used terms in the literature of mechanics are chosen for precise definition.

12-56   Dictionary of Refrigeration and Air Conditioning. By K. M. Booth. New York: American Elsevier, 1970. 315p.
Includes the basic terminology of refrigeration and air conditioning. All entries are arranged in one alphabet, with many cross-references. It should be noted that all references to gallons and tons are to United Kingdom Imperial units and not to American units, unless otherwise stated.

12-57   Dictionary of Telecommunications. By R. A. Bones. New York: Philosophical Library, 1970. 200p. illus.
Of the commonly encountered terms in the fast expanding field of telecommunications, custom has determined which term is to be defined with cross-reference from the least common term. Many of the terms and definitions are either reproduced from, or based on, British Standard 204:1960, "Glossary of Terms Used in Telecommunications (including Radio) and Electronics." Terms consisting of more than one word are fully cross indexed. Definitions are brief and to the point.

12-58   Dictionary of Water and Sewage Engineering. By Fritz Meinck and Helmut Mohle. 2nd rev. and enlarged ed. Amsterdam, New York: Elsevier Scientific Publishing, 1977. 737p. index.
This is a two part multi-language dictionary with no definitions. Part 1 is arranged by the German term with the corresponding term in English, French, and Italian. Part 2 is an index arranged by English, French, and Italian terms.

12-59   Drake's Radio-Television Electronic Dictionary. Comp. by Harold P. Manly. New York: Drake, 1971. 1v. (unpaged). illus.
Includes terms in radio transmission and reception, monochrome and color television, transistors, audio systems, high fidelity, and electricity and magnetism. A revision of an older work published under the title *Electrical and Radio Dictionary* (Chicago: American Technical Society, 1950. 133p.).

12-60   Electronics Dictionary. By John Markus. 4th ed. New York: McGraw-Hill, 1978. 745p.
Defines over 17,000 terms in the language of electronics as it is written and spoken today.

12-61   Elsevier's Telecommunication Dictionary in Six Languages: English/ American-French-Spanish-Italian-Dutch-German. Comp. by

W. E. Clason. 2nd ed. rev. New York: Elsevier Scientific Publishing, 1976. 604p.

A language dictionary arranged by the English term with the corresponding equivalent in French, Spanish, Italian, Dutch, and German. One of 33 descriptors is added to indicate what area of telecommunication each term applies to.

12-62   **Glossary: Water and Wastewater Control Engineering.** Prepared by Joint Editorial Board. New York: The Joint Editorial Board, 1969. 387p.

The Joint Editorial Board represents the American Public Health Association, the American Society of Civil Engineers, the American Water Works Association, and the Water Pollution Control Federation. The terms are arranged in straight alphabetical order. More than one definition is given if applicable.

12-63   **IEEE Standard Dictionary of Electrical and Electronics Terms.** Ed. by Frank Jay. 2nd ed. New York: Institute of Electrical and Electronics Engineers; distr. New York: Wiley-Interscience, 1977. 882p.

An excellent dictionary of terms and definitions standardized previously by IEEE, as well as many American National Standards and IEC terms. It is indispensable for engineers, technicians, teachers, students, editors, writers, and publishers.

12-64   **Modern Dictionary of Electronics.** By Rudolf F. Graf. 5th ed. Indianapolis, IN: Howard W. Sams, 1977. 832p. illus.

Over 20,000 electronic terms are listed in this comprehensive dictionary. The clear, concise definitions cover the fields of communications, microelectronics, fiberoptics, semiconductors, reliability, computers, and medical electronics.

12-65   **Semiconductors International Dictionary in Seven Languages.** Ed. by D. F. Dunster. Milan: Franco Angeli Editore; London: Business Books Ltd.; distr. Boston: Cahners, 1971. 2v. illus. index. (International Dictionaries of Science and Technology).

An impressive compilation covering all phases of semiconductors. Each term or concept is briefly defined with an added illustration or table, if necessary. All entries are in alphabetical order by the English term, with the corresponding term in German, French, Italian, Portuguese, Russian, and Spanish. The International System of Units has been used throughout. Four tables cover fundamental constants; units and quantities; signs, symbols, and abbreviations; and basic circuit symbols for semiconductor devices.

12-66   **Technical Dictionary of Heating, Ventilation, and Sanitary Engineering: English/German/French/Russian.** Ed. by Wolfgang Lindeke. New York: Pergamon, 1970. 182p.

Equivalents of about 5,300 terms are given in four languages. The English language section consists of an alphabetic list of English words and phrases, some with synonyms, preceded by a letter and number code for term identification and followed by the equivalent term in German, French, and Russian. The other three sections are alphabetic lists of German, French, and Russian terms with letter number code referring to the English equivalent.

Handbooks and Laboratory Guides

12-67 **ASHRAE Handbook of Fundamentals.** New York: American Society of Heating, Refrigerating and Air-Conditioning Engineers, 1972. 688p. illus. diagrams.

12-67a **ASHRAE Handbook and Product Directory.** New York: American Society of Heating, Refrigerating and Air-Conditioning Engineers, 1973-. illus. diagrams.

These two titles provide reference data pertaining to equipment for heating, refrigeration, ventilation, and air conditioning. The second title supplements the first, was formerly called *ASHRAE Guide and Data Book*, and is issued in alternate issues with the subtitles *Systems, Applications*, and *Equipment*. The papers and reports are based on transactions of the Society, ASHRAE research programs, and research and practices of the members of the Society and cooperating institutions. It covers basic theory, general engineering data, basic materials, basic tables, and terminology.

12-68 **ASME Handbook.** New York: McGraw-Hill, 1953-1958. 4v. illus.

12-68a **ASME Handbook.** v. 1- . 2nd ed. New York: McGraw-Hill, 1965- .

The American Society of Mechanical Engineers is in the process of revising this handbook. The volumes will cover metals engineering design, engineering tables, metals engineering process, and metals properties. Each volume is complete in itself and may be purchased separately.

12-69 **Electronic Components Handbook.** By Thomas H. Jones. Reston, VA: Reston Publishing, A Division of Prentice-Hall, 1978. 391p. illus. index.

Intended as a guide for engineers and technicians using military and commercial electronic components, this handbook covers capacitors, resistors, relays, connectors, semi-conductors, etc. Each chapter contains definitions, a reading list, a sample list of manufacturers, charts, photographs, component characteristics, and similar data.

12-70 **Electronic Design Data Book.** By Rudolf F. Graf. New York: Van Nostrand Reinhold, 1971. 312p. illus. index.

In order to bring together published and unpublished data that are needed by the engineer, technician, amateur, and student in the field of electronic design, the author has compiled an easy-to-use volume of nomograms, tables, charts, and formulas with as little theory as possible. The material is grouped into six sections: frequency data; communication; passive components and circuits; active components and circuits; mathematical data, formulas, and symbols; and physical data.

12-71 **Electronics Designers' Handbook.** Ed. by L. J. Giacoletto. 2nd ed. rev. New York: McGraw-Hill, 1977. 1v. (various paging). illus. index.

This excellent book for the professional engineer covers electronic components, circuit analysis, and circuit design. Each of the chapters covers a separate topic, such as filters, vacuum tubes, solid state devices, power amplifiers, computer-aided design and analysis, information theory, and reliability/availability of systems.

12-72 **Handbook for Electronics Engineering Technicians.** Ed. by Milton Kaufman and Arthur H. Seidman. New York: McGraw-Hill, 1976. 1v. (various paging). illus. index.

This is the first fully comprehensive handbook designed to meet the day-to-day needs of electronics technicians. Through the use of numerous worked-out practical examples, the reader is able to apply the information to his or her own needs. Each section follows a similar format: definitions of terms and parameters; breakdown of the types of characteristics of components; analysis of basic and special functions; detailed practical problems and clearly worked-out solutions; and various clarifying charts, tables, nomographs, and illustrations.

12-73 **Handbook of Applied Hydraulics.** Ed. by Calvin Victor Davis and Kenneth E. Sorensen. 3rd ed. New York: McGraw-Hill, 1969. 1v. (various paging). illus. index.

Clear and concise presentations of the principles basic to hydraulic engineering and demonstrations of their applications. In 42 sections, the 50 contributors treat, among other topics, reservoir hydraulics, natural channel, sewer diversion, massive buttress dows, tidal energy development.

12-74 **Handbook of Automation, Computation and Control.** Ed. by Eugene M. Grabbe and others. New York: Wiley, 1958-1961. 3v.

The three volumes cover Control Fundamentals, Computers and Data Processing, and Systems and Components. Dated but still not superseded. The discussions of process control and industrial automation are extremely thorough.

12-75 **Handbook of Chlorination for Potable Water, Waste Water, Cooling Water, Industrial Processes, and Swimming Pools.** By George Clifford White. New York: Van Nostrand Reinhold, 1972. 744p. illus. index.

This handbook, which brings together all the scattered information on chlorination, is directed to designers, operators, students, and regulatory agencies working with systems that use chlorine. Each chapter is a detailed treatise based on the author's many years of experience and a complete review of the literature. The 12 chapters cover history, facilities design, operation and maintenance, chemistry, determination of chlorine residuals in water treatment, chlorination of potable water, wastewater, swimming pools and cooling water, other applications, chlorine dioxide, hypochlorination, and other methods of disinfection.

12-76 **Handbook of Circuit Analysis Languages and Techniques.** Ed. by Randall W. Jensen and Lawrence P. McNamee. Englewood Cliffs, NJ: Prentice-Hall, 1976. 809p. illus. bibliog.

This is a supplement to the *Handbook of Electronic Circuit Designs* (12-78). It contains sets of instruction for applying ASTAP, BELAC, CIRC, and other programs to circuit analysis with emphasis on electrical and electronic applications.

12-77 **Handbook of Components for Electronics.** Charles A. Harper, ed. in chief. New York: McGraw-Hill, 1977. 1v. (various paging). illus. index.

This is a thorough and comprehensive sourcebook of practical data, guidelines, and information on the principles of integrated-circuit-component technology, digital integrated circuits, linear and special-purpose integrated circuits, discrete semiconductor devices, components for electro-optics, components for microwave systems, resistors and passive-parts standardization, capacitors, transformers and inductive devices, relays and switches, and connectors and connective devices. It includes property and performance data, comparison data, guidelines, extensive text and reliability data, detailed listings of important specifications and standards, and information on dimensions, configurations, and mechanical and environmental performance.

12-78   Handbook of Electronic Circuit Designs. By John D. Lenk. Englewood Cliffs, NJ: Prentice-Hall, 1976. 307p. illus. index.

A handbook for the experimenter who does not have advanced mathematics or theoretical background. It covers a wide range of circuit analysis including the design of filters, attenuators, amplifiers, wave generators, phototransistor circuits, switching, and oscillator circuits. Supplemented by *Handbook of Circuit Analysis and Languages and Techniques* (12-76).

12-79   Handbook of Electronic Formulas, Symbols and Definitions. By John R. Brand. New York: Van Nostrand Reinhold, 1979. 359p.

This is a portable ready reference source for engineers, technicians, hobbyists, and students. The text is arranged alphabetically by symbol within seven subject categories: passive circuits; English letters, passive circuits; Greek letters, transistors; static conditions, transistors; small signal conditions, operational amplifiers; symbols and definitions, and operational amplifiers; formulas and circuits.

12-80   Handbook of Hydraulics for the Solution of Hydraulic Engineering Problems. By Ernest F. Brater and Horace Williams King. 6th ed. New York: McGraw-Hill, 1976. 1v. (various paging).

Purpose is to present the fundamentals needed to solve hydraulics problems and to provide appropriate tables, graphs, and computer techniques to facilitate solutions.

12-81   Handbook of Modern Electronic Data. By Matthew Mandl. Reston, VA: Reston, 1973. 274p. illus. tables. index.

The material in this data compilation is arranged in a convenient topical grouping for use by electronic or electrical technicians and engineers. In addition to a clear and easily comprehensible description of the fundamental units and terms in electronics, chapter 1 offers some relevant mathematics and good tables of the basic SI units and symbols. Various series-parallel circuit combinations are presented in chapter 2, while the next chapter is entirely devoted to the basic parameters of tubes and field-effect transistors (FET). Transmission lines and antennas, including reflectors and flared waveguides, are covered in chapter 4. The next two chapters offer essential data tables, and basic circuitry is featured in chapter 7, while meter ranges, color codes, and symbols are listed and discussed in chapter 8. The last chapter is devoted to vectors and phase factors. A good index and cross-references within the text enhance the usefulness of this abundantly illustrated volume.

12-82   Handbook of Ocean and Underwater Engineering. By John J. Myers. New York: McGraw-Hill, 1969. 1070p. diagrams. tables.

Non-specialist engineers, scientists, or technicians involved with designing equipment, systems, or structures for the ocean, or with their installation, constitute the principal audience for this work, which is set forth in 12 sections ranging from basic oceanography to wind and wave loads.

12-83 **Handbook of Precision Engineering.** Ed. by A. Davidson. New York: McGraw-Hill, 1971-1974. 12v. illus. index. (Philips Technical Library). A multi-volume handbook intended to be the last word for precision engineers or, as some prefer, engineers of fine mechanisms. The volume titles are: 1, Fundamentals; 2, Materials; 3, Fabrication of Non-metals; 4, Physical and Chemical Fabrication Techniques; 5, Joining Techniques; 6, Mechanical Design Applications; 7, Electrical Design Applications; 8, Surface Treatment; 9, Machining Processes; 10, Forming Processes; 22, Production Engineering; and 12, Precision Measurement.

12-84 **Handbook of Semiconductor Electronics: A Practical Manual Covering the Physics, Technology, and Applications of Transistors, Diodes, and Other Semiconductor Devices in Conventional and Integrated Circuits.** Ed. by Lloyd R. Hunter. 3rd ed. New York: McGraw-Hill, 1970. 2v. illus. index.
Another outstanding handbook covering the field of semiconductor electronics. Treatment of integrated circuit techniques is new to this edition. Each section is developed in typical handbook format with historical and general considerations presented first, followed by the specifics of the topic. It is divided into four parts: Physics of Semiconductor Materials, Devices and Circuits; Technology of Semiconductor Materials, Devices, and Circuits; Circuit Design and Applications of Semiconductor Devices; and Measurement and Analytical Techniques.

12-85 **IC 1978 Master.** Garden City, NY: United Technical Publications, Inc., 1978. 2175p. illus.
A complete guide to integrated circuits. Includes an index by part number and a military parts directory. The most comprehensive guide available for integrated circuits.

12-86 **Illustrated Handbook of Electronic Tables, Symbols, Measurements and Values.** By Raymond H. Ludwig. West Nyack, NY: Parker Publishing Co., 1977. 352p. illus. index.
A good practical reference book that concentrates on standardized electronic data, including symbols, abbreviations, and other reference data.

12-87 **Instrument Engineers' Handbook.** Ed. by Bela G. Liptak. Philadelphia, PA: Chilton, 1969-1972. 3v.
A comprehensive work covering level, pressure, density, temperature, flow, viscosity, weight and analytical measurement, miscellaneous sensors, and safety instrumentation in volume 1; control valves, other types of final control elements, regulators, controllers and logic components, control boards and receiver displays, transmitters and telemetering, control theory, process computers, analog and hybrid computers, and process control systems in volume 2; and reports of the new developments in process sensors and computers and data transmission in volume 3.

12-88 **McGraw-Hill's Compilation of Data Communications Standards.** Ed. by Harold C. Folts and Harry R. Karp. New York: McGraw-Hill, 1978. 1133p. illus.
Five groups working in the communications field have their most often-used data communications standards reproduced in their entirety. The groups are the U.S.

federal government, the International Telegraph and Telephone Consultative Committee (CCITT), the International Organization for Standardization (ISO), the American National Standards Institute (ANSI), and the Electronic Industries Association (EIA). All of the standards are cross indexed.

12-89 **Marks' Standard Handbook for Mechanical Engineers.** Theodore Baumeister, ed. in chief. 8th ed. New York: McGraw-Hill, 1978. 1v. (various paging). index.

This noted and well-used handbook covers properties and handling of materials, machine elements, fuels and furnaces, power generation, pumps and compressors, shop processes, instrumentation, and environmental control with special reference to the impact of OSHA and EPA. It includes references through 1977.

12-90 **National Electrical Code Reference Book.** By J. D. Garland. Englewood Cliffs, NJ: Prentice-Hall, 1977. 561p. illus. index.

This book is for use as a code reference guide for electricians, electrical contractors and inspectors, as well as a reference for vocational students and individuals making a study of the *National Electrical Code*. An attempt is made to make over 4,000 code rules as clear and understandable as possible.

12-91 **Pump Handbook.** Ed. by Igor J. Karassik and others. New York: McGraw-Hill, 1976. 1v. (various paging). index.

This handbook deals first with the theory, construction details, and performance characteristics of all the major types of pumps—centrifugal, power, steam, screw and rotary, jet, and many variants. It also covers prime movers, couplings, controls, valves, and the instruments used in pumping systems, plus information on the selection, purchasing, testing, and maintenance of pumps.

12-92 **Radio Amateur's Handbook.** By A. Frederick Collins; rev. by Robert Hertzberg. 13th ed. New York: T. Y. Crowell, 1976. 378p. illus. index.

A standard work presenting fundamental theories in a manner suitable for comprehension by beginners. There is a good glossary.

12-93 **Shock and Vibration Handbook.** Ed. by Cyril M. Harris and Charles E. Crede. 2nd ed. New York: McGraw-Hill, 1976. 1v. (various paging). illus. index.

This handbook covering all aspects of shock and vibration tries to reflect current engineering practice and covers the theoretical basis for shock and vibration, instrumentation and measurements, standards, analysis and testing, concepts in the treatment of data, procedures for analyzing, isolation, damping, balancing, equipment design, packaging, and the effects of shock and vibration on man.

12-94 **Standard Handbook for Electrical Engineers.** Ed. by Donald Glen Fink and H. Wayne Beaty. 11th ed. New York: McGraw-Hill, 1978. 1v. (various paging). illus. index.

Contains in a single volume all pertinent data on generation, transmission, distribution, control, conversion, and application of power. It achieves a high degree of accuracy in its comprehensive technical treatment. Above all, it is oriented toward application with economic factors in mind.

12-95 **Standard Handbook for Mechanical Engineers.** Ed. by Theodore Baumeister. 8th ed. New York: American Society for Mechanical Engineers, 1978. illus. tables.

In 18 main sections, the book treats cryogenics, refrigeration, materials handling, and strength of materials, among others. An exhaustive analytic index.

12-96 **Standard Handbook of Lubrication Engineering.** Ed. by James J. O'Connor. New York: McGraw-Hill, 1968. 1060p. illus. (McGraw-Hill Handbooks).

Sponsored by the American Society of Lubrication Engineers, this work brings together scattered information on lubricating engineering for practicing engineers and scientists, machine designers, plant engineers, energy-systems engineers, lubricating engineers and technicians. Divided into four parts: lubrication principles, general lubrication engineering practices, lubrications of specific equipment, and lubrication in specific industries.

12-97 **Water Quality and Treatment. A Handbook of Public Water Supplies.** By the American Water Works Association. 3rd ed. New York: McGraw-Hill, 1971. 654p. index.

A one-volume desk reference book for the engineer and other professionals concerned with water quality and control. Contains 19 chapters written by subject experts on various treatment methods, saline water conversion, corrosion, fluorides, radioactivity, and other water-related topics. Most chapters contain extensive bibliographies.

Directories

12-98 **McGraw-Hill's Leaders in Electronics.** New York: McGraw-Hill, 1979. 651p. index.

This work includes some 5,200 current notables from private industry, government agencies, academia, and the consulting and military worlds. For each biographee, the usual biographical information is given. There are no photographs, but there is an index that lists all the names of persons affiliated with a given organization.

12-99 **Who's Who in Electronics: Electronic Industry Data in Depth.** Cleveland, OH: Harris, 1976- . illus.

This annual lists approximately 14,000 electronics firms, distributors, and representatives in alphabetical and geographic sequences and gives name, address, names of executives or key personnel, number of employees, products, major SIC code, annual sales volume, and the year the company was established.

PRODUCTION AND PROCESSING ENGINEERING

Guides to the Literature

12-100 **Brief Guide to Sources of Metals Information.** By Marjorie Hyslop. Washington: Information Resources press, 1973. 180p. index.

Addressed primarily to the metallurgist, this little volume contains useful information on publications and organizations of metallurgical interest. The text half of the book instructs the reader in the use of information sources, including the new computer-based services, and lists specifically the foremost libraries, societies, abstract services, journals, basic books, publishers, and search sources. The directory section (more than half the book) is arranged by organization; for each it provides address, purpose, publications, and services. The index refers to both the text and the directory. United States and Canadian sources are emphasized.

12-101 **Chemical and Process Engineering Unit Operations: A Bibliographical Guide.** By Kathleen Bourton. New York: IFI/Plenum, 1968. 534p. (Bibliographical Guide Series).
An annotated bibliography of bibliographies, with 4,586 entries on chemical engineering. It is divided into 38 sections with the first being a detailed survey of reference works (including bibliographies, biographies, directories, guides, handbooks, periodicals, and translations). Each entry is in complete bibliographical form and indicates the number of items in the bibliography and the years covered.

12-102 **Chemical Industries Information Sources.** By Theodore P. Peck. Detroit, MI: Gale Research Co., 1979. 595p. index. (Management Information Guide, 29).
A comprehensive guide to the sources of information for the chemical industries. The general information section is divided into two parts. The first covers governmental agencies; professional societies, trade associations, and scientific-technological related organizations; research institutes and industrial laboratories; educational institutions; and international organizations. The second part covers the literature of chemical engineering, including guides, handbooks, dictionaries, directories, encyclopedias, bibliographies, monographs, and indexing and abstracting services. The rest of the book is divided by subject giving associations and organizations and literature for each.

12-103 **How to Find Out about the Chemical Industry.** By Russel Brown and G. A. Campbell. Oxford: Pergamon, 1969. 219p. index.
Retains the format of other volumes in this series, treating in general organization of the literature, patents, research, safety measures, computers. Useful to the chemical engineer, primarily as a supplement to standard guides to the literature of chemistry.

12-104 **How to Find Out in Iron and Steel.** By D. White. New York: Pergamon, 1970. 184p. index.
Organized in 12 chapters, covering, among others, such topics as literature guides, bibliographies, and lists of periodicals; indexes and abstracting services; handbooks, dictionaries, and directories; standard textbooks and treatises; production statistics; education and visual aids; history.

Abstracts, Indexes, and Bibliographies

12-105 **IMM Abstracts.** v. 1- . London: Institution of Mining and Metallurgy, 1950- . bimonthly.
This is a survey of the world's literature on economic geology and mining of all minerals (except coal), mineral dressing, and non-ferrous extraction metallurgy.

12-106 **International Petroleum Abstracts.** v. 1- . Barking, England: Applied Science Publishing, 1973- . quarterly.

An abstracting service covering all aspects of petroleum and arranged in broad subject categories with author indexes.

12-107 **List of Publications Issued by the Bureau of Mines from July 1, 1910 to January 1, 1960: With Subject and Author Index.** Washington: U.S. Bureau of Mines; distr. Washington: GPO, 1960. 826p.

12-107a **List of Publications Issued by the Bureau of Mines from January 1, 1960 to December 31, 1964.** Washington: U.S. Bureau of Mines; distr. Washington: GPO, 1966. 197p.

12-107b **List of Publications Issued by the Bureau of Mines.** 1965- . monthly.

A basic bibliographic source on extractive engineering in the United States.

12-108 **Metals Abstracts and Index.** v. 1- . London: Institute of Metals and the American Society for Metals, 1968- . monthly.

Formed by the merger of *Review of Metal Literature* and *Metallurgical Abstracts.* International abstracting journal covering all aspects of the science and practice of metallurgy and related fields as published in primary and secondary forms of literature. The service yields about 25,000 descriptive abstracts a year.

12-109 **Theoretical Chemical Engineering Abstracts.** v. 1- . London: Technical Information Co., 1964- . bimonthly.

Principal headings are fluid dynamics, heat transfer, mass transfer, heat/mass transfer. Six yearly issues yield close to 2,000 abstracts. It has a classified arrangement.

Encyclopedias

12-110 **Chemical Engineering Practice.** By Herbert W. Cremer. New York: Academic Press, 1956-1965. 12v.

A definitive encyclopedic survey of chemical engineering. Each article is written by an authority.

12-111 **Chemical Technology: An Encyclopedic Treatment; The Economic Application of Modern Technological Development Based upon a Work Originally Devised by the Late Dr. J. F. Van Oss.** New York: Barnes and Noble, 1968-1975. 8v. illus. index.

Published in Great Britain and Europe under the title *Materials and Technology.* Written for the layman as well as the technologist, it encompasses all phases of chemical technology including sources, manufacture, and the processing and use of both natural and synthetic processes plus statistics on world trade, production, and prices with references for further reading. There are detailed bibliographies. The encyclopedia can be considered a successor to Dr. J. F. Van Oss's *Warenkennis en Technologie* (Amsterdam; New York: Elsevier, 1948-1950. 3v.).

12-112 **Corrosion and Corrosion Control: An Introduction to Corrosion Science and Engineering.** By Herbert H. Uhlig. 2nd ed. New York: Wiley, 1971. 1600p. index.

An excellent synthesis of today's state of the art.

12-113 Encyclopedia of Chemical Processing and Design. v. 1- . John J. McKetta, executive ed. New York: Marcel Dekker, 1976- .
This work is intended as a comprehensive source of information on the design of equipment, systems, and controls utilized in chemical processing. It stresses plant design and contains a large amount of graphic and illustrative material; extensive bibliographies accompany each article.

12-114 Encyclopedia of Chemical Technology. 3rd ed. New York: Wiley, 1978- .
Also known by its editors' names: Kirk-Othmer, this well-known encyclopedia is considered the authority in the chemical engineering field. Each volume is self-contained. In general it encompasses chemical technology, industrial processes, uses of materials, unit operations, processes and fundamentals of chemistry. While the third edition is being published, the second edition, which took ten years to complete, is still a useful reference source.

12-115 Encyclopedia of Industrial Chemical Analysis. Ed. by Foster D. Snell and others. New York: Wiley, 1966-1974. 20v.
A comprehensive encyclopedia intended to provide coverage of the methods and techniques used in industrial laboratories throughout the world for the analysis and evaluation of chemical products, including raw materials, intermediates, and finished products. The first three volumes cover general techniques; the main part of the encyclopedia begins with volume 4, covering individual compounds, elements and their compounds, compounds of similar chemical structure, compounds with the same end use, and groups of compounds by industry.

12-116 International Petroleum Encyclopedia. Ed. by George Weber. Tulsa, OK: Petroleum Publishing Co., 1970- . illus. (part col.). maps. index.
This yearbook takes the form of an annotated international atlas of regional maps, showing petroleum activity during the previous year. Appended is a statistical index, keyed to the atlas maps of oil and gas fields, refineries, and petrochemical plants; information on world tankers, off-shore operations, production, refining, and consumption of petroleum and natural gas; and directories of pipeline, drilling, and engineering contractors, and a list of government agencies.

12-117 Metals Reference Book. Ed. by Colin J. Smithells. 5th ed. Woburn, MA: Butterworths, 1976. 1566p. illus. index.
This is the world's authority on data relating to metallurgy. The sections covered are introductory tables; general physical and chemical constants; x-ray crystallography; crystallography; crystal chemistry; metallurgically important minerals; thermochemical data; physical properties of molten salts; metallography; equilibrium diagrams; gas-metal systems; diffusion in metals; general physical properties; elastic properties and damping capacity; temperature measurement and thermoelectric properties; radiating properties of metals; electron emission; electrical properties; steels and alloys with special magnetic properties; mechanical testing; mechanical propeties of metals and alloys; hard metals; lubricants; friction and wear; casting alloys and foundry data; refractory materials; fuels; controlled atmospheres for heat treatment; masers and lasers; guide to the corrosion resistance of metals; electroplating and metal finishing; welding; and solders and brazing alloys.

12-118 **Reservoir Engineering Manual.** By Frank W. Cole. 2nd ed. Houston, TX: Gulf Publishing, 1969. 385p. illus. index.
Covers drive reservoirs, pressure maintenance, secondary recovery, gas condensate reservoirs, and economics, with additional material on the application of hydrocarbon phase behavior to specific oil and gas reservoirs, the influence of wettability on oil recovery, recovery index data relating oil production to time in performance prediction techniques.

12-119 **Ullmanns Encyklopadie der Technischen Chemie.** 4th ed. completely rev. Berlin: Verlag Chemie, 1972- .
An outstanding encyclopedia encompassing all areas of chemical technology. When complete, the 24 volumes will be divided as follows: v. 1-6, thematically arranged: 1, General Principles of Process and Reaction Technology; 2, Process Engineering I (Unit Operations); 3, Process Engineering II and Reaction Plants; 4, Process Development (Cumulative Index v. 1-4); 5, Methods of Separation and Analysis; and 6, Measurement and Control, Environmental Conservation and Safety. Volumes 7 through 24 are alphabetically arranged by keywords from such fields as inorganic industry, dyes and pigments, fibroid textiles, metals and their compounds, plastics and rubber, and fluid dynamics. Each thematic volume is in itself complete. There will be a cumulated index in every third volume with a master index in volume 24. Index terms are in both German and English.

Dictionaries

12-120 **Chemical Synonyms and Trade Names: A Dictionary and Commercial Handbook Containing over 35,000 Definitions.** By William Gardner. 8th ed. Cleveland, OH: CRC Press, 1978. 769p. index.
An alphabetical list of over 32,500 chemical terms and proprietary trade names. Each entry is briefly described, all proprietary trade names are given a number, which refers to a list of manufacturers at the end of the book. It will be useful to industrial chemists, patent agents, sales directors and managers, and registrars of trademarks all over the world.

12-121 **Dictionary of Rubber Technology.** By Alexander S. Craig. New York: Philosophical Library, 1969. 228p. illus. bibliog.
A well-designed dictionary of encyclopedic scope. Bibliographical citations to leading expositions of the defined terms enhance its usefulness.

12-122 **Elsevier's Dictionary of Chemical Engineering.** By W. E. Clason. New York: American Elsevier, 1968. 2v. index.
A typical Elsevier multi-lingual dictionary arranged by English/American terms with corresponding French, Spanish, Italian, Dutch, and German terms. An index for each language refers the user to the English/American term by index number. Coverage is broad; some of the component fields are agricultural chemistry, inorganic industrial chemistry, bacteriological chemistry, chromatography, conserving industry, dairy industry, dyeing and printing, electrochemistry, fuels, and petroleum industry.

12-123 **Elsevier's Dictionary of Metallurgy and Metal Working in Six Languages: English/American, French, Spanish, Italian, Dutch and German.** Comp. by W. E. Clason. Amsterdam: Elsevier Scientific Publishing, 1978. 848p.

This dictionary is another attempt to standardize the nomenclature in metal working and the steel industry. It includes 8,406 entries listed by the English terms with the equivalents in the other languages.

Handbooks

12-124 **Chemical Engineer's Handbook.** Ed. by Robert H. Perry and Cecil H. Chilton. 5th ed. New York: McGraw-Hill, 1973. 1v. (various paging). illus. index. (McGraw-Hill Chemical Engineering Series).

The enormous amount of reference information is broken down into 25 sections, each with new or improved formulation of its subject matter. Each section begins with a detailed list of contents, and an overall alphabetical index is provided.

12-125 **The Chemical Formulary: Collection of Commercial Formulas for Making Thousands of Products in Many Fields.** Harry Bennett, ed. in chief. New York: Chemical Publishing, 1933-1977. 20v.

Contains formulas for chemical compounding and treatment of products from virtually every branch of the chemical industries. The introduction contains information on the apparatus, methods, cost calculations and elementary preparation of useful home products such as cleansing creams, hand lotions, and insect repellants. Other chapters are devoted to adhesives, coatings, cosmetics, detergents and disinfectants, drug products, emulsions and dispersions, food products, metals and their treatment, polish, rubber, resins, waxes, textile specialties, and miscellaneous products. There is no duplication of formulas included in previous volumes, and all formulas have been provided by reputable sources. Other useful information includes federal laws regulating foods, drugs, and cosmetics; a list of incompatible chemicals; tables of weights and measures; emergency first aid for chemical injuries; trademark chemicals; and list of suppliers.

12-126 **Dangerous Properties of Industrial Materials.** By N. Irving Sax; assisted by Marilyn C. Bracken and others. 5th ed. New York: Van Nostrand Reinhold, 1979. 1118p. bibliog.

A comprehensive guide with information on about 15,000 common industrial and laboratory materials. Each material is listed, followed by all the information that is needed to determine how hazardous it might be. This edition has attempted to supply actual toxicological data for humans as well as experimental animals. Data is also included on incompatibilities, e.g., "violent reaction on exposure to air"; "reacts violently with . . . "; and warnings against additional hazards that might be encountered in case of accident, fire, or earthquake.

12-127 **Handbook of Adhesives.** Ed. by Irving Skeist. 2nd ed. New York: Van Nostrand Reinhold, 1977. 921p. illus. index.

This handbook for chemists and engineers covers theory, economics, and properties of adhesives; descriptions of the important adhesive materials; and discussions of the industrial applications of adhesives. There are extensive bibliographies and a glossary of adhesive terminology.

12-128 **Handbook of Plastics and Elastomers.** Charles A. Harper, ed. in chief. New York: McGraw-Hill, 1975. 1v. (various paging). illus. index.

This is a comprehensive sourcebook, covering every aspect of polymer technology from design through manufacturing and end product use. It covers fundamentals of plastics and elastomers; electrical properties and tests; mechanical, physical, chemical, and environmental properties and tests; plastic and elastomeric product forms with extensive data and guidelines; specifications and standards; and design and fabrication of plastic products.

12-129 **Handbook of Pulp and Paper Technology.** Ed. by Kenneth W. Britt. 2nd ed. rev. and enlarged. New York: Van Nostrand Reinhold, 1970. 723p. illus. index.

Some of the topics treated are fiber science and technology, wood preparation, pulping, bleaching, stock preparation, paper manufacture, finishing, converting, and coating. In addition to the technology of paper manufacture, it includes information on pulpwood harvesting, printing, reprography, and aging of paper.

12-130 **Handbook of Reactive Chemical Hazards: An Indexed Guide to Published Data.** By L. Bretherick. London: Butterworths, 1975. 976p.

Gives researchers documented information to allow ready assessment of the likelihood of a reaction hazard potential associated with an existing or proposed chemical compound or reaction system. Arranged by compound.

12-131 **Industrial Solvents Handbook.** By Ibert Mellan. 2nd ed. Park Ridge, NJ: Noyes Data Corp., 1977. 567p.

Assembles over 800 tables of physical data, taken from manufacturers' literature and dealing with hydrocarbon solvents, halogenated hydrocarbons, nitroparaffins, organic sulfur compounds, monohydric alcohols, polyhydric alcohols, phenols, aldehydes, ethers, glycol ethers, ketones, acids, amines, and esters.

12-132 **Metals Handbook.** 8th ed. Metals Park, OH: American Society for Metals, 1961-1976. 11v. illus.

A comprehensive handbook whose first edition appeared in 1927. Covers all aspects of metals, with articles, graphs, tables, and illustrations.

12-133 **Petroleum Processing Handbook.** By William F. Bland and Robert L. Davidson. New York: McGraw-Hill, 1967. 1v. (various paging). illus. index.

Practical information on the refining of crude oil and its conversion to useful petroleum products, primarily fuels and lubricants. Within its scope is evaluation of crude oils, processes, equipment, materials of construction, chemicals and catalysts, maintenance and construction, offsite facilities and utilities, personal and plant safety, process control and instrumentation, petroleum products, and physical properties of hydrocarbons. Two final chapters deal with sources of information and a glossary of processing terms.

12-134 **Rare Metals Handbook.** Ed. by Clifford A. Hampel. 2nd ed. New York: Van Nostrand Reinhold, 1961; repr. Huntington, NY: Krieger, 1971. 715p. illus. index.

Encompasses the natural occurrence, available concentrations, unusual processing problems, and physical properties of those 55 metals classified as rare.

12-135 **Riegel's Handbook of Industrial Chemistry.** By Emil Raymond Riegel. Ed. by James A. Kent. 7th ed. New York: Van Nostrand Reinhold, 1974. 902p. illus. index.

Contrary to what the title implies, this reference work is not a handbook in the conventional sense, but rather a state-of-the-art review of the chemical process industry. The first through fifth editions were published under the title, *Industrial Chemistry*; the sixth edition appeared in 1962 as *Riegel's Industrial Chemistry*. This seventh edition of the review surveys 27 topics in industrial chemistry, maintaining much of the same format as the sixth edition. Two new topics have been added: air pollution control and wastewater technology.

12-136 **Toxic and Hazardous Industrial Chemical Safety Manual for Handling and Disposal with Toxicity and Hazard Data.** By the International Technical Information Institute. Tokyo: International Technical Information Institute, 1976. 591p.

A comprehensive manual that deals with the treatment and handling of chemicals. For each entry it gives: synonyms, use, properties, hazardous potential, toxicity, handling and storage, emergency treatment and measures, first aid, spills and leakage, and disposal and waste treatment.

CONSTRUCTION ENGINEERING

Guides to the Literature

12-137 **Guide to Literature on Civil Engineering.** Comp. by Rita McDonald. Washington: American Society for Engineering Education, 1972. 21p.

A brief listing of civil engineering bibliographies, indexes, abstracts, dictionaries, encyclopedias, almanacs, yearbooks, annuals, handbooks, manuals, directories, national associations, standards, specifications, serials, and information centers. Not annotated.

12-138 **Sources of Construction Information: An Annotated Guide to Reports, Books, Periodicals, Standards and Codes. Volume 1: Books.** By Jules B. Godel. Metuchen, NJ: Scarecrow Press, 1977. 661p. index.

This book is a description of existing books, reports, standards, and other publications of interest to the architect, planner, engineer, contractor and building official. Each entry gives author, title, publisher, date, pagination, and price, plus a brief annotation. Other volumes will cover government reports; professional, trade association, and planning organization reports, standards, and codes; and organizations, periodicals, and directories. It will not cover magazine and journal articles.

Abstracts, Indexes, and Bibliographies

12-139 **Bibliography on the Fatigue of Materials, Components and Structures, 1838-1950.** Comp. by J. Y. Mann. Elmsford, NY: Pergamon, 1970. 316p. index.

12-139a **Bibliography on the Fatigue of Materials, Components and Structures, 1951-1960.** Comp. by J. Y. Mann. Oxford; New York: Pergamon, 1978. 489p. index.

Published for the Royal Aeronautical Society, listing publications on the failure of materials and structures under repeated or cyclic loads. The bibliography is arranged chronologically and alphabetically within each year. Each entry gives full author and title, periodical, years, volumes, and pagination. A detailed author and subject index helps to locate specific articles.

12-140 Earthquake Engineering Reference Index. London: International Association for Earthquake Engineering, British National Section, 1963. 48p.

Despite its age, still the best bibliographic survey of significant articles and studies of tectonic descriptions of the earth's crust and of the ways they can be dealt with by civil and geo-engineers.

12-141 Geothechnical Abstracts. v. 1- . Essen, Deutsche Gesellschaft dur Erd- und Grundbau, (for) International Society for Soil Mechanics and Foundation Engineering, 1970- . monthly.

This is the English edition of *Dokumentation Bodenmechanik und Grundbau.* Sections include: General; Engineering Geology; Site Investigations; Soil Properties, Laboratory and Field Determinations; Analysis of Soil Engineering Problems; Rock Properties; Laboratory and Field Determinations; Analysis of Rock-Engineering Problems; Design, Construction and Behaviour of Engineering Works; Construction Methods and Equipment; Snow and Ice Mechanics and Engineering. Each monthly issue contains approximately 150 abstracts "written in English irrespective of the language of the original article." The abstracts are available in either sheet or card form. Annual author/subject indexes are also available. This is a continuation of Geodex Soil Mechanics Information Service. Geodex International developed and Deutsche Gesellschaft is continuing the Geodex Retrieval System, a coordinate indexing service for geo-technology based on a controlled vocabulary of 347 terms. The system is available from Geodex International.

Encyclopedias

12-142 Builders Encyclopedia. By Harry F. Ulrey. Indianapolis, IN: Audel, 1970. 593p. illus.

Designed to give the builder insight to terminology with which he or she is often confronted. Not primarily a list of trade names, but some are included if they are the standard accepted terms.

12-143 Composite Materials. By L. J. Broutman and others. New York: Academic Press, 1974. 7v.

Contents of this comprehensive treatise are: v. 1, Interfaces in Metal Matrix; v. 2, Mechanics of Composite Materials; v. 3, Engineering Applications; v. 4, Metallic Matrix Composites; v. 5, Fracture and Fatigue; v. 6, Interfaces in Polymer Matrix Composites; v. 7, Structural Design and Analysis.

12-144 Foundation Engineering. Ed. by G. A. Leonards. New York: McGraw-Hill, 1962. 1136p. illus. diagrams. maps.

Concentrates on foundation design and construction, omitting important related topics such as embankments, cut and natural slopes, and tunnels.

12-145 **Illustrated Encyclopedic Dictionary of Building and Construction Terms.** By Hugh Brooks. Englewood Cliffs, NJ: Prentice-Hall, 1976. 366p. index. illus.

A good, concise, easy-to-understand dictionary of building and construction terms plus commonly used formulas, tables, and charts.

12-146 **Planning and Estimating Underground Construction.** By Albert D. Parker. New York: McGraw-Hill, 1970. 300p. illus. index.

Invaluable in describing the methods and equipment that are used in tunnel construction, such as the modern drilling and blasting techniques and the tunnel-boring machines. Tunnel lining and shaft construction are discussed in detail. The major portion of the book deals with all aspects of estimating costs in tunnel construction. Exemplary estimates are used as an instructional aid.

Dictionaries

12-147 **Dictionary of Civil Engineering.** By John S. Scott. 2nd ed. rev. Harmondsworth, Middlesex, England: Penguin Books, 1965. 348p.

A wide but representative range of technical words in soil mechanics, heavy construction, and mining explained for the layman and non-specialist engineer. Differing applications of terms are listed and described in the contexts of specific engineering subfields.

12-148 **Plumbing Dictionary.** Ed. by I. D. Jacobson. 2nd ed. Cleveland, OH: American Society of Sanitary Engineering, 1975. 142p. illus. bibliog.

This excellent dictionary describes words and combinations of words used in plumbing and gives the definitions and meanings of these words and terms. Over 2,500 terms are defined with trade names and brand names omitted.

Handbooks

12-149 **American Metric Construction Handbook.** By R. J. Lytle. Farmington, MI: Structures Publishing, 1976. 305p. illus. index.

This book is a guide to the metric system as used in the construction industry. It covers the basic modern metric system; conversion factors; planning and managing metrification; use of metrics by civil engineers, architects, and surveyors, and in land title descriptions; history of measurement; metric legislation; and use of standard metric units and dimensions for masonry, concrete, lumber, and other building products.

12-150 **Building Construction Handbook.** By Frederick S. Merritt. 3rd ed. New York: McGraw-hill, 1975. 1v. (various paging). illus. (McGraw-Hill Handbooks).

This is a comprehensive and practical handbook for those working with building design and construction. Twenty-eight sections cover topics from insurance and bonds to water supply and purification. Frequent reference is made to other more comprehensive works.

12-151 **Civil Engineering Handbook.** Ed. by Leonard C. Urquhart. 4th ed. New York: McGraw-Hill, 1959. 1184p. illus. diagrams. tables.

The standard handbook for American practice of civil engineering. Arranged in ten sections as follows: surveying; railway, highway, and airport engineering; mechanics of materials; hydraulics; stresses in framed structures; steel design; cement and concrete; soil mechanics and foundations; sewerage and sewage disposal; water supply and purification. The product of specialists who supplement their contributions with bibliographies and footnotes.

12-152 **Data Book for Civil Engineers: Design.** By Elwyn Seelye. 3rd ed. New York: Wiley, 1957-1960. 3v. illus.
A set for the designer rather than for the student, based on the assumption that once data have been given and modus operandi described, readers will proceed to make their own calulations.

12-153 **Handbook of Composite Construction Engineering.** Ed. by Gajanan M. Sabnis. New York: Van Nostrand Reinhold, 1979. 380p. illus. index.
Composite construction utilizes a mixture of building materials, such as steel, concrete, wood, and masonry, to achieve improved structural performance at some economic advantage. This book covers fundamentals of composite construction; steel-concrete composite construction; the application of light gauge steel in this field; the use of prestressed concrete; the application of composite construction to bridges and buildings; and composite construction in wood and timber.

12-154 **Handbook of Heavy Construction.** Ed. by John A. Havers and Frank W. Stubbs, Jr. 2nd ed. New York: McGraw-Hill, 1971. 1v. (various paging). illus. index.
An authoritative reference work for persons involved in heavy construction, presenting the best of current U.S. practice. The 31 chapters are divided into three parts—construction management, construction equipment, and construction applications. Each chapter is supplied with supporting charts, illustrations, and diagrams.

12-155 **IES Lighting Handbook.** Ed. by John E. Kaufman and Jack F. Christensen. 5th ed. New York: Illuminating Engineering Society, 1972.
The 26 sections systematically treat lighting terms, physics of light, light and vision, measurement, color, light control, daylighting, light sources, lighting calculation, interior lighting design, offices and schools, institutions and public buildings, lighting for merchandising, industries, residential lighting, lighting system design factors, outdoor lights, light projection equipment and protective lighting, sports lighting, roadway lighting, aviation lighting, transportation lighting, lighting for advertising, theater, television and photographic lighting, miscellaneous and underwater lighting.

12-156 **Linemans and Cableman's Handbook.** By Edwin Kurtz and Thomas M. Shoemaker. 5th ed. New York: McGraw-Hill, 1976. 1v. (various paging). illus.
This handbook is intended for the apprentice, lineman, cableman, foreman, supervisor, and other associated individuals. The sections cover general understanding of electricity, electrical terms, and electric power systems, construction of overhead and underground distribution and transmission lines and the maintenance procedures. There is also one section on safety applications and a chapter of self-testing questions and exercises.

12-157 **SPI Handbook of Technology and Engineering of Reinforced Plastics/Composites.** Ed. by J. Gilbert Mohr and others. 2nd ed. New York: Van Nostrand Reinhold, 1973. 405p. illus. bibliog. index.

The sections of this handbook from the Society of the Plastics Industry cover low-temperature, intermediate, and high-temperature cures, thermoset molding, reinforced thermoplastics, unique composite materials, and the future of the reinforced plastics/composites industry. Each method within the sections on processing is treated with an introduction, materials and forms of materials required, equipment and tooling needed, description of processing methods, and product properties and performance data.

12-158 **Standard Handbook for Civil Engineers.** Ed. by Frederick S. Merritt. 2nd ed. New York: McGraw-Hill, 1976. 1v. (various paging). illus. index.

Civil engineering as conceived in this book includes planning, design, and construction for environmental control, development of natural resources, buildings, transportation facilities, and other structures required for the health, welfare, safety, employment, and pleasure of mankind, including highways, bridge, airport, rail-transportation, tunnels, water, and harbor engineers. Some of the treated topics are: computer operations; design management; construction management; structural theory; concrete design and construction; lightweight steel design and construction; surveying.

12-159 **Standards in Building Codes.** 15th ed. Philadelphia, PA: American Society for Testing and Materials, 1980, 2v.

Fulfillment of ASTM Standards is required in building codes of states, cities, and municipalities (e.g., The Basic Building Code, Uniform Building Code, Southern Standard Building Code, National Building Code, and the National Building Code of Canada). This book lists all standards and indicates which are approved by the U.S.A. Standards and the U.S.A. Standards Institute.

12-160 **Structural Engineering Handbook.** Ed. by Edwin H. Gaylord, Jr., and Charles N. Gaylord. 2nd ed. New York: McGraw-Hill, 1979. 1v. (various paging). illus. maps.

This comprehensive volume covers the entire field of structural engineering including concrete, masonry, steel, and timber. All of the text has been updated and new topics include finite element analysis, reinforced concrete silos and bunkers, and steel poles for transmission lines. There are numerous worked-out examples and commentaries.

12-161 **Structural Steel Designers' Handbook.** Ed. by Frederick Merritt. New York: McGraw-Hill, 1972. 1v. (various paging). illus. index.

Fourteen outstanding structural engineers have written a handbook that is a valuable addition to the reference shelf of any designer who must consider such troublesome problems as site conditions, esthetics, and functional requirements as well as economy. The sections cover properties of structural steels, fabrication and erection, general structural theory, special structural theories, connections, criteria for building design, design of building members, floor and roof systems, lateral-force design, criteria for bridge design, beam and girder bridges, truss bridges, arch bridges, and cable-supported bridges.

12-162 **Welding Handbook.** 7th ed. New York: American Welding Society, 1976- .

This is regarded as the most authoritative source available on welding. It is published in five sections, each complete in itself and each revised separately at different times. Information has been taken from private company files, technical reports, and articles.

TRANSPORTATION ENGINEERING

Abstracts, Indexes, and Bibliographies

12-163 **Catalog of the Transportation Center, Northwestern University.** Boston: G. K. Hall, 1972. 12v.

Divided into author and subject/geographic sections, this catalog contains records of 79,000 books and reports and 1,000 serials, of which 40% have been issued since 1965. Although its main emphasis is on transportation management (operations, planning, economics, regulation), monographs and reports in transportation engineering form a significant part of the total corpus.

12-164 **HRIS Abstracts.** Washington: Highway Research Information Service, Highway Research Board, Division of Engineering, National Research Council, National Academy of Sciences, National Academy of Engineering, 1968- . quarterly.

Each issue has about 1,000 indicative abstracts and references. It is arranged in broad subject categories covering transportation administration, personal management, photogrammetry, bituminous substances and mixes, construction and maintenance equipment, highway safety, exploration, legal studies, urban transportation, and research.

12-165 **Information Sources in Transportation, Material Management, and Physical Distribution: An Annotated Bibliography and Guide.** Comp. and ed. by Bob J. Davis. Westport, CT: Greenwood Press, 1976. 715p. index.

This bibliography covers books, government publications, organizations, educational materials and activities, statistical publications, and atlases and maps on transportation, materials management, and physical distribution. It includes brief annotations.

12-166 **International Aerospace Abstracts.** v. 1- . New York: American Institute of Aeronautics and Astronautics, 1961- . semi-monthly.

An abstracting service issued alternately with *STAR* (12-167). *STAR* and *IAA* cover the world's literature of aeronautics, space science, and technology. Divided into broad subject fields, *IAA* surveys periodicals, serials, books, proceedings, transactions, and translations. There are quarterly, semiannual, and annual indexes to subject, personal author, report number, and accession number.

12-167 **Scientific and Technical Aerospace Reports.** v. 1- . Washington: U.S. National Aeronautics and Space Administration, 1963- . semi-monthly.

This service is commonly called *STAR* and was preceded by *Technical Publication Announcement* in two volumes for 1958-1962. A semi-monthly abstracting service covering the worldwide literature on the science and technology of space and aeronautics as contained in the technical reports commissioned by the United States and other national governments. Some dissertations, translations, and patents also appear. Each issue contains an abstract section arranged by subject categories and an index of subjects, corporate authors, personal authors, report numbers, and accession numbers. Each report is assigned an accession number upon receipt by NASA and each issue of *STAR* contains a consecutive group of accession numbers.

12-168 **Transportation Research Abstracts.** v. 1- . Washington: Transportation Research Board, National Research Council, 1931- . monthly.
This is an abstracting service that covers articles, books, reports, etc. in all areas of transportation. Until 1974 it was called *Highway Research Abstracts.*

Encyclopedias

12-169 **Complete Illustrated Encyclopedia of the World's Aircraft.** Ed. by David Mondey. New York: A & W Publishers, 1978. 320p. illus. (part col.). index. (A Quarto Book).
"This encyclopedia is the first fully international guide to list all the world's manufacturers of production aircraft, both military and civil, from the beginning of powered flight to the present day." It is, in fact, a directory of manufacturers rather than a complete encyclopedia of aircraft.

12-170 **Encyclopedia of Aviation.** New York: Scribner's, 1977. 218p. illus. index.
This is a general encyclopedia that includes short histories of airlines, biographical data on aviation personalities, explanations of technical terms, brief descriptions of civil and military aircraft and of important air forces.

12-171 **International Encyclopedia of Aviation.** New York: Crown, 1977. 480p. illus. (part col.).
A good general encyclopedia that covers all aspects of aviation, including airlines, air sport, flying boats, balloons, airships, rotocraft, V/STOL, rockets, guided missiles, and space exploration. Also includes a section covering facts, feats, and records.

12-172 **Observer's Spaceflight Directory.** By Reginald Turnill. New York: Frederick Warne, 1978. 384p. illus. (part col.). index.
This is a comprehensive, authoritative survey of space flights through 1977. Following a short introductory section, space programs are listed alphabetically by sponsoring country or international agency, subarranged by name of program or vehicle. The descriptions range from two or three lines to four or five pages.

Dictionaries

12-173 **Astronautical Multilingual Dictionary.** Prepared by the International Academy of Astronautics. New York: American Elsevier, 1969. 936p. index.
Contains about 60,000 technical terms in English, Russian, German, French, Italian, Spanish, and Czech, as well as an additional 8,000 space law terms. The dictionary proper consists of almost 5,000 English scientific and technical terms

selected from the vocabulary of definitions of the National Aeronautics and Space Administration and their equivalents in the remaining six languages. The legal section contains 900 entries with corresponding translations. Alphabetical indexes are included to facilitate translations from Russian, German, French, Italian, Spanish, and Czech.

12-174 **Aviation and Space Dictionary.** Ed. by Ernest J. Gentle and Lawrence W. Reithmaier. 5th ed. Fallbrook, CA: Aero Publishers, 1974. 272p. illus.

6,000 terms in aerodynamics, air traffic control, avionics, astronomy, computer technology, geophysics, nucleonics, civil and military aviation, meteorology, navigation, and space flight. Basic sources consulted were the dictionaries and glossaries issued by the Air Force, Atomic Energy Commission, Federal Aviation Administration, National Weather Service, and National Aeronautics and Space Administration; additional entries were prepared by the editors.

12-175 **Language of Space: A Dictionary of Astronautics.** By Reginald Turnill. New York: John Day, 1971. 165p.

A brief introduction, a projection of forthcoming space endeavors for the next 20 years, a phonetic alphabet of space communications, and a list of abbreviations precede the main body of the volume. In the main section of the glossary are offered brief definitions of 1,100 terms created or adopted by the astronauts, engineers, and ground crews of the U.S. space program.

12-176 **NASA Thesaurus.** Springfield, VA: National Technical Information Service, 1976. 2v. (NASA SP-7050).

This is the authorized subject term list by which documents in the NASA scientific and technical information system are indexed and retrieved. Volume 1 is an alphabetical listing of all subject terms approved for use and volume 2 is an access vocabulary containing an alphabetical listing of all thesaurus terms and permutations of all multiword and pseudo-multiword terms.

12-177 **Road and Track Illustrated Dictionary.** By John Dinkel. New York: W. W. Norton, 1977. 92p. illus.

This is a handy dictionary to major automotive terms and concepts. It gives the basic mechanical concepts of the modern internal combustion engine.

12-178 **Space-Age Acronyms: Abbreviations and Designations.** By Reta C. Moser. 2nd ed. New York: IFI/Plenum, 1969. 534p.

An acronymic guide incorporating all MIL-STD-12B abbreviations used for engineering drawings and publications throughout the industry, most official U.S. Air Force, Army, and NASA acronyms. Many foreign acronyms of military and commerical aircraft manufacturers and airlines have been added to this edition, keyed to the country of origin.

Handbooks

12-179 **S.A.E. Handbook.** New York: Society of Automotive Engineers, 1924- . annual.

A collection of technical reports and standards as they pertain to the automotive industry. Each year's volume contains some new materials.

12-180 **Transportation and Traffic Engineering Handbook.** Ed. by John E. Baerwald. Englewood Cliffs, NJ: Prentice-Hall, 1976. 1080p. illus. bibliog.

This is a comprehensive handbook that emphasizes traffic engineering more than mass transportation. As with all Prentice-Hall books it is well designed, readable and up-to-date.

Directories

12-181 **Directory of Transportation Libraries in the United States and Canada.** Comp. by the Transportation Division, Special Libraries Association. New York: Special Libraries Association, 1973. 122p. index.

Lists 106 transportation libraries in classified arrangement (e.g., general, air, highway, motor vehicle, water, rail, etc.). The following information is provided: name of the library, address, full-time equivalent staff (professional and clerical), number of volumes (books, periodicals, reports, microforms), and a descriptive paragraph on the nature of the collection and the services provided.

12-182 **Interavia ABC: World Directory of Aviation and Astronautics 1975.** 23rd ed. Geneva: Interavia S.A.; distr. New York: International Publications Service, 1975. 3121p. index.

Systematically lists all companies, organizations, and institutes that are directly and indirectly related to the aeronautics industry. The 40,000 entries are arranged geographically, by continent and country.

12-183 **NASA Factbook: Guide to National Aeronautics and Space Administration Programs and Activities. 1974-1975.** 2nd ed. Chicago: Marquis, 1975. 613p.

A review of all the activities of NASA including grants and research contracts; personnel and facilities; statistics; legislation; applications of NASA Development; and glossary.

12-184 **Urban Mass Transit: A Guide to Organizations and Information Resources.** Ed. by Thomas N. Trzyna and Joseph R. Beck. Claremont, CA: Center for California Public Affairs, an Affiliate of the Claremont Colleges, 1979. 148p. index. (Who's Doing What Series, 5).

This new guide lists professionals and citizens who wish to become involved in any aspect of urban mass transit. It provides the names of key officials, addresses and telephone numbers, together with a short description of the agency and its responsibilities. Two sections to be noted are "Professional, Research, and Advocacy Organizations" and "Consultants and Manufacturers."

12-185 **World Aviation Directory.** v. 1- . Washington: Ziff-Davis, 1940- . twice a year.

A directory of airlines, major aerospace manufacturers and distributors, U.S. airports, repair stations and schools, aviation/aerospace publications, organizations, and government agencies. It lists executive, administrative, and operating personnel and provides addresses and telephone numbers.

13

Biomedical Sciences

Medicine is the science of diagnosing, treating, and preventing disease that impairs the health or threatens the existence of organisms. Because medicine is an applied science, its history is closely tied to the theoretical sciences. Harvey's discovery of the circulation of blood in 1628 was perhaps the most significant breakthrough in the history of medicine. From that time on research and experimentation grew at an accelerated rate. Louis Pasteur (1822-1895) proved that fermentation, or putrefaction, is caused by microscopic living organisms (bacteria) and not by chemical processes. Joseph Lister (1827-1912) adapted Pasteur's demonstration of the reality of germs as agents of disease to develop an antiseptic barrier between the wound and the germ-carrying atmosphere. Robert Koch (1843-1910) demonstrated how bacteria could be cultivated and how their specificity could be established. In 1897, Ross discovered the carrier of malaria to be the Anopheles mosquito. In 1902, Landsteiner discovered the A, B, and O blood groups, thus making transfusion a practical therapeutic option. Nicolle, in 1909, established the body louse as the vector of epidemic typhus.

There are many specialties of modern medicine. These specialties frequently are divided into three areas: diagnosis, therapy, and prevention. Diagnostic literature overlaps with the biological sciences, especially with anatomy, biochemistry, biophysics, embryology, endocrinology, genetics, microbiology, physiology, and psychology, and in this text are found mostly in the chapter on biological sciences. The present chapter covers the sources dealing with therapy and prevention. Therapeutic specialties include anesthesiology, cardiology, dentistry, dermatology, gastroenterology, general pathology, geriatrics and gerontology, gynecology and obstetrics, immunology, internal medicine, neurology and neurosurgery, nursing, ophthalmology, orthopedics, otorhinolarynology, pediatrics, pharmacology and toxicology, plastic surgery, psychiatry, radiology, rehabilitation, serology, surgery, thoracic surgery, urology, and venereology. Specialties within preventive medicine are environmental health, epidemiology, industrial medicine, occupational health, public health, and social medicine.

Due to the variety of medical specialties there have been a proliferation of specialized handbooks, encyclopedias, dictionaries, atlases, and directories in the medical literature. Serial indexing and abstracting is also a dominant force. Much of the indexing is sponsored by the federal government at the National Library of Medicine. In this chapter the many services of the National Library of Medicine are recognized.

GUIDES TO THE LITERATURE

13-1 **Bibliography of Pharmaceutical Reference Literature.** By Magda Pasztor and Jenny Hopkins. Philadelphia, PA: Rittenhouse, 1968. 167p. Contains about 570 annotated citations, abstracts, indexes, lists of periodicals, bibliographies, guides, encyclopedias, and directories, all relating to drugs and pharmaceuticals. Good for retrospective coverage.

13-2   **Current Medical References.** By Paul J. Sanazaro and Milton John Chatton. 6th ed. Los Altos, CA: Lange, 1970. 673p.

The most satisfactory bibliography of modern medical reference sources up to 1969 available in English.

13-3   **Guide to Drug Information.** By Winifred Sewell. Hamilton, IL: Drug Intelligence Publications, 1976. 218p. index.

This work is in four parts. Part 1 deals with books that list drugs and supply tables that compare features of drug information sources. Part 2 lists and describes treatises, perodicals, other primary sources, and reference works. Part 3 discusses resources such as index and abstract publications, card and microfiche files, reviews and encyclopedic treatises, computer searches, and network services. Part 4 is a look into the future, with suggestions on how to make the best use of the new information-handling technology.

13-4   **Guide to the Literature in Psychiatry.** By Bernice Ennis. Los Angeles, CA: Partridge Press, 1971. 127p. index. bibliog.

Organizes sources of information for psychiatrists into the following sections: journal literature; information sources (e.g., indexes, abstracts, directories, reviews, bibliographies, etc.); psychiatric monographs; pamphlets, reprints, exhibits; government documents; controlled circulation publications; translators and translations; psychiatric libraries; and a list of publishers.

13-5   **Handbook of Medical Library Practice.** By Gertrude L. Annan and J. W. Felter. 3rd ed. Chicago: Medical Library Association, 1970. 411p. index. (Medical Library Association Publication, no. 4).

A well-documented survey of medical library practice covering the emergence of the modern medical library; the medical librarian; administration; technical processing; readers' services; automation in medical libraries; audiovisual materials; research; rare books, archives, and history of medicine; library planning, furniture, and equipment; National Library of Medicine; the library and its public; identification and communication; professional associations; and British medical libraries. The chapter entitled "Research," a substantial part of the book, consists of a classified bibliography of medical reference works, which is not, however, as comprehensive as *Current Medical References* (13-2).

13-6   **Health Care Administration: A Guide to Information Sources.** By Dwight A. Morris and Lynne Darby Morris. Detroit, MI: Gale Research Co., 1978. 264p. index. (Health Affairs Information Guide Series, v. 1; Gale Information Guide Library).

This book contains over 1,100 citations, mostly annotated, on health care administration issues concerning mental health facilities, long-term care institutions, hospitals, and amulatory care facilities. References are drawn from journals, manuals, bibliographies, indexes, abstracts, proceedings, series, directories, and encyclopedias.

13-7   **Health Sciences and Services: A Guide to Information Sources.** By Lois F. Lunin. Detroit, MI: Gale Research Co., 1979. 614p. index. (Management Information Guide, 36).

This book is an annotated inventory of information resources including publications, data bases, and organizations of the disciplines that constitute the health sciences. For each of the subject sections the following items are covered:

definition or scope; publications arranged by form of literature; data bases, including files, services, and user aids; and organizations, including libraries and special collections.

13-8   **Health Sciences Librarianship: A Guide to Information Sources.** By Beatrice K. Basler and Thomas G. Basler. Detroit, MI: Gale Research Co., 1977. 186p. index. (Books, Publishing and Libraries Information Guide Series, v. 1; Gale Information Guide Library).

This is an annotated bibliography of 550 non-journal entries that were published before 1975. It basically covers books but does include pamphlets, vertical file materials, annual reports, technical reports, library serials lists, journal series, library or professional society newsletters, videotapes, and cassettes.

13-9   **How to Find Out in Pharmacy: A Guide to Sources of Pharmaceutical Information.** By Alice Lefler Brunn. Oxford; New York: Pergamon, 1969. 130p. illus.

Following the pattern of other volumes in this series, this one includes sections on indexing and abstracting services, bibliographies and bibliographical guides, and other reference material, as well as material on specific topics (e.g., law, crude drugs, and toxicology). Emphasis is on British imprints.

13-10   **Librarian and the Patient: An Introduction to Library Services for Patients in Health Care Institutions.** Ed. by Eleanor Phinney. Chicago: American Library Association, 1977. 352p. illus. index.

Background chapters of this text describe the patient, the institution, and the library as a therapeutic environment. Planning, organizing, equipping, staffing, operating, and supplementing library services are discussed from the viewpoint of principles to be applied rather than required techniques.

13-11   **Sourcebook on Health Sciences Librarianship.** By Ching-chih Chen. Metuchen, NJ: Scarecrow Press, 1977. 207p. illus. index.

Part 1 of this unique book is a systematic study of the articles published in the *Bulletin of the Medical Library Association* from 1966 through 1975. Subject scope, output of authors, frequency and dates of articles cited, relationship to trends in the field, and other data of professional interest are carefully documented statistically. Part 2 is a citation bibliography, arranged by broad subjects, of the *Bulletin* articles. These 3,000 citations refer to scientific or other books and journals that are related to librarians' interests but not readily retrievable from other sources.

13-12   **Use of Medical Literature.** Ed. by L. T. Morton. 2nd ed. Woburn, MA: Butterworths, 1977. 462p. index. (Information Sources for Research and Development).

Each chapter of this book is designed to be a self-contained guide to information sources within a broad range of preclinical, clinical, and applied health fields. While primarily oriented toward British library users, this is a useful collection of a broad range of source material for those interested in an in-depth introduction to research and information sources in the field of medicine.

ABSTRACTS, INDEXES, AND BIBLIOGRAPHIES

13-13 **Abridged Index Medicus.** v. 1- . Chicago: American Medical Association in cooperation with the National Library of Medicine, 1970- . monthly.
Designed for the use of the practicing physician whose daily needs for guidance in diagnosis and therapy can be satisfied by the information from a carefully selected list of major journals. The abridged index is based on the journals so identified and is a subset of *Index Medicus. New Series* (13-42).

13-14 **Acupuncture: A Research Bibliography.** By Allen Y. Liao. New York: New York University Medical Center Library, 1975. 66p. index.
This publication covers books, journal articles, and audiovisual materials that have been published from 1960 to early 1975. The purpose of the bibliography is to assist health professionals and others interested in acupuncture research.

13-15 **Aerospace Medicine and Biology: A Continuing Bibliography with Indexes.** Washington: NASA, 1964- . quarterly.
A selective annotated bibliography of unclassified reports and journal articles initially reported in *Scientific and Technical Aerospace Reports* (12-167) and *International Aerospace Abstracts* (12-166). Supersedes *Aerospace Medicine and Biology, 1952-1963*, published by the Library of Congress.

13-16 **American Hospital Association Guide to the Health Care Field.** Chicago: American Hospital Association, 1976. 618p. adv. index. (AHA Catalog no. 1476).
The main sections of this excellent guide are health care institutions; AHA membership; health organizations, agencies, and educational programs; and the buyer's guide. The list of hospitals is arranged alphabetically by state, then by city; information includes addresses, telephone numbers, association memberships or accreditations, facilities and services, bed size, expenses, etc.

13-17 **Annotated Bibliography on Diving and Submarine Medicine.** By Charles W. Shilling and Margaret F. Werts. New York: Gordon and Breach, 1971. 611p. index.
A comprehensive international bibliography covering the literature on submarine and deep sea diving published from January 1, 1962, through September 30, 1969. It contains some 1,900 citations and abstracts. Continues the two-volume work by Hoff and Greenbaum, *A Bibliographical Sourcebook of Compressed Air, Diving and Submarine Medicine* (Washington: U.S. Navy, 1948-1966. 2v.).

13-18 **Author Catalog of the Library of the New York Academy of Medicine.** Boston: G. K. Hall, 1969. 43v.

13-18a **Supplement.** 1st- . 1974- .

13-18b **Subject Catalog of the Library of the New York Academy of Medicine.** Boston: G. K. Hall, 1970. 34v.

13-18c **Supplement.** 1st- . 1974- .

13-18d **Illustration Catalog of the Library of the New York Academy of Medicine.** 3rd ed. enlarged. Boston: G. K. Hall, 1976. 264p.

13-18e Portrait Catalog of the Library of the New York Academy of Medicine. Boston: G. K. Hall, 1960. 5v.

13-18f Supplement. 1st- . 1959- .

13-18g Catalog of Biographies in the Library of the New York Academy of Medicine. Boston: G. K. Hall, 1960. 165p.
These are the catalogs to the collection of the New York Academy of Medicine that includes 365,000 bound volumes, 165,000 pamphlets, 10,784 separate portraits and 151,792 portraits appearing in books and journals. It is one of the most important medical libraries in the country, if not the world.

13-19 Bibliography of Bioethics. v. 1- . Detroit, MI: Gale Research Co., 1975- .
The subject matter covered is of growing interest covering such topics as proper conduct for physicians, value problems in biomedical and behavioral research, and the quest to develop public policy for medical care and human experimentation. It covers journal and newspaper articles, monographs, essays in books, court decisions, bills, and unpublished documents.

13-20 Bibliography of Drug Abuse, Including Alcohol and Tobacco. By Theodora Andrews. Littleton, CO: Libraries Unlimited, 1977. 306p. index.
The 725 titles listed in the work represent all aspects of the subject. Part 1 covers general reference sources and describes bibliographies, dictionaries, and periodicals. Part 2 covers source material arranged by 15 subject areas, such as histories and personal narratives; incidence and prevalence; prevention and rehabilitation; education; pharmacology and medical aspects; hallucinogens; stimulants; alcohol; and tobacco. Each entry gives a complete bibliographic description.

13-21 Bowker's Medical Books in Print: Subject Index, Author Index, Title Index. New York: R. R. Bowker, 1972- . annual.
Intended primarily for medical collections, this new member of Bowker's in-print series covers medicine, dentistry, psychiatry, psychology, veterinary medicine, nursing, and a number of paraprofessional areas such as medical technology. Some 30,000 books are listed in three separate sections by author, title, and subject, subdivided by some 5,000 subject headings, with a list of publishers in an appendix.

13-22 Cancer Therapy Abstracts. Bethesda, MD: National Cancer Institute, 1975- . quarterly.
Abstracts the significant cancer therapy articles collected from the current major biomedical sources of the world's literature. Author and subject indexes are included. Supersedes the institute's *Cancer Chemotherapy Abstracts, 1960-1974.*

13-23 Catalog of the F. B. Power Pharmaceutical Library, University of Wisconsin at Madison School of Pharmacy. Boston: G. K. Hall, 1976. 4v.
A catalog of 22,964 volumes covering pharmacy; pharmacology; pharmacognosy; pharmaceutics; pharmaceutical chemistry and biochemistry; pharmaceutical manufacturing; history of pharmacy; pharmacy jurisprudence; drug

literature and evaluation; hospital, retail and clinical pharmacy; pharmacy and public health; continuing education in pharmacy; radiopharmaceuticals; pharmacotherapeutics; therapeutics; toxicology; and therapeutic drug interactions.

13-24   **Catalog of the Menninger Clinic Library, The Menninger Foundation (Topeka, KS).** Boston: G. K. Hall, 1972. 4v.

13-24a **Supplement.** 1st- . 1978- .
A specialized collection of materials on treatment, research, and professional education in psychiatry. The catalog is divided into two parts: author-title and subject.

13-25   **Catalog of the Sophia F. Palmer Memorial Library of the American Journal of Nursing Company (New York).** Boston: G. K. Hall, 1973. 2v.
Although rich in historical data, the collection of this special library is also strong in contemporary materials related to nursing. The catalog is arranged in dictionary format.

13-26   **Coordinate Index Reference Guide to Community Mental Health.** By Stuart E. Golann. New York: Behavioral Publications, 1969. 237p. index. (Community Mental Health Series).
Offers 1,510 references on American community mental health studies published between the years 1960 and 1967. The coordinate index approach enables the user to choose items by geographical location, type of health situation, and related function. While not exhaustive, the index is an excellent starting point for those seeking information on community mental health literature.

13-27   **Cumulative Index to Nursing and Allied Health Literature.** Glendale, CA: Glendale Adventist Medical Center Publications Service, 1961- . quarterly with annual cumulations.
A subject-author index to approximately 120 worldwide journals of professional nursing. Formerly *Cumulative Index to Nursing Literature* until 1977.

13-28   **Current List of Medical Literature.** Washington: Army Medical Library, 1941-1959. 36v. index.
Not a true index but a table of contents listing of 1,200 then current medical journals. There were author and subject indexes in each issue.

13-29   **Current Therapy: Latest Approved Methods of Treatment for Practicing Physicians.** v. 1- . Philadelphia, PA: Saunders, 1949- . annual.
A useful compendium of the current state of the art arranged by broad medical subject headings. Includes a subject index.

13-30   **Death: A Bibliographical Guide.** By Albert Jay Miller and Michael James Acri. Metuchen, NJ: Scarecrow Press, 1977. 420p. index.
A partially annotated bibliographical guide of 3,848 entries from periodicals, books, essays, miscellaneous works, and audiovisual materials.

13-31   **Death, Grief and Bereavement: A Bibliography 1845-1975.** Comp. by Robert Fulton. New York: Arno Press, 1977. 253p. index. (The Literature of Death and Dying).

Consists of over 3,800 terms, arranged alphabetically by author with a subject index. No journalistic, literary, or theological works are included.

13-32 **Dental Abstracts.** v. 1- . Chicago: American Dental Association, 1956- . monthly.
A comprehensive abstracting service, divided into broad topics with descriptive abstracts. Annual subject and author indexes.

13-33 **Diabetes Literature Index.** v. 1- . Washington: National Institute of Arthritis and Metabolic Diseases, 1966- . monthly.
A product of the MEDLARS data base. Arranged by authors and title keywords.

13-34 **Dictionary Catalogue of the London School of Hygiene and Tropical Medicine, University of London.** Boston: G. K. Hall, 1965. 6v.

13-34a **Serials Catalogue.** 1965. 286p.

13-34b **Supplement.** 1st- . 1971- .
A specialized catalog covering public health reports of all tropical countries. Preventive medicine of temperate and tropical climates is especially well covered.

13-35 **Drug Abuse Bibliography.** 1st- . By Jean Cameron Advena. Troy, NY: Whitston, 1971- . annual.
A serial supplement to Menditto's *Drugs of Addiction and Non-Addiction 1960-1969* (13-36). Coverage is comprehensive, even to the extent of including (as indicated in the Preface) "trivia . . . on the grounds that it is of some, if only minor, immediate interest."

13-36 **Drugs of Addiction and Non-Addiction, Their Use and Abuse; A Comprehensive Bibliography, 1960-1969.** By Joseph Menditto. Troy, NY: Whitston, 1970. 315p.
Some 6,000 entries cover all areas of addiction. It is arranged under eleven broad subjects: amphetamines and stimulatns, barbiturates and tranquilizing drugs, lysergic acid diethylamide, marijuana, narcotic addiction, narcotic rehabilitation, narcotic trade, narcotics, narcotics control, narcotics laws and legislation, and narcotics and crime. Within each category, entries are arranged by form of publication: books and essays, doctoral dissertations, and periodical articles. Annual supplements are being published as *Drug Abuse Bibliography* (13-35).

13-37 **Dying and Death: An Annotated Bibliography.** By Irene L. Sell. New York: Tiresias Press, 1977. 144p. index.
This 506-item annotated bibliography is intended to provide citations to articles, books, and audiovisual materials that will enable nurses to understand emotional and other problems that dying patients and their families face.

13-38 **Endocrinology Index.** v. 1- . Washington: National Institutes of Health, 1968- . bimonthly.
Each issue is produced by the MEDLARS data base; however, this index contains additional subject headings not included in *Index Medicus* (13-42). Arranged in seven sections: table of contents; subject entries arranged under eight broad categories; reviews; methods; author (provides the user with a list of keywords or

index terms assigned to each author); subject index; and author index. No abstracts.

13-39 **Excerpta Medica.** Amsterdam: Excerpta Medica Foundation, 1947- .
A major international abstracting service screening 3,500 medical journals in the health sciences. The abstracts are divided into 44 numbered subject sections, each beginning at a different time. Abstracts, all in English, are descriptive or indicative, with author and subject indexes to each section. Sections of the classified arrangement are:

1) *Anatomy, Anthropology, Embryology, and Histology*, v. 1- . 1947- .
2) *Physiology*, v. 1- . 1948- .
3) *Endocrinology*, v. 1- . 1947- .
4) *Microbiology: Bacteriology, Mycology and Parasitology*, v. 1- . 1945- .
5) *General Pathology and Pathological Anatomy*, v. 1- . 1948- .
6) *Internal Medicine*, v. 1- . 1947- .
7) *Pediatrics and Pediatric Surgery*, v. 1- . 1947- .
8) *Neurology and Neurosurgery*, v. 1- . 1948- .
9) *Surgery*, v. 1- . 1947- .
10) *Obstetrics and Gynecology*, v. 1- . 1948- .
11) *Oto-, Rhino-, Laryngology*, v. 1- . 1948- .
12) *Ophthalmology*, v. 1- . 1947- .
13) *Dermatology and Venereology*, v. 1- . 1947- .
14) *Radiology*, v. 1- . 1947- .
15) *Chest Diseases, Thoracic Surgery, and Tuberculosis*, v. 1- . 1948- .
16) *Cancer*, v. 1- . 1953- .
17) *Public Health, Social Medicine, and Hygiene*, v. 1- . 1955- .
18) *Cardiovascular Diseases and Cardiovascular Surgery*, v. 1- . 1957- .
19) *Rehabilitation and Physical Medicine*, v. 1- . 1958- .
20) *Gerontology and Geriatrics*, v. 1- . 1958- .
21) *Developmental Biology and Teratology*, v. 1- . 1961- .
22) *Human Genetics*, v. 1- . 1963- .
23) *Nuclear Medicine*, v. 1- . 1964- .
24) *Anesthesiology*, v. 1- . 1966- .
25) *Hematology*, v. 1- . 1967- .
26) *Immunology, Serology, and Transplantation*, v. 1- . 1967- .
27) *Biophysics, Bio-Engineering and Medical Instrumentation*, v. 1- . 1967-.
28) *Urology and Nephrology*, v. 1- . 1967- .
29) *Biochemistry*, v. 18- . 1965- .
30) *Pharmacology and Toxicology*, v. 18- . 1965- .
31) *Arthritis and Rheumatism*, v. 1- . 1965- .
32) *Psychiatry*, v. 1- . 1948- .
33) *Orthopedic Surgery*, v. 1- . 1956- .
34) *Plastic Surgery*, v. 1- . 1970- .
35) *Occupational Health and Industrial Medicine*, v. 1- . 1971- .
36) *Health Economics and Hospital Management*, v. 1- . 1971- .
37) *Drug Literature Index*, v. 1- . 1969- .
38) *Adverse Reactions Title*, v. 1- . 1966- .
40) *Drug Dependence*, v. 1- . 1972- .
46) *Environmental Health and Pollution Control*, v. 1- . 1971- .
47) *Virology*, v. 1- . 1971- .
48) *Gastroenterology*, v. 1- . 1971- .

49) *Forensic Science Abstracts*, v. 1- . 1975- .
50) *Epilepsy Abstracts*, v. 1- . 1971- .

13-40 Hospital Literature Index. v. 1- . Chicago: American Hospital Association, 1945- .
This is an author-subject index of literature about the administration, planning, and financing of hospitals and related health care institutions. Includes the administrative aspects of the medical, paramedical, and prepayment fields.

13-41 Index-Catalogue of the Library of the Surgeon General's Office, United States Army (Army Medical Library), Authors and Subjects. Washington: GPO, 1880-1961. 61v. (Ser. 1-4 and Public Health Service Series).
The Library of the Surgeon General's Office, the predecessor and the collection nucleus of the NLM, was, during its day, one of the largest medical libraries in the world. The *Index-Catalogue*, as its name implies, was more than a book catalog. It included thousands of analytic entries for articles and parts of serials. It is particularly adapted for biographical reference work because of the extent to which it indexed biographical and obituary notices. The five series cover: Ser. 1, A-Z, 1880-1895. 16v.; ser. 2, A-Z, 1896-1915. 21v.; ser. 3, A-Z, 1918-1932. 10v.; ser. 4, A-Mn, 1936-1955. 11v.; and Public Health Service Series, 1959-1961. 3v.

13-42 Index Medicus. New Series. v. 1- . Washington: National Library of Medicine, 1960- . monthly with annual cumulations.
This *Index Medicus*, which supersedes the *Current List of Medical Literature* (13-28) indexes several hundred periodicals. Articles are analyzed by specialists who assign appropriate subject descriptors listed in NLM's Medical Subject Headings (MeSH). Close familiarity with MeSH is recommended for all users of *Index Medicus* since it is also used by MEDLARS. In the January issue appears the "List of Journals Indexed."

13-43 Index Medicus, or Quarterly Classified Record of the Current Medical Literature of the World. New York, Boston and Washington: Carnegie Institute, 1879-1899. 21v.; 2nd series. 1903-1920. 18v.; 3rd series. 1921-1927. 6v.
This first *Index Medicus* was a subject-arranged list of medical literature with author and subject indexes, covering periodical articles, books, theses, and parts of books. No volumes published between 1899 and 1903.

13-44 Index to Dental Literature. v. 1- . Chicago: American Dental Association, 1921- . quarterly.
A cumulative index to dental literature. It is produced from the MEDLARS of the National Library of Medicine and is arranged by subject and name sections. There is also a list of new dental books, dissertations, and a list of publications indexed.

13-45 International Nursing Index. v. 1- . New York: American Journal of Nursing Co., 1966- .
This index of nursing literature is based on the National Library of Medicine's MEDLARS retrieval system. In addition to the indexed publications, there is a list of serials indexed, publications of organizations and agencies, new nursing books, and dissertations.

13-46   **Katalog der Sammlung Schoenlein.** (Catalog of the Schoenlein Collection on Epidemics). Würzburg Library. Boston: G. K. Hall, 1972. 543p.
This is a specialized alphabetical catalog that covers about 8,000 epidemiological works of the fifteenth through the nineteenth centuries.

13-47   **Medical Book Guide.** v. 1- . Boston: G. K. Hall, 1974- . monthly with annual cumulations.
A bibliography of all English and French language medical monographs and serial titles cataloged by the Library of Congress and the National Library of Medicine.

13-48   **Medical Literature Analysis and Retrieval System (MEDLARS).**
This is the National Library of Medicine's data base and is one of the most important in the world today. This data base has been operational since 1964. A multipurpose system, its two primary purposes are to: 1) build the *Index Medicus—Current Catalog* data complex; and 2) to extract from it, on demand, bibliographic citations, ranging from single citations to complex subject searches. Within MEDLARS are several subordinate computer systems, each with its own distinguishing acronym. They are as follows:

AVLINE (AudioVisual Catalog On-Line) represents bibliographic and review data for nonprint materials in the health sciences.

BIOETHICS provides citations from the *Bibliography of Bioethics* since 1973.

CANCERLIT provides literature discussing all aspects of cancer since 1963.

CANCERPROJ contains summaries of ongoing cancer research projects, which have been provided by cancer scientists around the world since 1976.

CATLINE (Catalog-on-Line) is available for cataloging data on items cataloged by NLM since 1965.

CHEMLINE is an online chemical dictionary of chemical substance names or name surrogates.

CLINPROT contains summaries of clinical investigations of new anticancer agents and treatment modalities.

EPILEPSY contains citations to abstracts in *Epilepsy Abstracts* from 1945.

HEALTH is the Health Planning and Administration file and contains citations to the literature dealing with the non-clinical aspects of health care delivery from 1975.

HISTLINE (History of Medicine On-Line) provides citations from the *Bibliography of the History of Medicine* since 1978.

MEDLINE contains citations that appear in *Index Medicus*. Backfiles are available from 1966.

MESH VOC is NLM's controlled vocabulary online. It contains all descriptors and qualifiers used to index and catalog materials.

NAME AUTH contains authority records in all corporate, conference, and series names used in NLM cataloging.

RTECS (Registry of Toxic Effects of Chemical Substances) contains toxicity data for some 36,000 substances since 1978.

SDILINE (Selective Dissemination of Information On-Line) makes possible searches from the current month's portion of MEDLINE.

TDB (Toxicology Data Bank) contains facts and data for approximately 2,500 substances.

TOXLINE (Toxicology Information On-Line) contains information from a variety of subfiles in toxicology. Older information is in the TOXBACK file.

13-49 **National Library of Medicine.**
The National Library of Medicine is the keystone of medical bibliography in the United States. Its services are so extensive and so pervasive in influence that all other systems of medical bibliographical control must establish a practical working relationship with the NLM if they are to be successful. Founded in 1836 as the Library of the Surgeon General's Office (Department of War), the library was first conceived as an adjunct to military medicine. Gradually its focus and collecting interests broadened until it became what it is today, a national medical library whose collections embrace human medicine and health, veterinary medicine, dentistry, nursing, pharmacy, and allied or ancillary subjects in the physical, biological, behavioral, and engineering sciences. To document the point about the centrality of the NLM in American medicine, one needs only to list the following regional medical libraries, all of which are directly tied to NLM:

1) New England Region (CT, ME, MA, NH, RI, VT)
Francis A. Countway Library of Medicine
10 Shattuck Street
Boston, MA 02115

2) New York and Northern New Jersey Region
(New York and the 11 northern counties of New Jersey)
New York Academy of Medicine Library
2 East 103rd Street
New York, NY 10029

3) Mid-Eastern Region (PA, DE, and the ten southern counties of New Jersey)
Library of the College of Physicians
19 South 22nd Street
Philadelphia, PA 19103

4) Mid-Atlantic Region (VA, WV, MD, DC, NC)
National Library of Medicine
8600 Rockville Pike
Bethesda, MD 20014

5) East Central Region (KY, MI, OH)
Wayne State University Medical Library
4325 Brush Street
Detroit, MI 48201

6) Southeastern Region (AL, FL, GA, MS, SC, TN, Puerto Rico)
A. W. Calhoun Medical Library
Emory University
Atlanta, GA 30322

7) Midwest Region (IL, IN, IA, MN, ND, WI)
University of Illinois at the Medical Center
1750 West Polk Street
Chicago, IL 60612

8) Midcontinental Region (CO, KS, MO, NE, SD, UT, WY)
   University of Nebraska Medical Center
   42nd Street and Dewey Avenue
   Omaha, NE 68105
9) South Central Region (AR, LA, NM, OK, TX)
   University of Texas Southwestern Medical School at Dallas
   5323 Harry Hines Boulevard
   Dallas, TX 75235
10) Pacific Northwest Region (AK, ID, MT, OR, WA)
   University of Washington Health Sciences Library
   Seattle, WA 98105
11) Pacific Southwest Region (AZ, CA, HI, NV)
   University of California
   Center for the Health Sciences
   Los Angeles, CA 90024

13-50 **National Library of Medicine Current Catalog.** Jan. 1/14, 1966- .
Washington: National Library of Medicine; distr. Washington: GPO,
1966- . quarterly with annual and quinquennial cumulations.
Supersedes the U.S. Library of Congress, *Catalog*. Contains citations to all
publications cataloged by the National Library of Medicine. Issues are arranged
in four sections: 1) Monographs Subject, 2) Monographs Name, 3) Serials Subject, and 4) Serials Name. Citations appear in full only under the main entry in
the Name Sections; subject headings used are from MeSH. This catalog is
computer-produced and is the companion series to *Index Medicus. New Series*
(13-42), the index to serial literature. Both publications are part of the
MEDLARS data base.

13-51 **Nutrition Abstracts and Reviews.** v. 1- . Aberdeen, Scotland: Commonwealth Bureau of Nutrition, 1931- . monthly.
An abstracting service, issued under the auspices of the Commonwealth Bureau
Council and Medical Research Council, covering areas such as chemical composition of foodstuffs, vitamins, physiology of nutrition, human diet in relation to
health and disease, and feeding of animals. Abstracts are short and indicative. In
1977 the title divided into two sections: A—Human and Experimental, and
B—Livestock Feeds and Feeding.

13-52 **Quarterly Cumulative Index Medicus.** Chicago: American Medical
Association, 1927-1956. 60v.
More than 1,000 separate journal titles and selected new books were indexed each
year. In dictionary arrangement, complete bibliographic information appeared
under the author's name with the title in the original language if it was English,
French, German, Spanish, Italian, or Portuguese. Titles in all other languages
were translated. Under subject headings, all titles were translated with a note
identifying the original language. There was no subject authority list or thesaurus
of terms as is now available in MeSH. Usually subjects were simple or common
names—e.g., cancer, eyes, liver, and heart. Included also were lists of publishers
and of journals indexed. Overlapping in time but extending three years beyond
the 1956 termination date of this work was the *Current List of Medical Literature*
(13-28).

13-53   **Quarterly Cumulative Index to Current Medical Literature, 1916-1926.**
Chicago: American Medical Association, 1917-1927. 12v.
A brief subject and author index to a selection of current medical periodicals. The coverage was not complete but, used with the old *Index Medicus* (13-43), medical literature of the period was adequately analyzed. Each annual volume included an appended bibliography of important medical books. In 1927 both the old *Index Medicus* and the *Quarterly Cumulative Index to Current Medical Literature* merged to form the *Quarterly Cumulative Index Medicus* (13-52).

13-54   **Selective Bibliography of Orthopaedic Surgery: Including a Basic Science Supplement on the Musculoskeletal System.** By the American Academy of Orthopaedic Surgeons. 3rd ed. St. Louis, MO: C. V. Mosby, 1975.
Citations to articles and monographic works are arranged in three sections: clinical orthopaedics; basic science and related subjects; and miscellaneous. The first two sections are subdivided by appropriate categories—e.g., surgery, trauma, rehabilitation. Entries are not annotated but include adequate bibliographic descriptions. There is no index.

13-55   **Toxicity Bibliography.** Washington: U.S. National Library of Medicine; distr. Washington: GPO, 1968-1977. 10v.
Extracted from *Index Medicus* (13-42) are all documents dealing with adverse drug reactions and poisoning in animals and people up to 1977.

13-56   **Underwater Medicine and Related Sciences: A Guide to the Literature; An Annotated Bibliography, Key Word Index, and Microthesaurus.** By Charles W. Shilling and Margaret F. Werts. New York: IFI/Plenum, 1973-1975. 2v. indexes.
This title succeeds and updates *An Annotated Bibliography on Diving and Submarine Medicine* (New York: Gordon and Breach, 1971. 622p.). Each volume includes almost 1,800 citations, primarily from 1970 to 1974; each item provides an abstract, many of which were prepared by the authors of the book. Other features include a permuted subject index (manual), an author index, and a microthesaurus. The citations are arranged alphabetically by author.

13-57   **Venereal Disease Bibliography. 1966-1975.** Comp. by Stephen H. Goode. Troy, NY: Whitston, 1972-1977. 6v.
A comprehensive bibliography of world literature on venereal disease from 1966 to 1975. Each citation is given twice, once in the alphabetical section by title and again in a subject listing. Includes an author index.

ENCYCLOPEDIAS

13-58   **Encyclopaedia of Antibiotics.** By John S. Glasby. 2nd ed. New York: Wiley, 1979. 467p. (A Wiley-Interscience Publication).
This is an alphabetically arranged encyclopedia of antibiotics. For each entry the following is given: formula, structure, melting point, elaborating organism, methods of preparation and purification, those organisms against which the antibiotic is effective, toxicity, and literature and/or patent references.

13-59   **Encyclopedia and Dictionary of Medicine, Nursing, and Allied Health.** By Benjamin Frank Miller and Claire Brackman Keane. 2nd ed. Philadelphia, PA: W. B. Saunders, 1978. 1148p. illus.

Definitions of a large number of terms, with cross-references and indications of pronunciation. The definitions are brief but adequate. Its encyclopedic function manifests itself in the number of concepts receiving an extended discussion, often up to several pages. There are no literature references and eponymic syndromes are not identified. Useful for nurses, paramedical professionals, and laypersons.

13-60 **Encyclopedia of Bioethics.** Ed. by Warren T. Reich. New York: The Free Press, 1978. 4v.
Contains 315 signed articles on ethical and legal problems; basic concepts and principles; ethical theories; religious traditions; historical perspectives; and disciplines bearing on bioethics. Written for the layperson.

13-61 **Encyclopedia of Common Disease.** By the Staff of Prevention Magazine. Emmaus, PA: Rodale Press, 1976. 1296p. index.
An easy-to-read nontechnical book that covers the major diseases and disorders that afflict men, women, and children.

13-62 **Encyclopedia of Sport Sciences and Medicine.** Ed. by Leonard A. Larson; under the sponsorship of the American College of Sports Medicine, the University of Wisconsin, and in cooperation with other organizations. New York: Macmillan, 1971. 1707p.
Contains over 1,000 articles on the variables that affect the human organism before, during, and after participation in sports. Includes such areas as the environment, emotions, the intellect, and drugs.

13-63 **Merck Index: An Encyclopedia of Chemicals and Drugs.** 9th ed. Rahway, NJ: Merck, 1976. 1835p.
A dictionary of chemicals, drugs, and allied products, with a detailed cross index of names and trademarks. For each entry the variant names or trademarks are listed plus the properties, formulas, medical uses, and veterinary uses. It is necessary to use the cross index, since the main work lists the drugs or chemicals under only one name. There are also numerous tables that include calories in food, Greek and Russian alphabets, alchemical symbols used in biology and botany, conversion formulas, and logarithms.

13-64 **Penguin Medical Encyclopedia.** By Peter Wingate. 2nd ed. Baltimore, MD: Penguin Books, 1976. 489p. illus.
Definitions in this encyclopedia are brief, consisting of about 30 lines on the average, but they are adequate for the layperson and paraprofessional. Included are some biographical sketches of famous physicians.

13-65 **Potter's New Cyclopaedia of Botanical Drugs and Preparations.** By R. C. Wren; re-edited and enlarged by R. W. Wren. Hengiscote, Bradford, N. Devon: Health Sciences Press, 1975. 400p. illus. index.
This classic offers information on herbal compounds (continental), forms of medical preparation, and the plants from which the herbs are derived. It was originally published in England in 1907.

13-66 **Reston Encyclopedia of Biomedical Engineering Terms.** By Rudolf F. Graf and George J. Whalen. Reston, VA: Reston, a Prentice-Hall Co., 1977. 415p.

Biomedical and engineering terms (particularly in mechanical, electrical, and chemical engineering, and in physics and materials science) have been merged in this encyclopedia. The definitions are short, adequate, and nontechnical.

13-67 **Understanding Medical Terminology.** By Sister Agnes Clare Frenay. 6th ed. St. Louis, MO: Catholic Hospital Association, 1977. 393p.

Designed for students in allied health fields, this tool is organized into chapters that are devoted to single types of disorders—e.g., cardiovascular, digestive, etc. A second part treats terms used in the subdisciplines of medicine—e.g., physical medicine and nuclear medicine.

DICTIONARIES

13-68 **Abbreviations and Acronyms in Medicine and Nursing.** By Solomon Garb, Eleanor Krakauer, and Carson Justice. New York: Springer, 1976. 122p.

Over 4,000 abbreviations and acronyms are listed with an identification consisting of a few words. The virtue of this work is its list of abbreviations that are currently in use or that have been in use; recommends that for the sake of clarity abbreviations should be used much less frequently than they are.

13-69 **Abbreviations in Medicine.** By Albrecht Schertel. 2nd ed. Munich: Verlag Dokumentation; distr. New York: K. G. Saur, 1977. 204p.

This book is a compilation of medical abbreviations, initialisms, and acronyms from the German, English, and French languages. The meaning of an abbreviation is only explained if its definition is incomprehensible out of context.

13-70 **Anatomical Dictionary with Nomenclature and Explanatory Notes.** By Tibor Donath. Oxford: Pergamon, 1968. 634p. index.

A comprehensive dictionary for students in anatomy. Covers in preliminary chapters a comparison of the nomenclature systems of Basle, Jena, and Paris.

13-71 **Bailliere's Nurses Dictionary.** 18th ed. London: Bailliere, Tindall, 1974. 479p.

A well-known authoritative dictionary for nurses that gives clear concise definitions.

13-72 **Biomedical Thesaurus and Guide to Classification.** By Michael S. Koch. New York: CCM, 1972. 181p. bibliog.

This thesaurus of about 5,000 headings was developed by a medical librarian over a period of many years. It is of value to both practitioners and researchers in information handling. The terms listed were drawn from three major sources of medical subject headings: The National Library of Medicine printed catalogs from 1950 to 1959, the Library of Congress list of subject headings, and the National Library of Medicine Medical Subject Heading List (MeSH). In addition, appropriate classification numbers from both the NLM and LC classification schedules are given.

13-73 **Black's Medical Dictionary.** Ed. by William A. R. Thomson. 31st ed. New York: Barnes and Noble, 1976. 950p. illus.

One of the best known dictionaries in medicine, first published in Great Britain in 1906. Designed as a handy quick reference for the medical profession it provides brief definitions of all important medical terms. The text has been thoroughly revised in this edition.

13-74 **Blakiston's Gould Medical Dictionary: A Modern Comprehensive Dictionary of the Terms Used in All Branches of Medicine and Allied Sciences.** 3rd ed. New York: McGraw-Hill, 1972. 1828p. illus.

This is an outstanding medical dictionary covering about 75,000 words in all branches of medicine and its allied sciences.

13-75 **Butterworths Medical Dictionary.** Ed. by Macdonald Critchles. 2nd ed. Woburn, MA: Butterworths, 1978. 1942p.

Initially published as the *British Medical Dictionary* (London: Caxton, 1961, 1680p.), its 60,000 entries are arranged alphabetically and cover basic medical terminology, chemical nomenclature, anatomical vocabulary, pharmaceuticals, and eponymous terms. In addition to the basic definition, pronunciation, etymological derivation, variant definitions, individuals, and many subentries are included.

13-76 **Computer Glossary for Medical and Health Sciences.** By William T. Blessum and Charles J. Sippl. New York: Funk and Wagnalls; distr. New York: T. Y. Crowell, 1973. 262p. illus. index.

On the argument that all doctors and other health professionals should learn about computers and their applications, the authors have designed this book to meet that educational goal. The glossary is printed in dictionary form. There are two appendixes: a 13-page essay on "The Impact of Computers on Medicine and the Health Sciences" and a 3-page "Glossary of Acronyms."

13-77 **Current Procedural Terminology.** By Burgess L. Gordon, William R. Barclay, and Charlotte Fanta. 4th ed. Chicago: AMA, 1970. 368p.

A computer-produced dictionary of terms and codes for the naming and designation of diagnostic and therapeutic procedures in surgery, medicine, and the specialties.

13-78 **Dictionary of Abbreviations in Medicine and the Health Sciences.** By Harold K. Hughes. Lexington, MA: Lexington Books, a division of D. C. Heath, 1977. 313p.

This book contains more than 12,000 entries with 20,000 meanings taken from the areas of clinical, research, and production activities of professional care; food and energy resources; remedial education; veterinary science; and safety. Professional and industrial associations, boards, and commissions; government agencies; widely circulated journals; Latin and Greek combining forms and terms used in prescriptions and nursing procedures; anatomy and physiology; drugs and chemicals; analytical and therapeutic instruments; financial terms; and funding agencies may be found among the entries.

13-79 **Dictionary of Immunology.** Ed. by W. J. Herbert and P. C. Wilkinson. Oxford, England: Blackwell Scientific Publication, 1971; distr. Philadelphia, PA: F. A. Davis, 1971. 195p. illus.

This dictionary defines over 1,500 words and meets the needs of the undergraduate student, biologist, clinician, and biochemist. The definitions are simple enough to be understood by one with minimal biological knowledge. Included are short descriptions of diseases with immunological features, human and animal vaccines, abbreviations in current use, and some obsolete terms that will help in the interpretation of older literature.

13-80   **Dictionary of Medical Ethics.** Ed. by A. S. Duncan, G. R. Dunstan, and R. B. Welbourn. London: Darton, Longman and Todd; distr. Atlantic Highlands, NJ: Humanities Press, 1977. 335p.

This outstanding reference tool surveys the main ethical issues found in contemporary biomedical sciences. Provides excellent treatment of major religious ethical viewpoints toward medicine, as well as the complete texts of the leading twentieth century medical ethics codes.

13-81   **Dictionary of Medical Syndromes.** By Sergio Magalini. Philadelphia, PA: Lippincott, 1970. 591p.

Defines some 1,800 individual syndromes or "symptom complexes" from every branch of medicine. Each entry lists synonyms, symptoms and signs, etiology, diagnostic procedures, therapy, prognosis, and bibliography. The data was compiled from 10,000 medical articles and more than 50 textbooks.

13-82   **Dictionary of Nutrition.** By Richard Ashley and Heidi Duggal. New York: St. Martin's Press, 1975. 236p.

Intended primarily for the layperson, this dictionary provides common and technical terms in nutrition. The functions; sources; requirements; results of deficiencies; and the potential toxicity of the various vitamins, minerals, and other nutrients are presented.

13-83   **Dictionary of Pharmaceutical Science and Techniques.** By A Sliosberg. New York: American Elsevier, 1968. 2v.

A polyglot lexicon of equivalent pharmaceutical terms in English, French, Italian, Spanish, and German.

13-84   **Dictionary of Speech and Hearing Anatomy and Physiology.** By Joseph F. Brown. Sacramento, CA: Speech and Hearing Service, 1976. 322p. illus.

Defines several thousand terms covering the wide spectrum of anatomy and physiology of speech and hearing organs. Speech disorders and diseases directly related to speech are not represented. No pronunciation guide is given.

13-85   **Dorland's Illustrated Medical Dictionary.** 25th ed. Philadelphia, PA: Saunders, 1974. 1748p. illus.

A standard dictionary in the health sciences that is frequently revised. Contains more than 700 illustrations and 50 plates.

13-86   **Dorland's Pocket Medical Dictionary.** 22nd ed. Philadelphia, PA: Saunders, 1977. 741p. illus. (col.).

This is an abridgement of *Dorland's Illustrated Medical Dictionary* (13-85). It presents current terminology of medicine and allied sciences in a compact and convenient form together with the essentials of spelling, punctuation, and meaning. Definitions are brief, accurate, and clear.

13-87   **Duncan's Dictionary for Nurses.** By Helen Duncan. New York: Springer, 1971. 386p. illus.

This dictionary defines approximately 11,000 terms from nursing, medicine, and related disciplines. The definitions are accompanied by a pronunciation guide. Variant definitions, cross-references, and synonyms are included, as well as prefixes, suffixes, and some common abbreviations. Etymologies and chemical formulas are omitted.

13-88   **Eight-Language Dictionary of Medical Technology: English, German, French, Russian, Spanish, Polish, Hungarian, Slovak.** Ed. by Roald Albert and Harry Hahnewald. New York: Pergamon; Berlin: VEB Verlag Technik, 1978. 595p.

This dictionary lists nearly 8,000 English words in the interdisciplinary field of medical technology with comparable terms in seven other languages.

13-89   **Elsevier's Dictionary of Public Health: In Six Languages, English-French-Spanish-Italian-Dutch and German.** Comp. by Nic J. I. Deblock. New York: Elsevier Scientific, 1976. 196p. index.

Provides translations of leading public health words and phrases in six basic European languages.

13-90   **Glossary of Words and Phrases Used in Radiology, Nuclear Medicine and Ultrasound.** By Lewis E. Etter. 2nd ed. Springfield, IL: Charles C. Thomas, 1970. 355p.

Prepared for medical secretaries, x-ray technicians, medical students, and residents in radiology. Preceding the specialized glossary are chapters on radiological symbols, eponyms, confusing words, terminology and preparation of radiological reports, and an index to radiological and ultrasonic reports.

13-91   **Illustrated Dictionary of Eponymic Syndromes, Diseases and Their Synonyms.** By Stanley Jablonski. Philadelphia, PA: Saunders, 1969. 335p.

"The purpose in this dictionary is to gather together in one volume the profusion of eponyms and descriptive synonyms used to designate syndromes and diseases." Synonymous eponyms are cross-referenced.

13-92   **Inverted Medical Dictionary: A Method of Finding Medical Terms Quickly.** By Waldo A. Rigal. Westport, CT: Technomic, 1976. 261p.

Each interpretation of terms has been reduced to a key phrase. Phrases are alphabetically arranged by their most identifiable subject. Provisions are made for synonomous terms.

13-93   **Medical Abbreviations.** By Edwin Benzel Steen. 3rd ed. Philadelphia, PA: F. A. Davis, 1971. 102p.

A compact listing of the more common medical abbreviations that appear in major reference tools.

13-94   **Medical Abbreviations and Acronyms.** By Peter Roody, Robert E. Forman, and Howard B. Schweitzer. New York: McGraw-Hill, 1977. 255p. (A Blakiston Publication).

About 14,000 health care-related abbreviations and acronyms in their preferred forms arranged alphabetically followed by one or more definitions.

13-95 Medical and Health Sciences Word Book. Comp. by Ann Ehrlish. Boston: Houghton Mifflin, 1977. 448p.

This is an alphabetical listing of 60,000 medical, nursing, and allied health terms that the author has gathered from medical dictionaries, MeSH terminology found in *Index Medicus*, medical journals, and professional consultants from health-related fields. These terms are divided into syllables with accent marks so the reader should be able to pronounce and spell each entry.

13-96 Medical Dictionary, Medizinisches Worterbuch, Dictionnaire Medical. By Albert Nobel. 5th ed. rev. and enlarged. New York: Springer-Verlag, 1969. 1329p.

A three-language medical dictionary in which the first section lists the English term followed by German and French terms running in parallel columns. A German and French index refers to the consecutive serial numbers in section 1.

13-97 Medical Secretary Medi-Speller: A Transcription Aid. By Harriette L. Carlin. Springfield, IL: Charles C. Thomas, 1973. 244p. bibliog.

This source is a spelling book for medical secretaries. There are few definitions, although the words are categorized into the main fields of transcription. Definitions are given when word pronunciations are very similar but meanings are different. The sections are arranged as follows: 1) General Terminology, 2) Anatomy, 3) Surgical Pathology, 4) Radiology, 5) Diseases/Syndromes, 6) Surgical Procedures, 7) Instruments, 8) Clinical Pathology, and 9) Commonly Misspelled English Words.

13-98 Medical Specialty Terminology. v. 1- . By Clara Gene Young and James D. Barger. St. Louis, MO: Mosby, 1971- . illus. index.

This is a projected multi-volume dictionary-treatise. Following the model of the first volume, each chapter of the succeeding volumes will be devoted to a single topic; the discussion takes the form of a narrative exposition, followed by definitions of the terms relevant to the topic.

13-99 Medical Terminology from Greek and Latin. By Sandra R. Patterson and Lawrence S. Thompson. Troy, NY: Whitston Publishing, 1978. 275p. index.

A text-preference that permits the user to find readily exact meanings of the elements of medical terms. The following sections are covered: 1, Greek prefixes and suffixes; 2, Latin prefixes and suffixes; 3, bases; 4, Greek inflection and plural formation; 5, Latin inflection and plural formation; 6, scientific words from Greek mythology; and 7, words of non-European origin in science.

13-100 Medical Terms: Their Origin and Construction. By Francon Roberts. 5th ed. London: William Heinemann Medical Books, 1971. 102p. index.

Part 1 of this guide gives origins of words and the principles of their construction. Part 2 brings together synonyms and antonyms demonstrating how words are related by resemblance, contrast, and shades of meaning.

13-101 Medical Word Finder. Comp. by George Willeford. 2nd ed. West Nyack, NY: Parker, 1976. 490p.

This book is intended to provide medical, dental, and allied personnel with a single easy-to-use guide to the spelling, syllabication, and pronunciation of frequently used medical terms. The various sections cover prefixes and suffixes; phonetic spelling of problem words; a listing of anatomical organs and tissues; abbreviations; frequently prescribed pharmaceuticals; and troublesome medical words. The longest and most comprehensive section is the alphabetical list of current medical and paramedical terminology.

13-102 **Mellon's Illustrated Medical Dictionary.** By Ida Dox, Biagio John Melloni, and Gilbert M. Eisner. Baltimore, MD: Williams & Wilkins, 1979. 530p. illus. (col.).

This new dictionary provides clear and concise definitions to some 25,000 terms in the health sciences. Also includes approximately 2,500 illustrations, which help visualize the meaning of the definition. An attempt has been made to use terms in the definitions that do not require additional reference. Not as comprehensive as some other dictionaries, but more inclusive than the abridged medical dictionaries.

13-103 This number has not been used.

13-104 **Merck Manual of Diagnosis and Therapy.** 13th ed. Rahway, NJ: Merck and Co., 1977. 2165p. index.

A standard work, frequently revised, emphasizing medical diagnosis and treatment. Divided into 24 broad subject sections. Includes an extensive subject index.

13-105 **New American Pocket Medical Dictionary.** By Nancy Roper; adapted from the 13th British Edition by Jane Clark Jackson. New York: Churchill Livingstone; distr. New York: Longman, 1978. 380p. illus.

Lists about 6,500 entries. Some sections do not pertain to the United States, such as the section on snakebite, which deals exclusively with the adder, the only poisonous snake found in Great Britain. British and American spellings are provided.

13-106 **Obstetric-Gynecologic Terminology with Section on Neonatology and Glossary of Congenital Anomalies.** Ed. by Edward C. Hughes. Philadelphia, PA: Davis, 1972. 731p. index.

Seeks to standardize usage so that variability of definition may be minimized. It is arranged by the following sections: anatomy; anatomy of generative organs; disease and conditions of generative organs; benign and malignant neoplasms; physiology of reproduction; obstetrics; neonatology; and congenital anomalies.

13-107 **Psychiatric Dictionary.** By Leland E. Hinsie and Robert Jean Campbell. 4th ed. New York: Oxford University Press, 1970. 816p.

Lists approximately 9,700 entries ranging from community and social psychiatry to the neurophysiology of sleep. Its broad scope embraces every form of explanatory hypothesis found in psychiatric practice today: functional, behavioral, physical, chemical, and pharmaceutical.

13-108 **Stedman's Medical Dictionary: A Vocabulary of Medicine and Its Allied Sciences, with Pronunciations and Derivations.** 23rd ed. Baltimore, MD: Williams and Wilkins, 1976. 1678p. illus. index.

This outstanding dictionary of over 28,000 entries groups definitions by category. An appendix serves as the central cross-referencing system for adjectival or descriptive terms in the vocabulary; cross-references are not found in the text. Eponymic terms are cross-referenced to diseases and syndromes. Brief biographical information is also included. Special sections include instructions for use of the work, medical etymology, blood groups, a glossary of common Latin terms used in prescription writing, symbols and abbreviations, and results of laboratory analysis. It offers a variant approach to Black's and Blakiston's.

13-109 **Structural Units of Medical and Biological Terms.** By Jacob Edward Schmidt. Springfield, IL: Charles C. Thomas, 1969. 172p.
A convenient guide to English roots, stems, prefixes, suffixes, and other combining forms that are the building blocks of medical and related scientific words. Each of the approximately 1,000 main entries begins with the meaning of the combining form in English, followed by the Greek or Latin root and its definition, the combining stem in its available forms, and finally a selection of defined examples showing the usage.

13-110 **Taber's Cyclopedic Medical Dictionary.** Ed. by Clayton L. Thomas. 13th ed. Philadelphia, PA: F. A. Davis. 1977. 1v. (various paging). illus.
This dictionary is primarily for nurses and persons in allied health areas. It is arranged alphabetically, letter-by-letter, with phonetic spellings for most entries as well as abbreviations where appropriate. Brief first aid information is included plus biographies. An appendix gives information such as Latin and Greek nomenclature; lists of muscles, joints, arteries, veins, etc.; basic dictionary information; and lists of common health-related questions and answers translated into five languages.

HANDBOOKS AND LABORATORY GUIDES

13-111 **APhA Drug Names: An Index to Drug Names.** Washington: American Pharmaceutical Association, 1976. 192p.
This is an expansion of an earlier publication, *Proprietary Names (Trade Names) of Official Drugs* (Washington: American Pharmaceutical Association, 1965, 525p.). It lists approximately 7,200 drug names in a cross-referenced alphabetical list. Proprietary names and non-proprietary names only are listed. The manufacturers of the products and the combinations in which each drug appears are indicated. It also contains a list of trade names of discontinued products and a directory of manufacturers whose products appear in the drug names section.

13-112 **American Drug Index.** 1956- . Philadelphia, PA: Lippincott, 1956- .
This book lists, identifies, gives brief information about, and correlates the many pharmaceuticals available to the medical, pharmaceutical, and allied health professions. The main section lists drugs by generic name (also called non-proprietary, public, or common name), brand name (also called trademark, proprietary, or specialty name), and chemical name. Other sections of the book contain common abbreviations used in medical orders, common systems of weights and measure, approximate practical equivalents, pharmaceutical company labeler code index, and pharmaceutical manufacturers and/or drug distributors addresses.

13-113 CRC Handbook of Clinical Laboratory Data. By Willard R. Faulkner. 2nd ed. Cleveland, OH: Chemical Rubber Co., 1968. 710p. illus.

In authoritative articles, accompanied by tables, charts and diagrams, this CRC handbook comprehensively deals with the principles and processes of biochemistry, blood banking, hematology, microbiology, and pathology.

13-114 Clinical Guide to Undesirable Drug Interactions and Interferences. By Solomon Garb. New York: Springer, 1971. 491p.

The interaction of drugs with foods, chemicals, and other drugs can have serious and fatal consequences. Most of this volume consists of table 1, the basic entry, which lists a pair of drugs or foods followed by diagnostic tests with code letters designating the kind of undesirable interaction that may result when used together. Numbers are included referring to 984 literature references used by the author. There is no index, but the codification of comparatively elusive information is quite useful.

13-115 Clinical Handbook of Psychopharmacology. By Alberto Dimascio and Richard I. Shader. New York: Science House, 1970. 395p. illus.

A collection of specialized papers written by noted authorities. In 24 chapters, it covers clinical drug actions, mechanisms of drug action, side-effects or adverse reactions, drug use in selected populations, and general clinical considerations.

13-116 Clinical Toxicology of Commercial Products: Acute Poisoning. By Robert E. Gosselin and others. 4th ed. Baltimore, MD: Williams and Wilkins, 1976. 1v. (various paging). illus. index.

A manual designed to help physicians deal with acute chemical poisonings from suicidal or accidental ingestion of products used in homes, on farms, or in business and industry. The seven sections cover: first aid and general emergency treatment; ingredients index; therapeutics index; supportive treatment; trade name index; general formulations; and manufacturers' names and addresses.

13-117 Common Symptom Guide: A Guide to the Evaluation of 100 Common and Pediatric Symptoms. By John Wasson and others. New York: McGraw-Hill, 1975. 353p. index.

Intended primarily for the resident, medical student, and physicians's assistant, the guide arranges pertinent information for evaluating the patient's symptoms under his or her presented complaint.

13-118 Complete Handbook for Medical Secretaries and Assistants. By Jean Monty Doyle and Robert Lee Dennis. 2nd ed. Boston: Little, Brown, 1978. 619p. illus. index.

This is a greatly revised and expanded new edition covering new material of medical procedures with illustrations, first aid techniques, laboratory workup of diseases, and a complete list of laboratory tests and their diagnostic significance. Simple medical and surgical reports are included to help with the structure and form of these reports.

13-119 Contemporary Nursing Practice: A Guide for the Returning Nurse. By Signe Skott Cooper. New York: McGraw-Hill, 1970. 348p. illus. bibliog. index.

A textbook for inactive nurses enrolled in refresher courses. Covers overview of nursing, concepts of nursing care, skills and equipment, and reponsibilities and opportunities for the professional nurse.

13-120 **Current Diagnosis and Treatment.** v. 1- . Los Altos, CA: Lange Medical, 1962- . annual.
Describes new techniques for the diagnosis and treatment of internal disorders, malfunctions, and diseases.

13-121 **Current Diagnosis: Clinical Diagnostic Methods.** v. 1- . Philadelphia, PA: Saunders, 1966- . biennial.
Contains more detailed descriptions of diagnostic methods than *Current Diagnosis and Treatment* (13-120), but omits consideration of therapy.

13-122 **Current Drug Handbook 1976-78.** By Mary W. Falconer, H. Robert Patterson, and Edward A. Gustafson. Philadelphia, PA: Saunders, 1976. 279p. index.
Presents in tabular form information on approximately 1,500 drugs with the following information included: name, source, synonyms, preparations; dosage and administration; uses; action and fate; side effects and contraindications; and remarks.

13-123 **Emergency Medical Guide.** By John Henderson. 4th ed. New York: McGraw-Hill, 1978. 681p. illus. bibliog. index.
Nontechnical discussion of medical emergencies. Under each emergency situation, signs, symptoms, and treatment are given.

13-124 **Essential Guide to Prescription Drugs: What You Need to Know for Safe Drug Use.** By James W. Long. New York: Harper and Row, 1977. 768p. bibliog. index.
The first comprehensive source of information on drugs for the patient. Over 200 drugs are described mentioning 1,370 brand names for drugs marketed in the United States and Canada.

13-125 **Facts and Comparisons.** 1979- . St. Louis, MO: Facts and Comparisons, 1978- .
Provides quick comparison of the therapeutic aspects of similar products, as well as the relative prices of similar or identical products from its cost index figure. It is organized by therapeutic classes of drugs. Also published in looseleaf and microfiche editions. Prior to 1979, there were monthly updates but these have been discontinued.

13-126 **French's Index of Differential Diagnosis.** Ed. by F. Dudley Hart. 10th ed. Baltimore, MD: Williams and Wilkins, 1973. 972p. illus. (part col.). index.
This work remains a classic and standard treatise on the application of differential diagnosis. All signs and symptoms are collected in the index under headings of the various diseases in which they occur. The symptomologies range from the basic to the curious and the arcane in all the major medical and surgical specialties. Contains 805 illustrations, including 212 in color. The 160-page index contains over 40,000 entries.

13-127 **Guide to Medical Mathematics.** By D. A. Franklin and G. B. Newman. New York: Wiley, 1973. 453p. illus. index.
A sound introduction to mathematics for the medical research worker. It assumes only a knowledge of simple arithmetic and geometry. Graph construction, trigonometric functions, calculus, series, and differential equations are the principal mathematics topics considered. Particularly helpful are the examples from medical literature and practice used to illustrate various mathematical techniques.

13-128 **Handbook of Analytical Toxicology.** Cleveland, OH: Chemical Rubber Co., 1969. 1081p. index.
Supplies essential data to scientists working with the analysis of drugs, environmental hazards, economic poisoning, and industrial chemicals. It summarizes published works on current methods for the detection of toxicities in biological specimens. The first section is an alphabetically arranged index of drugs, poisons, and chemicals giving all data in tabular form; the second provides a sequential tabulation of physical properties; and the third includes a brief discussion of physical methods of instrumental analysis and their application to toxic substances.

13-129 **Handbook of Clinical Neurology.** v. 1- . Ed. by P. J. Vinken and G. W. Bruyn. North Holland; distr. New York: Wiley-Interscience, 1969- .
A comprehensive work covering all facets of clinical neurology. It replaces the Bumke-Foerster series on clinical neurology published in 1936.

13-130 **Handbook of Medical Specialties.** By Henry Wechsler. New York: Human Sciences Press, 1976. 315p. illus. index. (Health Services Series).
The information presented in this guide is designed to help the medical student choose a particular field of medicine. Part 1 discusses medical specialties in general, outlining certification requirements and the distribution of specialists in the United States. Part 2 summarizes each of 20 specialties, providing the following information: certifying board and year of its incorporation, total number of active certificates, certification requirements, distribution, and selected characteristics of the practice.

13-131 **Handbook of Medical Treatment.** Ed. by Milton J. Chatton. 16th ed. Greenbrae, CA: Jones Medical Publication, 1979. 643p. index.
A classic reference tool designed to assist the medical worker in the day-to-day care of patients. The author provides concise, pertinent, and timely data on a broad range of diseases from disturbances of the respiratory tract to psychiatric disorders. For greater depth, the reader will obviously need to consult more specialized works, but this handbook is an excellent starting point.

13-132 **Handbooks of Neurochemistry.** Ed. by Abel Lajtha. New York: Plenum, 1969-1972. 7v. in 1. illus.
This comprehensive handbook condenses the field of neurochemistry into seven volumes. Each volume and each chapter is written by an authority in the field covering a specific topic: v. 1, Chemical Architecture of the Nervous System; v. 2, Structural Neurochemistry; v. 3, Metabolic Reactions in the Nervous System; v. 4, Control Mechanisms in the Nervous System; v. 5, Metabolic Turnover in the Nervous System; v. 6, Alterations of Chemical Equilibrium in the Nervous System; and v. 7, Pathological Chemistry of the Nervous System.

13-133 **Handbook of Non-Prescription Drugs.** 5th ed. Washington: American Pharmaceutical Association, 1977. 388p. illus. index.

This handbook is composed of 32 chapters, each on a different type of product (such as antacid products, acne products, cold and allergy products, etc.). Information has been included on product formulas, indications/contraindications, safety, and appropriate use of specific non-prescription drugs. More than 1,500 drug products are covered.

13-134 **Handbook of Obstetrics and Gynecology.** By Ralph C. Benson. 6th ed. Los Altos, CA: Lange Medical, 1977. 772p. illus. index.

A practical handbook intended primarily for use by the physician and medical student, though it can also serve the needs of nurses, midwives, and paramedical workers. It emphasizes the clinical application of knowledge derived from anatomy, physiology, pathology, and pharmacology. Although not a substitute for standard textbooks on obstetrics and gynecology, it does supplement them by its sound, practical orientation.

13-135 **Handbook of Pediatrics.** By Henry K. Silver, C. Henry Kempe, and Henry B. Bruyn. 11th ed. Los Altos, CA: Lange Medical, 1975. 705p. illus. index.

A pocket-sized handbook, intended for the practicing physician and medical student, focusing on information necessary for the diagnosis and treatment of pediatric disorders. Clinical aspects of the subjects are stressed; however, physiologic principles and recent advances are also treated.

13-136 **Handbook of Physical Medicine and Rehabilitation.** By Frank Hammond Krusen. 2nd ed. Philadelphia, PA: Saunders, 1971. 920p. illus. index.

A detailed text on physical medicine covering principles of evaluation and management, techniques of management, evaluation and management of specific disorders, and basic considerations of importance in total rehabilitation. One of the best texts available.

13-137 **Handbook of Poisoning: Diagnosis and Treatment.** By Robert H. Dreisbach. 10th ed. Los Altos, CA: Lange Medical, 1980. 578p. illus. index.

A concise summary of the diagnosis and treatment of clinically important poisons, with information on a number of non-poisonous but hazardous substances. There are discussions of the prevention of poisoning and on emergency treatment. The handbook focuses on an analysis of agricultural, industrial, and medicinal poisons; and of cosmetics, foods, bleaches, soaps, and detergents as poisons.

13-138 **Handbook of Psychiatry.** Ed. by Philip Solomon and Vernon D. Patch. 3rd ed. Los Altos, CA: Lange Medical, 1974. 706p. index.

Although not a handbook of facts for ready reference, it does serve as a well-indexed entry point for the person interested in exploring the field of psychiatry. Its primary audience consists of general practitioners, students, and mental health trainees whose work brings them into contact with psychiatric problems.

13-139 **International Handbook of Medical Science.** Oxford: Medical and Technical Publishing Co., 1972. 832p. bibliog. index.

A handy tool discussing cardiac medicine, surgery, psychiatry, orthopedic surgery, anaesthetics, dermatology, pediatrics, obstetrics and gynecology, otolaryngology, physiology, immunology, genetics, drugs, therapy of common diseases, common emergencies, and laboratory tests.

13-140 **Measurement and Classification of Psychiatric symptoms: An Instruction Manual for the PSE and Catego Program.** By J. K. Wing, J. E. Cooper, and N. Sartorius. New York: Cambridge University Press, 1974. 233p. illus. charts. bibliog. index.

The Present State Examination (PSE) is an interview technique that allows symptoms to be elicited, recorded, and precisely defined. According to the authors, once rated, symptoms can be grouped into syndromes, thus giving a profile of the mental state. The components of the PSE-Catego system, some of which can be used independently of the others, are described in detail. They include: 1) the PSE interview and glossary of definitions, 2) the PSE checklist of symptoms, 3) the syndrome checklist (SCL), and 4) the Catego computer program. The PSE interview is an abbreviated textbook of functional psychopathology. In addition to monographic and journal citations, there are symptom, subject, and name indexes. The full version of the PSE (9th ed., May 1973) is contained in the appendix.

13-141 **Medical Mycology Manual.** By Everett Smith Beneke and Alvin Lee Rogers. 3rd ed. Minneapolis, MN: Burgess, 1971. 226p. illus. (part col.).

Intended for use by laboratory technicians, microbiologists, and researchers in medical mycology. For each fungus the defining characteristics are given, followed by etiological agents, occurrence, and selected references to the literature.

13-142 **National Formulary.** 14th ed. Washington: American Pharmaceutical Association; distr. Easton, PA: Mack, 1975. 1123p. illus. index.

"A compendium of standards designed to ensure the safety, effectiveness, quality, and purity of drugs. Such standards shall include the titles, definitions, descriptions, and specifications . . . appropriately developed and established to ensure adequate and appropriate characteristics of quality, strength, purity, and identity of the articles to which the respective standards apply" (Preface). The 14th edition of the NF and the first supplement to the *Pharmacopoeia* (13-145) are a composite listing, reflecting the attempt to unify the two works.

13-143 **Nurse's Drug Handbook.** By Suzanne Loebl, George Spratto, and Andrew Wit. New York: Wiley, 1977. 803p. illus. index. (A Wiley Medical Publication).

A useful handbook for the nursing student and the practicing nurse, providing not only a broad overview of the drug groups but in-depth descriptions of individual drugs.

13-144 **Nurse's Guide to Drugs.** Barbara E. Reed, pharmacy ed.; Jean Robinson and others, eds. Horsham, PA: Intermed Communications, 1979. 1355p. illus. index. (Nurse's Reference Library; Nursing 79 Books).

This guide presents complete, accurate material on drug contents, dosages, and possible bad side effects. It covers over 1,000 drugs, giving both generic and trade names, regular dosages, interactions with other drugs, possible side effects, and nursing considerations, which also pertain to all health professionals and the public in the correct use of drugs.

13-145 **Pharmacopoeia of the USA.** Easton, PA: U.S. Pharmacopoeial Convention, 1820- . Rev. every five years.

A pharmacopoeia is an official compendium of drugs, chemicals, and medicinal preparations; descriptions and tests for identification, purity, and strength; and formulas for preparation. The first *U.S. Pharmacopoeia* was published in 1820 and was adopted as the authority for drugs by the U.S. Food and Drug Administration in 1906. It is now issued under the authority of the United States Pharmacopoeial Convention. The early editions cited drugs by Latin names, but later editions use the English nomenclature.

13-146 **Physicians Desk Reference to Pharmaceutical Specialties and Biologicals.** 1st ed.- . Ordell, NY: Medical Economics, 1947- . annual.

A valuable working tool supplementing other standard compendia by addressing itself to pharmaceutical concerns of specific interest to physicians: dosage, contraindications, precautions, side effects, and undesirable interactions.

13-147 **Physicians' Handbook.** By Marcus A. Krupp and others. 19th ed. Los Altos, CA: Lange Medical, 1978. 754p. illus. index.

A pocket-sized handbook of diagnostic and therapeutic information and procedures that are of value to practicing physicians and medical students. Diagnosis is stressed. Data from texts, laboratory examinations, and procedures are included.

13-148 **Practical Handbook of Psychiatry.** Ed. by Joseph R. Novello. Springfield, IL: Charles C. Thomas, 1974. 621p. index. (American Lecture Series, no. 907).

Combines in a single volume data on psychiatric nasology, evaluation, diagnosis and therapy; a basic reading list on continuing education in psychiatry; a directory of professional organizations and institutions; and tables of laboratory indicators in clinical medicine. Other guides, checklists and tables in the book include the "suicide caller" and "differential" diagnosis, major types of drug abuse, acute schizophrenia, acute mania, and hallocinosis.

13-149 **Prescription Drugs and Their Side Effects.** By Edward L. Stern. New York: Grosset and Dunlap, 1975. 96p. index.

For each of the 150 most frequently prescribed drugs the following information is provided: brand name, manufacturer, generic name, common dosage, form, strength, and route of administration. In addition, one finds the reason a particular drug is usually prescribed, the precautions and warnings to observe when taking the drug, as well as the possible side effects and adverse reactions that may occur.

13-150 **Student's Guide to Geriatrics.** By Trevor H. Howell. 2nd ed. Springfield, IL: Charles C. Thomas, 1970. 212p. illus.

An introduction to the diseases that are common among the aged. It covers anatomical and physiological changes, pathology of old age, psychology of old age, diseases of the central nervous system, cardiovascular disease, respiratory disease, digestive disorders, rheumatic disorders, diseases of the blood, endocrine disorders, surgery in old age, special problems, mental disorders, rehabilitation, and social medicine in old age.

13-151 **U.S. Dispensatory.** Ed. by Arthur Osol, Robertson Pratt, and Alfonso R. Genuaro. 27th ed. Philadelphia, PA: Lippincott, 1973. 1292p.

A collection of articles, alphabetically arranged, about individual drugs. The articles supply chemical, generic, and brand name nomenclature; chemical structure; preparation summaries; pharmacologic actions; therapeutic uses; contraindications; warnings and precautions; drug interactions; dosage for adults and children, including variations of dosage in different diseases; and dosage forms. The *Dispensatory* contains general articles on classes of drugs and a classification of drugs and related substances based on principal pharmacologic actions and/or therapeutic uses of drugs. This is an indispensable reference work for the pharmacologist.

POPULAR MEDICAL GUIDES

The medical librarian is frequently asked to recommend simple but technically sound popular medical books that can circulate. Following is a representative sampling of such works, which public librarians have identified as meeting these criteria.

13-152 **Advocate Guide to Gay Health.** By R. D. Fenwick. New York: E. P. Dutton, 1978. 240p. illus. bibliog. index.

13-152a **Compleat Herbal: Being a Description of the Origins, the Lore, the Characteristics, the Types, and the Prescribed Uses of Medicinal Herbs, Including an Alphabetical Guide to All Common Medicinal Plants.** By Ben Charles Harris. Barre, MA: Barre Publishers, 1972. 243p. bibliog.

13-152b **Complete Medical Guide.** By Benjamin F. Miller, with Lawrence Galton. 4th ed. rev. and updated. New York: Simon and Schuster, 1978. 639p. illus. index.

13-152c **Dictionary of Drugs: The Medicines You Use.** By Richard B. Fisher and George A. Christie. updated and rev. ed. New York: Schocken Books, 1976. 256p. index.

13-152d **Dr. Fishbein's Popular Illustrated Medical Encyclopedia.** By Morris Fishbein. New York: Doubleday, 1979. 773p. illus.

13-152e **Drugs from A to Z: A Dictionary.** By Richard R. Lingeman. 2nd ed. rev. New York: McGraw-Hill, 1974. 310p.

13-152f **Encyclopedia of Alternative Medicine and Self-Help.** Ed. by Malcolm Hulke. New York: Schocken Books, 1979. 243p. bibliog. index.

13-152g **Parents' Guide to Allergy in Children.** By Claude A. Frazier. New York: Grosset & Dunlap, 1978. 338p. illus. index.

13-152h **Patients' Guide to Medicine, from the Drugstore through the Hospital.** By Warren J. Brown. 8th ed. Largo, FL: Aero-Medical Consultants, Inc., 1978. 244p. illus. index.

13-152i **Pocket Medical Encyclopedia and First Aide Guide.** By James Bevan. New York: Simon and Schuster, 1979. 133p. illus. (part col.). (Fireside Book).

ATLASES

13-153 **Atlas of Clinical Haematology.** By Herbert Begemann and Johann Rastetter. 3rd ed. New York: Springer-Verlag, 1979. 324p. illus. (part col.). index.
This atlas depicts the cytology of peripheral blood, bone marrow, lymph nodes, and the spleen. Individual cells are described in detail, and their changes in different diseases are discussed. Cyto-chemical methods, panoptic staining, and blood parasites are also included. The fine illustrations include color reproductions of the paintings used in the first edition, plus new photomicrographs.

13-154 **Atlas of Haematological Cytology.** By F. G. J. Hayhoe and R. J. Flemans. New York: Wiley-Interscience, 1970. 320p. illus.
A reference collection of photomicrographs in color, showing examples of cell types most likely to be encountered.

13-155 **Atlas of Neuroanatomy and Special Sense Organs.** By Jean Bossy. Philadelphia, PA: Saunders, 1970. 348p. illus. index.
Graphically correlates the nervous system with the special sense organs for the purpose of identifying the clinical and pathological manifestations of sensory-related disorders and dysfunctions.

13-156 **Atlas of Orthodontic Principles.** By Raymond C. Thurow. 2nd ed. St. Louis, MO: C. V. Mosby, 1977. 231p. illus. index.
An overview of orthodontic principles, practices, and procedures susceptible of pictorial representation, including dental forces and movements, functional anatomy of dental mechanisms, and cephalometric radiography.

13-157 **Atlas of the Anatomy of the Ear.** By Branislav Vidic and Ronan O'Rahilly. Philadelphia, PA: Saunders, 1971. 1v. (unpaged). illus. index.
Fifty drawings and slides of the ear from the outer portion to the inner sections showing stapes and the incudostapedial joint. Each drawing is accompanied by a descriptive paragraph and references to additional readings.

13-158 **Grant's Atlas of Anatomy.** By James E. Anderson. 7th ed. Baltimore, MD: Williams and Wilkins, 1978. 1v. (various paging). illus. (part col.). index.
The parts of the book correspond to the regions of the body: upper limb, abdomen, perineum, pelvis, lower limb, vertebrae, vertibral column, thorax, head and neck, cranial nerves, and dermatomes. Detailed cross-sections and dissections of every part of the body follow the same order that a student would find in laboratory dissecting classes. The text is minimal. Excellent plates are distinguished for their clarity of definition.

DIRECTORIES

13-159 **American Dental Directory.** 1st- . Chicago: American Dental Association, 1947- . annual.

Arranged in five lists: geographical, alphabetical, honorary and associate members, geographical list of affiliate members, and geographical list of dentists describing type of practice.

13-160 **American Medical Directory.** 1st- . Chicago: American Medical Association, 1906- . annual.
The most complete directory of physicians on the North American continent available, listing all legally qualified physicians in alphabetical order, and by state and city within each state. Includes Isthmian Canal Zone, Puerto Rico, Virgin Islands, certain Pacific Islands, and U.S. physicians temporarily located in foreign countries. Very brief sketches, primarily education and subsequent practice.

13-161 **Directory of Health Sciences Libraries, 1973.** Comp. and ed. by Susan Crawford and Gary Dandurand. Chicago: American Medical Association, 1975. 196p. illus. index.
Identifies some 3,000 health science libraries in the United States. Each entry is arranged alphabetically by state, then by city within the state. The following information is included: chief librarian, address and telephone number, type of institution, number of volumes and currently received serial titles, number of staff, and primary occupations of each library's users.

13-162 **Directory of Medical Specialists Certified by American Boards.** Chicago: Marquis Who's Who, 1939- . biennial.
The authorized publication of the American Specialty Boards, which certify physicians as medical specialists. Includes the requirements for certification. Lists diplomates first by specialty and then geographically by state and city. Name and certification, birth date and place, education, career, teaching positions, military record, professional memberships, and office address are provided for each entry.

13-163 **Directory of Pathology Training Programs.** 1st- . Bethesda, MD: Intersociety Committee on Pathology Information, 1969/70- . annual.
Designed as an aid to young physicians seeking the pathology training program best suited to their individual talents, preferences, and needs. It covers some 235 hospitals, medical schools, and laboratories located in 37 states, 6 Canadian provinces, and Puerto Rico.

13-164 **Health: A Multimedia Source Guide.** By Joan Ash and Michael Stevenson. New York: R. R. Bowker, 1976. 185p. index.
This is an annotated guide to organizations that deal with health related matters. Information about the 700 sources is presented in eight chapters: Associations, Societies, and Foundations; Audiovisual Producers and Distributors; Book Dealers; City, State, and Federal Agencies; Libraries and Information Services; Pharmaceutical and Other Companies; Publishers; and Research Institutes. In addition to address, phone number, and the name of the executive director, the following information is included with most entries: purpose, membership, special services offered, and publications.

13-165 **Pharmaceutical Manufacturers of the United States.** Park Ridge, NJ: Noyes Data, 1977. 266p. index.

This directory provides detailed information to more than 300 pharmaceutical and health care product manufacturing corporations in the United States. When available, the following information is given: address, telephone number; whether or not it is a division or subsidiary of another company; product category; annual sales; number of employees; names and titles of the principal executives; a listing of products, including trade names and pharmacological classification; locations of facilities and research laboratories; domestic subsidiaries and affiliates; and either a listing of foreign subsidiaries by name or, if the list would be too long, a listing of the cities and countries where the subsidiaries are located.

14

Bibliography

The following is a selected classified list of recent articles, books, and reports on various topics of special relevance to science librarians.

GENERAL

Cawkell, A. E. "Connections between Engineering and Science," *IEEE Transactions on Professional Communications* PC 18 (June, 1975):71-73.

Cawkell, A. E. "Understanding Science by Analyzing Its Literature," *Information Scientist* 10 (March, 1976):3-10.

Flanagon, Dennis. "The Future of Scientific Communication," *Federation Proceedings* 33 (June, 1974):1721-23.

Frame, J. Davidson, and Francis Narin. "The International Distribution of Biomedical Publications," *Federation Proceedings* 36 (1977):1790-95.

Garfield, Eugene. "Significant Journals of Science," *Nature* (London) 264 (1976):609-615.

Griffith, Belver C. "Mapping the Scientific Literature," *Nato Advanced Study Institutes Series, Series E. Applied Science* 10 (1975):457-81.

Houghton, Bernard. "Scientific Journals and Natural Selection," *New Scientist* 68 (Dec. 18/25, 1975):727-28.

Hyman, Richard J. "From Cutter to MARC: Access to the Unit Record," *Queens College Studies in Librarianship* 1 (1977):40pp.

Lea, Peter W. "Trends in Scientific and Technical Primary Journal Publishing in the USA," *British Library. Research and Development Department. BLRD Report* 5272 (1976):29pp.

Line, Maurice B. "National Libraries," *British Librarianship and Information Science* 1971-1975:132-45.

Van Styvendale, J. H. "University Scientists as Seekers of Information: Sources of Reference to Periodical Literature," *Journal of Librarianship* 9 (Oct., 1977):270-77.

BIBLIOGRAPHIC CONTROL

Cohen, Diana B. "Unesco's Bibliographic Services," *International Library Review* 9 (April, 1977):127-60.

Fasana, Paul. "Serials Data Control: Current Problems and Prospects," *Journal of Library Automation* 9 (March, 1976):19-33.

Gorman, Michael. "A Modest Proposal for a Future National Bibliographic System," *American Libraries* 10 (March, 1979):147-49.

Kennedy, H. E., and D. A. Fisher. "Information Support for the Biomedical Sciences," *Federation Proceedings* 33 (June, 1974):1714-17.

Lucker, Jay K. "Library Resources and Bibliographic Control," *College and Research Libraries* 40 (March, 1979):141-53.

Markuson, Barbara Evans. "Bibliographic Systems, 1945-1976," *Library Trends* 25 (July, 1976):311-28.

"National Planning for Bibliographic Control. Minutes of the Ninety-Fourth Meeting" (May 10-11, 1979 at Cambridge, Massachusetts). Washington, DC: Association of Research Libraries, 1979. (ED176795).

Wellisch, Hans H. "Script Conversion and Bibliographic Control of Documents in Dissimilar Scripts: Problems and Alternatives," *International Library Review* 10 (Jan., 1978):3-22.

NETWORKS

Avram, Henriette D. "The Role of the Library of Congress in the National Bibliographic Network," *Journal of Library Automation* 10 (June, 1977): 154-62.

Davies, G. W. P. "EURONET, The European Information Network," *Second European Congress on Information Systems and Networks* (Luxembourg) May 27-30, 1975:55-63.

Institute on the National Bibliographic Network, sponsored by the American Library Association's Information Science and Automated Division, Chicago, February, 1977. *Journal of Library Automation* 10 (June, 1977): 101-180.

Knapp, John F. "Requirements for the National Library Network: A View from the Local Network," *Journal of Library Automation* 10 (June, 1977):131-41.

Little, John L. "A Computer Network Protocol at the Application Level for Libraries and Other Information Science Services," *Journal of Library Automation* 11 (Sept., 1978):239-45.

Lorenz, John G. "The National Bibliographic Network: The View from the Research Library," *Journal of Library Automation* 10 (June, 1977):114-19.

Martin, Susan K. "The Quest for a National Bibliographic Network," *Library Journal* 103 (Jan. 1, 1978):19-22.

Nyren, Karl. "Conference Report: ALA/ISAD Institute on the National Network," *Library Journal* 102 (April 1, 1977):761-63.

Rather, Lucia J., and Peter J. de la Garza. "Getting It All Together: International Cataloguing Cooperation and Networks," *Journal of Library Automation* 10 (June, 1977):163-69.

Rouse, Sandra H., and William B. Rouse. "Design of a Model-Based Online Management Information System for Interlibrary Loan Networks," *Information Processing and Management* 15 (1979):109-122.

Silberstein, Stephan M. "Standards in a National Bibliographic Network," *Journal of Library Automation* 10 (June, 1971):142-53.

Trezza, Alphonse F. "The NCLIS View—A Full Service Network," *Journal of Library Automation* 10 (June, 1977):170-76.

Wright, Gordon H. "The Canadian Mosaic-Planning for Shared Partnership in a National Network," *Aslib Proceedings* 30 (Feb., 1978):88-102.

INDEXING (including Citation Indexing)

Avemey, Brian, and Rod Slade. "Indexing of Popular Periodicals: The State of the Art," *Library Journal* 103 (Oct. 1, 1978):1915-23.

Cawkell, A. E. "Science Perceived through the Science Citation Index," *Endeavour* 1 (April, 1977):57-62.

Culnan, Mary J. "An Analysis of the Information Usage Patterns of Academics and Practitioners in the Computer Field: A Citation Analysis of a National Conference Proceedings," *Information Processing and Management* 14 (1978):395-404.

Hirst, Graeme, and Nadia Talent. "Computer Science Journals—An Iterated Citation Analysis," *IEEE Transactions on Professional Communication* PC-20 (Dec., 1977):233-38.

LaBorie, Tim, and Michael Halperin. "Citation Patterns in Library Science Dissertations," *Journal of Education for Librarianship* 16 (Spring, 1976):271-83.

MacCafferty, Maxine. "Thesauri and Thesauri Construction," *Aslib Bibliography Series* 7 (1977):191pp.

Murugesan, Poovanalingam, and Michael J. Moravcsik. "Variation of the Nature of Citation Measures with Journals and Scientific Specialities," *Journal of the American Society for Information Science* 29 (May, 1978):41-47.

LIBRARY AUTOMATION

Butcher, Roger. "National Data on Microfiche: Bibliographic and MARC Based Use of COM," *Reprographics Quarterly* 9 (Summer, 1976):104-107.

Coward, R. E. "Computers in National Libraries: Where They Have Made Their Greatest Impact," *Use of Computers in Libraries and Information Centres: Proceedings of a Conference Held by Aslib with the Cooperation of the Aslib Computer Applications Group in London on 19-20 May, 1975.* p. 46-50.

Daniels, Mary Kay. "Automated Serials Control: National and International Considerations," *Journal of Library Automation* 8 (June, 1975):127-46.

Ganning, Mary Kay Daniels. "The Catalog: Its Nature and Prospects," *Journal of Library Automation* 9 (March, 1976):48-66.

Malinconico, S. Michael. "Bibliographic Data Base Organization and Authority File Control," *Wilson Library Bulletin* 54 (Sept., 1979):36-45.

Sauer, Mary. "Automated Serials Control: Cataloguing Considerations," *Journal of Library Automation* 9 (March, 1976):8-18.

ABSTRACTS AND INDEXES—PRINTED AND COMPUTERIZED

Ball, A. W., and Molly A. Wolfe. "Excerpta Medica On-Line by Informatics, Inc.," *Drug Information Journal* 9 (May-Sept., 1975):142-44.

Burylo, Michelle A. "EIS Plants," *Online* 1 (July, 1977):53-58.

Butcher, William S. "Water Resources Information Services," *Library Science with a Slant to Documentation* 14 (June, 1977):95-98.

Collier, H. R. "ISI and the Science Citation Index," *Seminar Wholesalers of Documentary Information: Proceedings* (1976):101-106.

Dayton, David L., et al. "CASSI: File for Document Access," *Special Libraries* 69 (Sept., 1978):337-47.

Dow, Ronald F. "Inspec on BRS," *Database* 1 (Sept., 1978):70-81.

Frazier, John P. "An Introduction to DOE/RECON and NASA/RECON," *Sci-Tech News* 32 (Oct., 1978):85-87.

Garfield, E., M. Koenig, and T. DiRenzo. "ISI Data-Base-Produced Information Services," *IEEE Transactions on Professional Communication* PC-20 (Sept., 1977):95-99.

Gilreath, Charles L. "AGRICOLA: Multipurpose Data Base for Agricultural and Life Sciences Libraries," *Serials Librarian* 3 (Fall, 1978):89-95.

Kenton, C., and Y. B. Scott. "Medline Searching and Retrieval," *Medical Informatics* 3 (1978):225-36.

Metcalfe, John R. "The CAB World Agricultural Information Service," *Aslib Proceedings* 31 (March, 1979):110-17.

O'Dette, R. "The CAS Data Base," *Pure and Applied Chemistry* 49 (Dec., 1977):1781-92.

Rowlett, R. "Symposium on User Reactions to CAS Data and Bibliographic Services. Concluding Remarks." *Journal of Chemical Information and Computer Sciences* 15 (1975):186-89.

Savage, Gretchen S., and Jeff Pemberton. "SCI Search on Dialog," *Database* 1 (Sept., 1978):50-67.

Sumkins, M. A. "A Comparison of Data Bases for Retrieving References to the Literature on Drugs," *Information Processing and Management* 13 (1977):141-54.

Wilson, J. P., and M. J. Wood. "An SDI Experiment Using COMPENDEX: An Evaluation," *New Zealand Libraries* 41 (1978):4-10.

ONLINE SEARCHING

Antony, Arthur, et al. "An Online Component in an Interdisciplinary Course on Information Resources for Science and Engineering Students," *Online Review* 2 (Dec., 1978):337-44.

Bellando, Trudi, Gail Kennedy, and Gretchen Temoulet. "On-line Bibliographic System Instruction," *Journal of Education for Librarianship* 19 (Summer, 1978):21-31.

Calkins, Mary L. "On-line Services and Operational Costs," *Special Libraries* 68 (Jan., 1977):13-17.

Hawkins, Donald T. "Bibliometrics on the Online Information Retrieval Literature," *Online Review* 2 (Dec., 1978):345-52.

Hawkins, Donald T. "On-line Data Base Coverage of the On-line Information-Retrieval Literature," *Online Review* 1 (March, 1977):59-64.

Hawkins, Donald T. "On-line Information Retrieval Bibliography: Second Update," *Online Review* 3 (March, 1979):37-73.

Kidd, J. S. "On-line Bibliographic Services: Selected British Experiences," *College and Research Libraries* 38 (July, 1977):285-90.

Marshall, Doris B. "A Survey of the Use of On-line Computer-Based Scientific Search Services by Academic Libraries," *Journal of Chemical Information and Computer Sciences* 15 (Nov., 1975):247-49.

Miller, Betty. "Coverage of Library and Information Science Literature by On-line Data Bases," *Information 5 Politics: Proceedings of the 39th ASIS Annual Meeting, San Francisco, October 4-9, 1976.* v. 13. 66p.

Pope, Noland F. "Bibliographic Data Base Use," *Sci-Tech News* 32 (Oct., 1978):91-93.

Ruhl, Mary Jane, and Elizabeth J. Yeates. "Introducing and Implementing On-line Bibliographic Retrieval Services in a Scientific Research and Development Organization," *Journal of Chemical Information and Computer Sciences* 16 (August, 1976):147-50.

Tedd, Lucy A. "Education, Training and Marketing for Online Information Retrieval Systems," *Online Review* 3 (June, 1979):205-212.

Tomberg, A. "Data Banks: Survey." First International On-Line Information Meeting, London, 13-15 December 1977, organized by *On-Line Review*, 159-68 (Paper E19).

Williams, Martha E. "Online Retrieval—Today and Tomorrow," *Online Review* 2 (Dec., 1978):353-66.

Wolters, Peter H. "CAN/OLE: A Canadian System for the On-line, Interactive Searching of Large Bibliographical Reference Files," *Third Open Conference on Information Science in Canada, Quebec, 8-10 May, 1975: Proceedings*, 221-29.

Woodford, Susan E., and Mary E. Pensyl. "Continuing User Education to Promote the Effective Use of an Established On-line Search Service in a University Community," *Information Management in the 1980s: Proceedings of the 40th ASIS Annual Meeting, Chicago, September 26-October 1, 1977.* v. 14. 50p.

INFORMATION SCIENCE

Becker, Joseph. "The Rich Heritage of Information Science," *Bulletin of the American Society for Information Science* 2 (March, 1976):9-13.

Kent, Allen. "Information Science," *Journal of Education for Librarianship* 17 (Winter, 1977):131-39.

Salton, Gerard. "Computers and Information Science," *Bulletin of the American Society for Information Science* 2 (March, 1976):19-21.

Zurkowski, Paul G. "Perspectives on the Information Industry," *Bulletin of the American Society for Information Science* 2 (March, 1976):28-29.

MICROGRAPHICS

Carroll, C. Edward. "Bibliographic Control of Microforms: Where Do We Go from Here?," *Microform Review* 7 (Nov.-Dec., 1978):321-26.

Kneitel, A. M. "New Approach to Microfiche Indexing," *Journal of Micrographics* 7 (March, 1974):159-61.

Lee, L. S. "Speak Softly But Carry a Big Fiche," *Journal of Micrographics* 7 (March, 1974):185-87.

Saffady, William. *Micrographics*, Littleton, CO: Libraries Unlimited, 1978. 238pp.

Author/Title/Subject Index

References in this index are to item number. Those titles mentioned only in annotations to other items are designated by "n."

Guide to (cont'd) . . . *Medical Mathematics*, 13-127; *Microforms in Print*, 3-34; *Reference Books*, 3-2; *Reference Material*, 3-3; *Reference Sources in the Computer Sciences*, 5-4; *Reprints*, 3-35; *Scientific and Technical Journals in Translation*, 3-154; *Special Issues and Indexes of Periodicals*, 3-88; *Tables in Mathematical Statistics*, 5-5; *the Applications of the Laplace and 8-Transforms*, 5-67; *the History of Science*, 4-10; *the Literature in Psychiatry*, 13-4; *the Literature of Astronomy*, 6-4; *the Literature of Botany*, 4-50; *the Literature of Chemistry*, 8-3; *the Literature of Mathematics and Physics Including Related Works on Engineering Science*, 5-6; *the Literature of the Life Sciences*, 9-1; *the Literature of the Zoological Sciences*, 9-1n; *the Research Collections of the New York Public Library*, 3-4; *the Study of Lichens*, 9-131; *the Taxonomic Literature of Vertebrates*, 9-172; *the Vocabulary of Biological Literature*, 9-4n; *Theses and Dissertations*, 3-148; *U.S. Government Maps*, 10-3

Guidebook to Biochemistry, 9-58

Guidebook to Microscopical Methods, 9-59

Guides to the literature — astronomy, 6-1 — 6-4; biology, 9-1 — 9-3; biomedical, 13-1 — 13-12; chemistry, 8-1 — 8-6; energy, 11-2; engineering, 12-1 — 12-3, 12-38 — 12-40, 12-100 — 12-104, 12-137, 12-138; environment, 11-1, 11-3 — 11-6; geoscience, 10-1 — 10-7; mathematics, 5-1 — 5-9; multidisciplinary, 3-1 — 3-11; physics, 7-1 — 7-3

Gupta, S. D., *Bibliography of Multivariate Statistical Analysis*, 5-11

Gurr, E., *Encyclopedia of Microscopic Stains*, 9-48

Gustafson, E. A., *Current Drug Handbook*, 13-122

HEALTH, 13-48n

HISTLINE, 13-48n

HRIS Abstracts, 12-164

Hackh's Chemical Dictionary, 8-40

Haensch, G., *Dictionary of Biology*, 9-23

Hahnewald, H., *Eight-Language Dictionary of Medical Technology*, 13-88

Hale, M. E., *How to Know the Lichens*, 9-34n; *Lichens* (1973), 9-131n; *Lichens* (1969), 9-131n

Hall, E. R., *Mammals of North America*, 9-225

Halstead, B. W., *Poisonous and Venomous Marine Animals of the World*, 9-199

Hamilton, W. R., *Larousse Guide to Minerals, Rocks, and Fossils*, 10-77

Hammond, K. A., *Sourcebook on the Environment*, 11-4

Hampel, C. A., *Encyclopedia of Chemical Reactions*, 8-20; *Encyclopedia of Chemistry*, 8-22; *Glossary of Chemical Terms*, 8-39; *Rare Metals Handbook*, 12-134

Hanchey, M. M., *Abstracts and Indexes in Science and Technology*, 3-81

Handbook for Electronics Engineering Technicians, 12-72

Handbook of . . . Adhesives, 12-127; *Analytical Toxicology*, 13-128; *Applied Hydraulics*, 12-73; *Applied Mathematics*, 5-68; *Automation, Computation and Control*, 12-74; *Biochemistry and Molecular Biology*, 9-60; *British Birds*, 9-211n; *Chemical Technicians*, 8-48; *Chemistry and Physics*, 3-76, 8-49; *Chlorination for Potable Water, Waste Water, Cooling Water, Industrial Processes, and Swimming Pools*, 12-75; *Chromatography*, 8-50; *Circuit Analysis Languages and Techniques*, 12-76; *Clinical Neurology*, 13-129; *Components for Electronics*, 12-77; *Composite Construction Engineering*, 12-153; *Data Processing for Libraries*, 3-77; *Dimensional Measurement*, 12-24; *Electronic Circuit Designs*, 12-78; *Electronic Formulas, Symbols and Definitions*, 12-79; *Engineering Fundamentals*, 12-25; *Environmental Control*, 11-49; *Environmental Data on Organic Chemicals*, 11-50; *Freshwater Fishes Biology*, 9-216; *Heavy Construction*, 12-154; *Hydraulics for the Solution of Hydraulic Engineering Problems*, 12-80; *Industrial Noise Control*, 11-51; *Intermediary Metabolism of Aromatic Compounds*, 9-61; *Living Primates*, 9-223; (cont'd)

World Patents Index—WPI, 3-166

World Space Directory, Including Oceanology, 12-37

World Who's Who in Science, 3-138

World Wildlife Guide, 9-231

Wren, R. C., Potter's New Cyclopaedia of Medicinal Herbs and Preparations, 13-65

Wren, R. W., Potter's New Cyclopaedia of Medicinal Herbs and Preparations, 13-65

Wyatt, H. V., Use of Biological Literature, 9-3

Wycoff, R. W. G., Crystal Structures, 8-46

Wynar, B. S., American Reference Books Annual, 3-23

Wynter, H., Scientific Instruments, 4-67

Yamashita, Y., Atlas of Representative Stellar Spectra, 6-56

Yannerella, P. A., U.S. Government Scientific and Technical Periodicals, 3-180

Yates, B., How to Find out about Physics, 7-1

Yates, F., Statistical Tables for Biological, Agricultural and Medical Research, 9-37

Yearbook of . . . Astronomy, 6-27; International Congress Proceedings, 3-103

Yeasts, 9-149

Youden, W. W., Computer Literature Bibliography, 5-17

Young, C. G., Medical Specialty Terminology, 13-98

Young, H. C., Directory of Special Libraries and Information Centers, 3-109; Subject Directory of Special Libraries and Information Centers, 3-121

Young, M. L., Directory of Special Libraries and Information Centers, 3-109; Subject Directory of Special Libraries and Information Centers, 3-121

Yudkin, M., Guidebook to Biochemistry, 9-58

Zahlenwerte und Funktionen aus Naturwissenschaften und Technik-Neue Serie, 7-22a

Zahlenwerte und Funktionen aus Physik, Chemie, Astronomie, Geophysik und Technik, 7-22

Zentralblatt fuer . . . Bakteriologie, Parasitenkunde, Infectionskrankheiten und Hygiene, 9-46; Geologie und Palaeontologie, 10-37, 10-37a; Mathematik und ihre Grenzgebiete, 5-27; Mineralogie, 10-37b, 10-37c; Mineralogie, Geologie und Palaeontologie, 10-37n

Ziegler, E. N., Encyclopedia of Environmental Science and Engineering, 11-33

Zimmerman, E., Bibliographische Berichte, im Auftrag des deutschen bibliographischen Kuratoriums, 3-27

Zimmermann, H., Concise Encyclopedia of Astronomy, 6-16

Zoobooks, 9-160

Zoological Record, 9-161

Zoology—abstracts, 9-162, 9-165; atlases, 9-176, 9-177; bibliographies, 4-15, 4-19, 9-155, 9-158, 9-160, 9-161, 9-164, 9-166—9-169, 9-172—9-175; dictionaries, 9-186—9-190, 9-192, 9-193; dictionaries—foreign language, 9-191; directories, 9-227—9-232; encyclopedias, 9-178—9-185; handbooks, 9-194—9-226; library catalogs, 9-156, 9-159, 9-163, 9-170

Zoos and Aquariums in the Americas, 9-232

Zweig, G., Handbook of Chromatography, 8-50